UNITEXT for Physics

UNITEXT for Physics series, formerly UNITEXT Collana di Fisica e Astronomia, publishes textbooks and monographs in Physics and Astronomy, mainly in English language, characterized of a didactic style and comprehensiveness. The books published in UNITEXT for Physics series are addressed to upper undergraduate and graduate students, but also to scientists and researchers as important resources for their education, knowledge and teaching.

More information about this series at http://www.springer.com/series/13351

Wladimir-Georges Boskoff ·
Salvatore Capozziello

A Mathematical Journey to Relativity

Deriving Special and General Relativity
with Basic Mathematics

 Springer

Wladimir-Georges Boskoff
Mathematics and Informatics
Ovidius University
Constanţa, Romania

Salvatore Capozziello
Complesso Universitario Monte
Sant'Angelo
University of Naples Federico II
Naples, Italy

ISSN 2198-7882   ISSN 2198-7890   (electronic)
UNITEXT for Physics
ISBN 978-3-030-47896-4   ISBN 978-3-030-47894-0   (eBook)
https://doi.org/10.1007/978-3-030-47894-0

This Springer imprint is published by the registered company Springer Nature Switzerland AG
The registered company address is: Gewerbestrasse 11, 6330 Cham, Switzerland

Wir müssen wissen, Wir werden wissen.

D. Hilbert

We are all in the gutter but some of us are looking at the stars.

O. Wilde

Dedicated to all those who do not like Mathematics with the hope that they will change their mind.

Preface I

This book is an approach to Special and General Relativity from a full mathematical point of view. When Physics is studied, there is the need for understanding its language, that is, Mathematics. Dirac's words describe very well what we want to do: *God used beautiful Mathematics in creating the world;* therefore, we present a part of this *divine plan*, the beautiful Mathematics of Special and General Relativity. We wrote a textbook which, we believe, can be easily used by students in Mathematics, Physics and Engineering studies; by teachers or by some other people who are interested in this subject. If someone already knows Mathematics, that is, both basic Geometry and Differential Geometry, this person can neglect the first six chapters. She/he can start from Gravity in Newtonian Mechanics. People who study Physics should start from the very beginning in order to understand the development of Geometry. The improvement of mathematical language, in more than two thousand and five hundred years, allowed to produce a common language for both Calculus and Linear Algebra; this approach ends up to a *dialect*, the Differential Geometry, which constitutes the basic tool of Relativity. Without the effort to understand the nature of the Non-Euclidean Geometry, the Differential Geometry could not occur. Without Differential Geometry, General Relativity could not exist. The first six chapters represent the adventure of Geometry from axioms until the Non-Euclidean Geometries through Differential Geometry. A lot of examples and solved exercises help the reader to understand the theory. Actually, the entire book, which is written in a unitary way, offers clear statements and proofs. About the proofs: it offers complete proofs; all computations are presented. In our opinion, this is the only way to understand the complicated computations which depend on the Differential Geometry language. Reading line by line, the reader can understand every single proof. The references which inspired us are mentioned not only at the beginning of each chapter but also in the text. Some proofs and some approaches of the theory are completely original. If somebody is reading from the beginning to the end of this book, it becomes understandable why each subject presented is important for the topic. We hope that our humble efforts are useful, first of all, for learning people to whom this book is mainly dedicated. We thank our colleagues; our teachers; our friends and, first of all, our students whose questions, discussions and remarks allowed us to enter the *perfect world* of Geometry towards its amazing realization which is General Relativity. We also want to thank Dr. Marina Forlizzi and the Springer staff for invaluable support in publishing this book.

As a final remark, we want to say that this book was conceived about two years ago during pleasant discussions on Mathematics and General Relativity in scientific congresses and meetings between the authors and was concluded during the severe period of the global Coronavirus disease. We hope that Science and its high values can be comforting even in difficult situations like the present one, as happened so many times in history.

Constanţa, Romania   Wladimir-Georges Boskoff
Naples, Italy   Salvatore Capozziello
March 2020

Preface II

What does a mathematical journey towards the general theory of Relativity look like? The authors propose an original itinerary moving from Euclidean and non-Euclidean Geometry created from axioms to models of geometric Euclidean and non-Euclidean worlds. Differential Geometry of surfaces and then abstract Differential Geometry are special stops for two reasons:

1. To understand non-Euclidean Geometry models from this point of view and
2. To create the language by which we can describe the General Relativity and its consequences.

The physical world allows both Euclidean and non-Euclidean descriptions. To have an image of this physical world, we need to continue the itinerary with supplementary stops: Newtonian and Lagrangian Mechanics, Special Relativity to reach, finally, General Relativity.

The content of the book is written to be self-contained. All the proofs are done with all the details presented for the reader. The problems are solved, or they have hints. Almost all the contents were presented to students at different university courses and, in our experience, they were well received.

In Chap. 1, we present, using a slightly modified Hilbert's axioms system, Euclidean and non-Euclidean geometries and what they mean. Here, the mathematical theory is built from a set of primary objects, which do not require definitions, together with a set of axioms. The collection of primary objects is chosen from the set theory. The axioms are stated in a formal form and the axiomatic theory is built as a collection of mathematically rigorous statements deduced from the axioms.

It exists a common part for Euclidean and Non-Euclidean Geometry, the so-called Absolute Geometry. Absolute Geometry consists in all the results that can be thought and proved using the axiomatic system before introducing a Parallelism Axiom. The main theorem in Absolute Geometry is the Legendre one, which states that the sum of measures of angles of a given triangle is less than or equal to two right angles.

The two consequences are the following:

1. The sum of angles in each triangle is equal to two right angles.
2. The sum of angles in each triangle is strictly less than two right angles.

A further axiom, the Parallelism Axiom, makes us to discover the Euclidean world, corresponding to the first case, i.e. the sum of angles is equal to two right angles.

The denial of the previous Parallelism Axiom leads us to the Non-Euclidean Geometry; here the sum of angles is strictly less than two right angles.

Euclidean Geometry and Non-Euclidean geometries are the frameworks which formulate Newtonian Mechanics and Relativity, respectively, as we will see later.

Chapter 2 highlights how the Euclidean Geometry, previously introduced, can be constructed and viewed using algebra and trigonometry. All happens in a two-dimensional vector space endowed with an inner product invariant with respect to the group of Euclidean rotations. Basic facts on Euclidean Geometry are presented, the most important being Pythagoras Theorem and the generalized Pythagoras theorems.

Even if it seems there is no connection between Minkowski Plane Geometry and the geometries created from the axiomatic point of view, we present in detail the Minkowski Plane Geometry. The construction is related to the same two-dimensional vector space used to describe Euclidean Geometry, but instead of the Euclidean inner product, we have a Minkowski product. There exists a group of hyperbolic rotations which leaves invariant the Minkowski product. Minkowski Geometry is not as simple as Euclidean Geometry. There are space-like vectors, null vectors and time-like vectors. Minkowski-Pythagoras Theorem has different statements with respect to the type of the involved side-vectors of the triangle. Even though in this chapter we construct an algebraic image of the Euclidean Geometry, we have not yet constructed images of the Non-Euclidean Geometry. The next chapters deal with this issue.

Chapter 3 is dedicated to the tools we need to construct the first model of Non-Euclidean Geometry. This model is constructed in the interior disk of a given circle.

To construct, we need to understand the geometric inversion and basic facts about Projective Geometry. A projective invariant of a special projective map, attached to the previously given circle, allows us to construct a distance inside the disk of the circle. The Poincaré disk model is highlighted.

The "lines", that is, the geodesic lines of this distance, are orthogonal arcs of circles to the given circle. It is easy to see that there are more than two non-intersecting "lines" through a given point with respect to a given "line". The sum of angles of a triangle in this Poincaré model is of course less than π.

Chapter 4 is related to the Differential Geometry of surfaces. In the first part, the surfaces are seen as subsets of a three-dimensional Euclidean space. In this context, we understand how the Euclidean inner product of the Euclidean space induces a way to measure lengths and angles for vectors belonging to the tangent planes of the surface. We can also measure length of curves that belong to surfaces, areas of regions, and all these using the metric attached to the surface.

The Differential Geometry of a surface continues by introducing a fundamental notion: the Gaussian Curvature of a surface at a point. If, at the beginning, the Gaussian Curvature seems to be dependent on the fact that the surface is seen in the Euclidean ambient space, after we prove Gauss' formulas, we step into the intrinsic theory of surfaces. There, Gauss' equations and *Theorema Egregium* offer another perspective: each surface can be seen as a piece of a plane endowed with a metric, and this metric only determines the curvature.

The study continues in Minkowski 3-spaces where we have to take care of the Minkowski-type nature of the normal vector to the surface. However, we have almost the same picture; the Minkowski product determines a non-Euclidean metric of a surface which allows us to conclude about the intrinsic Geometry of it. Therefore, in both cases the surface becomes irrelevant for our study. In fact, we study the Geometry of a metric and obtain relevant geometric aspects about the piece of plane endowed with that metric.

The covariant derivative introduced in the last part of the chapter allows us to define the parallel transport and the geodesics of surfaces. At the end of the chapter, we introduce a short story about a person embedded in a surface with the aim to reveal how the person can develop a theory about his universe, which is the surface where he lives. The study is continued in the next chapter, where we better understand the nature of geometric objects which appear in Differential Geometry.

Chapter 5 is fundamental for the book: the final image about the Non-Euclidean Geometries cannot be given without what we learn here. Basic Differential Geometry is about Differential Geometry when an extra-dimension does not exist. In the previous chapter, we claimed that we did not know yet the mathematical nature of the multi-index quantities which appear in the Differential Geometry of surfaces. In this chapter, we prove the tensor character of the metric coefficients, of the Riemann symbols, of the Ricci symbols and also for the geodesics equations. All these multi-index quantities remain invariant when we deal with a change of coordinates. Why this is important? The substance of General Relativity is related to these changes of coordinates. A change of coordinates may reflect an acceleration field which is equivalent to a gravitational field and, in the context described, a nice example developed later is about the constant gravitational field.

The covariant derivative for contravariant vectors, which appears as a geometric property, allows us to think of a general definition for the covariant derivative of tensors. How we parallel transport vectors along curves, how geodesic lines appear, and some other important properties of parallel transport of vectors along geodesic lines are also studied. At the end of the chapter, the covariant derivation of Einstein's tensor allows the reader to have the first image on Einstein's field equations.

Chapter 6 is devoted to Non-Euclidean Geometry models and their physical interpretation. It is worth stressing that we are dealing with *models* and not only with *a model*. In fact, we imagine some other models of Non-Euclidean Geometry.

There are some steps before to provide these models. Differential Geometry gives us the possibility to see:

- how distances of the models produce the metrics;
- how the geodesics of the distances are also the geodesics of the metrics; and
- how the models are related among themselves through their metrics according to convenient changes of coordinates.

It is important to stress that all these models are equivalent and contribute to the big picture. Specifically, the Poincaré disk model, the Poincaré half-plane model, the exterior disk model, the hemisphere model and the hyperboloid models are studied and presented.

The first three models are connected among themselves by geometric inversions. The remaining models need appropriate changes of coordinates to connect to the first three. In particular, the hyperboloid model is described by a Minkowski metric. At this point, we have the first connection between Non-Euclidean Geometry and Minkowski Geometry. Next, in the Eighth Chapter devoted to Special Relativity, a Minkowski-type metric appears giving a geometric image of the (so-called) *physical reality* [1].

The physical example, developed at the end of the chapter, is due to Poincaré. The question is: can we develop the Poincaré disk model starting from simple physical rules? The answer is yes and this is possible combining Physics and Geometry. Even if Poincaré developed the model stating that reality cannot be fully understood, after this example, it is easy to accept the fact that the Geometry is related to the Relativity description. In fact Non-Euclidean Geometry, seen through Differential Geometry, is needed to understand basic facts of General Relativity, as we see later.

Euclidean Geometry constitutes the framework of the Newtonian Mechanics. Chapter 7 is dedicated to understand how forces can explain what is happening in our surrounding world modeled into a three-dimensional Euclidean Space where only one clock gives the universal time. In this sense, the Newtonian Mechanics reveals an absolute space and an absolute time [2].

It is described as the gravitational force together with the gravitational field. A mathematical artifact, the gravitational potential, is involved in two fundamental results: the vacuum field equation and the gravitational field equation. Looking at these equations and how difficult we mathematically obtained them, somebody can think that this is the maximum we can say about the gravitational field. But the tidal forces and the tidal acceleration equations offer another perspective. The vacuum field equation is encapsulated in the trace of the Hessian matrix involved in tidal acceleration equations.

If we try to obtain the geometric equivalent of these equations in a curved space, that is, if we cancel out the Euclidean three-dimensional space, the Hessian matrix is replaced by a curvature-dependent tensor whose trace is the Ricci symbol. In the future, we prove via Fermi coordinates that this is a possible way to obtain

Einstein's vacuum field equations. This is the first geometric change of the Euclidean frame when one studies forces.

This situation may arouse another important change of perspective.

Suppose we have a force and the trajectories of a point subjected to this force. Is it possible to locally find a metric whose geodesics are the previous trajectories? The answer is yes. The Euler-Lagrange equations become the equations of the previous trajectories and the same Euler-Lagrange equations are the geodesic equations of a metric induced by another *mathematical artifact* called Lagrangian.

The study of the Lagrangian, starting from the mechanical one, is made through a function called action. If the first-order variation of the action vanishes, we obtain the Euler-Lagrange equations. Later, we prove how another action, the Hilbert action, allows us to derive the Einstein field equations in General Relativity.

Kepler's laws are also studied in this chapter with the aim to prepare the reader's understanding about planet trajectories in a metric, of course, later, in the part related to General Relativity.

Chapter 8 is devoted to Special Relativity. Reflection and refraction of light were explained in a satisfactory way by Newton who looks at light rays as trajectories of particles (after called photons).

In the middle of *XIXth* century, James Clark Maxwell offered another view: the light is an electromagnetic wave and it satisfies four equations, known as Maxwell's equations of the Electromagnetism. If we try to put them in accordance with Newton's theory, it appears the necessity of considering a medium in which the electromagnetic waves travel through space. It was called by physicists of that time the *ether*.

Ernst Mach did not agree with the idea of ether and observed the necessity of the revision of all fundamental concepts of Physics [2].

Michelson-Morley experiment, which initially was designed to reveal the ether, had a result completely different with respect to the expectations. Albert Einstein explained the result of the experiment in a theory where he revised in a fundamental way the ideas of space and time, and no place for ether remained. The absolute space and the absolute time of the Newtonian mechanics were replaced by the specific space and the specific time of each observer. Different observers mean different inertial frames of coordinates; each one having its time axis and its space axes [3].

Einstein formulated the Special Relativity starting from two main postulates:

1. The laws of Physics has to be the same in all inertial reference frames.
2. The speed of light in vacuum, denoted by c, is the same for all the observers and it is the maximum speed reached by a moving object.

Presenting the theory, we preferred to balance it starting from two important works, the book by Callahan [4] and a paper by Varićak [5], where we found the most possible geometric approach to Special Relativity.

Therefore we have adapted, in a new form, the basic ideas discussed there. The first idea is related to the consideration of two inertial frames of coordinates, one moving at a constant speed, another considered at rest, in which the two observers have to agree on the same laws of Physics [4]. In this way, the old Galilean transformations of coordinates are replaced by the Lorentz transformations. There are a lot of consequences: another formula for velocity addition, the time dilation, the length contraction, the covariance of Maxwell's law under Lorentz transformation, the rest-energy formula, the Doppler effect and so on. The second one is related to the geometric understanding of these facts: we have a sort of equivalence between the so-called "geometric coordinates" and the "physical coordinates" [5]. The entity called "physical spacetime" is understood through the geometric spacetime where the results are easier to be viewed. This idea can be originally seen in [3].

Then, when we introduce a constant gravitational field via the accelerated frames (see also [4] point of view), we can prove the bending of the light-rays; the interpretation of the Doppler gravitational effect shows that accelerated frames are not inertial ones. Further, a contradiction between the Minkowski flat spacetime of Special Relativity and the gravitational Doppler effect occurs. A physical theory containing the old Mechanics, including gravity, and the modern electromagnetic waves theory needs to integrate the accelerated frames. In this way, we can add another argument towards the General Relativity.

Chapter 9 is devoted to General Relativity and Relativistic Cosmology. There is no other better description of the subject than the sentence by John Archibald Wheeler: *Spacetime tells matter how to move; matter tells spacetime how to curve* [6, 34]. How the space is curved appears from the Einstein field equations

$$R_{ij} - \frac{1}{2}g_{ij}R = \frac{8\pi G}{c^4}T_{ij}.$$

In the left-hand side, we have "Geometry," a metric g_{ij} and its derivatives are involved; in the right-hand side, we have a tensor depending on matter, the so-called energy-momentum tensor. The energy-momentum tensor establishes the metric; the metric produces geodesics described by the equations

$$\frac{d^2x^r}{dt^2} + \Gamma^r_{pq}\frac{dx^p}{dt}\frac{dx^q}{dt} = 0.$$

They are trajectories of objects moving according to the Geometry of spacetime. Therefore, the geodesic equations are the equivalent equations of curves which satisfy $\overrightarrow{F} = m\overrightarrow{a}$ from Mechanics.

The equations of geodesics of an initial metric switch accordingly a change of coordinates into the equations of the geodesics of the newly obtained metric. Changes of coordinates may provide a new state of a given frame; therefore, the new state is described by a new metric provided by the old state metric via the coordinates change. The reader will understand how it works looking at the case of the constant gravitational field.

The chapter starts looking at the differences between the classical Newtonian Mechanics and Einstein's landscape of gravity described by Geometry. Einstein's field equations can be derived from the Hilbert action. A generalization of such an action, the so-called $f(R)$ gravity is also presented. The straightforward solution, in the case of vacuum field equations for spherical symmetry, is Schwarzschild's one. We present computations related to the orbits of planets and the bending of light rays.

Fermi's viewpoint on Einstein's vacuum field equations allows to obtain the classical counterparts of the relativistic equations in the case of weak gravitational field.

After, we analyze the Einstein static universe and the basic considerations on the cosmological constant, as a part of the classical approach to the General Relativity.

A "metric for the Universe" is obtained for the cosmic expansion. It is the Friedmann-Lemaître-Robertson-Walker metric. The way we obtain it is related to the energy-momentum tensor and the Cosmological Principle.

Black holes are an important prediction of General Relativity. We propose an introduction to their theory starting from the Rindler metric. The singularities which can be removed using Rindler's idea are geometric only. Schwarzschild metric is important in the study of black holes. The anomalies of black holes are explained via Kruskal-Szekeres coordinates, and light cones inside and outside black holes are presented.

Another important prediction, the existence of gravitational waves, is discussed in this chapter to give a more complete picture about the physical landscape of General Relativity. Furthermore, cosmic strings are presented as a hypothetical structure considered, until now, only as a possible solution of Einstein's field equations. Another important solution of Einstein's field equations is Gödel, which succeeded to prove that a homogeneous universe without a global time coordinate can theoretically exist. We present the above solutions with all the details necessary to be easily understood at undergraduate student level.

In Chap. 10, as a full geometric realization of Relativity, we present the so-called Affine Universe and the de Sitter spacetime. From a cosmological point of view, this solution is fundamental to discuss, at an elementary level, the problems of primordial inflation and the late accelerated behavior, often dubbed as dark energy. Starting from two different parameterizations, it is possible to describe the cosmological constant, the main ingredient of de Sitter solution. Essentially, it is possible to show that a curved universe can be achieved without a mass distribution. A possible explanation can be obtained starting from a Minkowski spacetime where gravitational field (without masses) is considered. In this sense, this is a full geometric realization of Relativity.

Constanţa, Romania   Wladimir-Georges Boskoff
Naples, Italy   Salvatore Capozziello

Contents

1 Euclidean and Non-Euclidean Geometries: How They Appear 1
  1.1   Absolute Geometry 2
  1.2   From Absolute Geometry to Euclidean Geometry
        Through Euclidean Parallelism Axiom 24
  1.3   From Absolute Geometry to Non-Euclidean Geometry
        Through Non-Euclidean Parallelism Axiom 26

2 Basic Facts in Euclidean and Minkowski Plane Geometry 29
  2.1   Pythagoras Theorems in Euclidean Plane 30
  2.2   Space-Like, Time-Like and Null Vectors in Minkowski
        Plane 35
  2.3   Minkowski-Pythagoras Theorems 37

**3 Geometric Inversion, Cross Ratio, Projective Geometry
    and Poincaré Disk Model** 39
  3.1   Geometric Inversion and Its Properties 39
  3.2   Cross Ratio and Projective Geometry 45
  3.3   Poincaré Disk Model 59

4 Surfaces in 3D-Spaces 65
  4.1   Geometry of Surfaces in a 3D-Euclidean Space 65
  4.2   Intrinsic Geometry of Surfaces 80
  4.3   Geometry of Surfaces in a 3D-Minkowski Space 94
  4.4   A Short Story of a Person Embedded in a Surface 104

5 Basic Differential Geometry 109
  5.1   Covariant and Contravariant Vectors and Tensors.
        The Christoffel Symbols 109
  5.2   Covariant Derivative for Vectors and Tensors. Geodesics 118
  5.3   Riemann Mixed, Riemann Curvature Covariant, Ricci
        and Einstein Tensors 123

6   Non-Euclidean Geometries and Their Physical Interpretation 139
   6.1   Poincaré Distance and Poincaré Metric of the Disk 142
   6.2   Poincaré Distance and Poincaré Metric of the Half-Plane 147
   6.3   Connections Between the Models of Non-Euclidean
       Geometries 149
   6.4   The Exterior Disk Model 155
   6.5   A Hemisphere Model for the Non-Euclidean Geometry 159
   6.6   A Minkowski Model for the Non-Euclidean Geometry:
       The Hyperboloid Model 161
   6.7   The Theoretical Minimum About Non-Euclidean Geometry
       Models. A Possible Shortcut 163
   6.8   A Physical Interpretation 166

7   Gravity in Newtonian Mechanics 169
   7.1   Gravity. The Vacuum Field Equation 171
   7.2   Divergence of a Vector Field in an Euclidean 3D-Space 176
   7.3   Covariant Divergence 178
   7.4   The General Newtonian Gravitational Field Equations 179
   7.5   Tidal Acceleration Equations 183
   7.6   Geometric Separation of Geodesics 186
   7.7   Kepler's Laws 189
   7.8   Circular Motion, Centripetal Force and Dark Matter
       Problem 196
   7.9   The Mechanical Lagrangian 199
   7.10 Geometry Induced by a Lagrangian 203

8   Special Relativity 217
   8.1   Principles of Special Relativity 218
   8.2   Lorentz Transformations in Geometric Coordinates
       and Consequences 225
      8.2.1   The Relativity of Simultaneity 225
      8.2.2   The Lorentz Transformations in Geometric
           Coordinates 226
      8.2.3   The Minkowski Geometry of Inertial Frames in
           Geometric Coordinates and Consequences: Time
           Dilation and Length Contraction 228
      8.2.4   Relativistic Mass, Rest Mass and Energy 230
   8.3   Consequences of Lorentz Physical Transformations: Time
       Dilation, Length Contraction, Relativistic Mass and Rest
       Energy 233
      8.3.1   The Minkowski Geometry of Inertial Frames in
           Physical Coordinates and Consequences: Time
           Dilation and Length Contraction 233
      8.3.2   Relativistic Mass, Rest Mass and Rest Energy in
           Physical Coordinates 236

8.4	Maxwell's Equations	238
8.5	Doppler's Effect in Special Relativity	244
8.6	Gravity in Special Relativity: The Case of the Constant Gravitational Field	246
	8.6.1 Doppler's Effect in Constant Gravitational Field and Consequences	251
	8.6.2 Bending of Light-Rays in a Constant Gravitational Field	254
	8.6.3 The Basic Incompatibility Between Gravity and Special Relativity	255

9   General Relativity and Relativistic Cosmology 257

9.1	What Is a Good Theory of Gravity?	259
	9.1.1 Metric or Connections?	261
	9.1.2 The Role of Equivalence Principle	261
9.2	Gravity Seen Through Geometry in General Relativity	264
	9.2.1 Einstein's Landscape for the Constant Gravitational Field	267
9.3	The Einstein–Hilbert Action and the Einstein Field Equations	273
9.4	A Short Introduction to $f(R)$ Gravity	279
9.5	The Energy-Momentum Tensor and Another Proof for Einstein's Field Equations	282
	9.5.1 The Covariant Derivative of the Energy-Momentum Tensor	283
	9.5.2 Another Proof for Einstein's Field Equations	285
9.6	Introducing the Cosmological Constant	286
9.7	The Schwarzschild Solution of Vacuum Field Equations	288
	9.7.1 Orbit of a Planet in the Schwarzschild Metric	290
	9.7.2 Relativistic Solution of the Mercury Perihelion Drift Problem	293
	9.7.3 Speed of Light in a Given Metric	297
	9.7.4 Bending of Light in Schwarzschild Metric	299
9.8	About Einstein's Metric: Einstein's Computations Related to Perihelion's Drift and Bending of the Light Rays	304
9.9	Solutions of General Einstein's Field Equations: The Friedmann–Lemaître–Robertson–Walker Models of Universe	309
	9.9.1 The Cosmological Expansion	317
9.10	The Fermi Coordinates	324
	9.10.1 Determining the Fermi Coordinates	326
	9.10.2 The Fermi Viewpoint on Einstein's Field Equations in Vacuum	329
	9.10.3 The Gravitational Coupling in Einstein's Field Equations: $K = \frac{8\pi G}{c^4}$	332

9.11   Weak Gravitational Field and the Classical Counterparts
       of the Relativistic Equations 335
9.12   The Einstein Static Universe and its Cosmological Constant ... 343
9.13   Cosmic Strings 345
9.14   The Case of Planar Gravitational Waves 348
9.15   The Gödel Universe 350
9.16   Black Holes: A Mathematical Introduction 355
       9.16.1   Escape Velocity and Black Holes 355
       9.16.2   Rindler's Metric and Pseudo-Singularities 357
       9.16.3   Black Holes Studied Through the Schwarzschild
                Metric 359
       9.16.4   The Light Cone in the Schwarzschild Metric 364

10 **A Geometric Realization of Relativity: The Affine Universe
    and de Sitter Spacetime** 367
    10.1   About the Minkowski Geometric Gravitational Force 367
    10.2   The de Sitter Spacetime and Its Cosmological Constant 369
    10.3   Some Physical Considerations 376
    10.4   A FLRW Metric for de Sitter Spacetime Given by the Flat
           Slicing Coordinates Attached to the Affine Sphere 379
    10.5   Deriving Cosmological Singularities in the Context of de Sitter
           Spacetime....................................... 384

11 **Conclusions** 387

References 389

Index 393

Chapter 1
Euclidean and Non-Euclidean Geometries: How They Appear

Omnibus ex nihilo ducendis sufficit unum.

G. W. von Leibniz

We intend to construct these geometries using a slightly modified Hilbert's axioms system in the same way as it is done in [7–10]. An interesting thing is related to the fact that it exists a common part for Euclidean and Non-Euclidean Geometry, the so called Absolute Geometry. Roughly speaking, the Absolute Geometry consists in all theorems that can be thought and proved using the axiomatic system before introducing a parallelism axiom.

In our vision, the most important theorem in Absolute Geometry is the Legendre one:

"The sum of angles of a triangle is less than or equal two right angles."

It allows us to prove that only two situations hold:

"The sum of angles in each triangle is equal to two right angles."

or, the other situation:

"The sum of angles in each triangle is strictly less than two right angles"

Choosing an appropriate parallelism axiom we discover the Euclidean world, corresponding to the first case, i.e. the sum of angles is equal to two right angles. The denial of the previous parallelism axiom leads us to the Non-Euclidean Geometry; here the sum of angles is strictly less than two right angles. We have used few figures to illustrate these concepts, because, the reader can remain with a false image about how lines look like. However, in Absolute Geometry the reader can think and draw images as in the Euclidean Geometry, because all the objects and all the theorems valid in Absolute Geometry are also valid in Euclidean Geometry. Here the lines are the ordinary straight lines of the plane. The images can be thought in a more complicated way if someone try to imagine them in a model for the Non-Euclidean

© The Editor(s) (if applicable) and The Author(s), under exclusive license
to Springer Nature Switzerland AG 2020
W. Boskoff and S. Capozziello, *A Mathematical Journey to Relativity*,
UNITEXT for Physics, https://doi.org/10.1007/978-3-030-47894-0_1

Geometry, because lines can look like arcs of circles, segments, etc. All the proofs are reported in such a way that they can be understood by reading them directly.

1.1 Absolute Geometry

The key idea of the axiomatic method is to build a theory from a set of primary objects, which do not require definitions, together with a set of axioms. Therefore the theory is built as a collection of mathematically rigorous statements deduced from the axioms.

The collection of primary objects of the Geometry are the following, inherited from set theory. The objects of the first collection are called *points*, and they are denoted by capital letters $A, B, C, \ldots$ The second collection contains *lines*, denoted by $l, l', \ldots$ The third collection contains *planes*, denoted by Greek letters $\alpha, \beta, \gamma, \ldots$ Finally, the last collection contains only one element called *space,* denoted by S.

The first important group of axioms is related to the incidence of the objects described above. They describe who belongs to who, which set of objects can be included in which, how many objects are necessary to create another object, etc. Let us introduce the so called "axioms of incidence".

The first axiom which helps us to construct a Geometry establishes the existence and uniqueness of a line and its connection with to two given distinct points.

Axiom $\mathbf{I}_1$: *For any two distinct points A and B there exists a unique line l which is incident with both A and B, i.e. $A \in l$ and $B \in l$.*

The unique line l of the previous axiom is often denoted by AB, indicating that it is the line that passes through the points A and B.

Axiom $\mathbf{I}_2$: *There exist at least two distinct points on any line. Moreover, there exist at least three distinct points which are not on the same line.*

In view of the axiom, it seems useful to be able to distinguish between points which are on a line from points which do not belong to the same line, therefore we introduce the following notion.

Definition 1.1.1 Any number of points are called *collinear* if there is a line which is incident to all of them. Otherwise, they are called *non-collinear*.

For example, axiom $\mathbf{I}_1$ asserts that every two distinct points are collinear, and axiom $\mathbf{I}_2$ guarantees the existence of at least three non-collinear points in the Geometry we are constructing. The next two axioms establish the relationship between points and planes.

Axiom $\mathbf{I}_3$: *For any three arbitrary non-collinear distinct points A, B and C, there exists an unique plane α which contains A, B and C.*

In general, such a plane is denoted by $\alpha := (ABC)$.

The following axiom establishes the relationship among points on a given line and a plane containing that line. This axiom plays a crucial role once we construct geometries with more number of points and lines.

Axiom $\mathbf{I_4}$: If two points A and B, which determine the line l, lie in the plane α, then every point of the line l lies in the plane α.

In this case, we write $l \subset \alpha$ (regarded as a subset of points). The following axiom states that the minimum number of points in an intersection of two planes is two.

Axiom $\mathbf{I_5}$: If two planes α and β have a common point A, then they have another common point B distinct from A.

An immediate consequence of axioms $\mathbf{I_4}$ and $\mathbf{I_5}$ is that if the planes α and β contain the two distinct points A and B, then they contain the whole line $l = AB$, and we write $\alpha \cap \beta = \{l\}$, again as an equality of sets of points.

The last axiom of incidence states the minimum number of points necessary to create the space.

Axiom $\mathbf{I_6}$: There exist at least four points which do not belong to the same plane.

In the view of this last axiom $\mathbf{I_6}$, we give the following.

Definition 1.1.2 Any number of points are called *coplanar* if there is a plane which passes through all of them. Otherwise, they are called *non-coplanar*.

Axioms $\mathbf{I_1}$–$\mathbf{I_6}$ give rise to a simple model of a space created only with 4 points, 6 lines and 4 planes.

The model described above can be written as follows. The distinct points are A, B, C, D, and the six lines are given by the following sets of points: $l_{AB} = \{A, B\}$, $l_{AC} = \{A, C\}, l_{BC} = \{B, C\}, l_{BD} = \{B, D\}, l_{CD} = \{C, D\}$, and $l_{AD} = \{A, D\}$. The four planes are $(ABC) = \{A, B, C\}, (ABD) = \{A, B, D\}, (ACD) = \{A, C, D\}$, $(BCD) = \{B, C, D\}$, and the space built by Axiom $\mathbf{I_6}$ is by definition $(ABCD)$.

We study below some immediate consequences of the group of six axioms of incidence. Notice that the results we prove below make sense even when applied to the simple model described above.

Theorem 1.1.3 *Two distinct lines have at most one common point.*

Proof Let l_1, l_2 two distinct lines. We distinguish the following two cases. If $l_1 \cap l_2 = \emptyset$, then they have no point in common, therefore the conclusion of the theorem is true.

If $l_1 \cap l_2 \neq \emptyset$, then let $A \in l_1 \cap l_2$ a point in their intersection. We assume, by contradiction, that there is another point $B \in l_1 \cap l_2$, $B \neq A$. In particular, $A, B \in l_1$, therefore $l_1 = AB$ (axiom $\mathbf{I_1}$). Similarly, $A, B \in l_2$, therefore $l_2 = AB$. Axiom $\mathbf{I_1}$ says then that $AB = l_1 = l_2$, in contradiction with the hypothesis that $l_1 \neq l_2$. Therefore, our assumption on the existence of a different point $B \in l_1 \cap l_2$ is false. In conclusion, A is the only common point of the two lines l_1 and l_2. □

The previous theorem motivates the following

Definition 1.1.4 Two distinct lines that intersect in exactly one point are called *secant* lines.

The "Axioms of Order" deal with the undefined yet relation of *betweenness*, i.e. of a point lying between two other points. Once the axioms of order appear, the previous very simple model of Geometry fail to exists. The axioms of order are formulated as follows.

Axiom $\mathbf{O}_1$: *If a point B is between A and C, then A, B, C are three distinct collinear points on a line l, and B is between C and A.*

Imagine the line as a circle. The previous axiom tells us that such an image is not possible. The line l has no predefined "orientation". The only correct concept of order among points is defined to be "between".

Axiom $\mathbf{O}_2$: *For every pair of distinct points A and B, there is at least another distinct point C such that B is between A and C.*

An immediate consequence of axiom $\mathbf{O}_2$, combined with axiom $\mathbf{I}_2$, is that a line contains at least three points. The axiom can be applied again to the pair $\{A, C\}$, so there exists another point D such that C is between A and D, etc.

Axiom $\mathbf{O}_3$: *Given three arbitrary points on a line, at most one of them is between the other two.*

Notice that the axiom $\mathbf{O}_2$ does not guarantee the existence of a point B between two given ones A and C. This will be proven below. Nevertheless, if we assume that there exists B between A and C, then the axiom $\mathbf{O}_3$ guarantees that A cannot be between B and C, and C cannot be between A and B. Another theorem will clarify the situation of three given points on a line.

Axiom $\mathbf{O}_4$*(Pasch): Let A, B, C be three non-collinear points, and l a line situated in the plane* (ABC) *which does not pass through any of the points A, B, C. If the line l contains a point which is between A and B, then the line l contains either a point between A and C or a point between B and C.*

We denote by $\overline{ABC}$ when B is on the line AC and B is between A and C, and we will refer to it as the *order* ABC. Note that by axiom $\mathbf{O}_1$, the order $\overline{ABC}$ is the same as the order $\overline{CBA}$.

An immediate consequence of the axioms of order is the following

Theorem 1.1.5 *Given two points A and B on a line l, there is a point $M \in l$ such that we have the order* $\overline{AMB}$.

Proof There exists a point C not on the line AB (axiom $\mathbf{I}_2$). Then there exists a point D such that we have the order $\overline{ACD}$ (axiom $\mathbf{O}_2$).

Similarly, there exists the point E with respect to the order $\overline{DBE}$ (axiom $\mathbf{O}_2$). Then we apply axiom $\mathbf{O}_4$ for the points C, D, E and the line AB, so there exists a point M on the line AB such that we have order $\overline{AMB}$. $\qquad\qquad$ □

The previous theorem suggests the following

Definition 1.1.6 The set of points M on the line AB with the property that M is between A and B is called a *segment*, and it is denoted by $[AB]$.

Formally we can write

$$[AB] = \{M \in AB \mid \overline{AMB}\} \cup \{A, B\}$$

The *interior* of the segment $[AB]$ is defined to be the set $[AB] - \{A, B\}$.

Note that the segment $[AB]$, seen as a set of points, is equal to the segment $[BA]$. Moreover, the order $\overline{AMB}$ is equivalent to $M \in [AB] - \{A, B\}$, so the previous theorem can be reformulated as follows: *the interior of every segment is non-empty.* We have also $[AA] = \{A\}$. Moreover, we can define now one of the most important object of any Geometry: the triangle.

Definition 1.1.7 A configuration of three distinct non-collinear points A, B, C is called a *triangle*, and it is denoted by $\triangle ABC$. Moreover, the points A, B, C are called the *vertexes* of the triangle, and the segments determined by each pair of two vertexes are called the *sides* of the triangle.

The next theorem guarantees the existence and uniqueness of ordering for three collinear points.

Theorem 1.1.8 *Let A, B, C three points on a line l. Then one and only one of the orders $\overline{ABC}, \overline{ACB}$ or $\overline{BAC}$ occurs.*

Proof We assume that we have neither the order $\overline{ACB}$, nor the order $\overline{BAC}$, and we prove that we must have the order $\overline{ABC}$. In our Euclidean intuition, we will prove that if B is not "to the left" of A and not "to the right" of C, then it must be between A and C.

There exists a point $D \notin AC$ (axiom $\mathbf{I}_2$). Then there exists a point $E \in DB$ with the order $\overline{EDB}$ (axiom $\mathbf{O}_2$). Looking at the triangle $\triangle BEC$ and the secant line AD, then there is a point F at the intersection of AD and EC, such that we have the order $\overline{EFC}$ (axiom $\mathbf{O}_4$). In the same way, there exists the point $\{G\} = CD \cap AE$, such that we have the order $\overline{AGE}$. The line CG is a secant line for the triangle $\triangle AEF$, as we have the order $\overline{ADF}$. Moreover, considering the triangle $\triangle AFC$ and the secant line DE, it follows the order $\overline{ABC}$.   □

The following theorems establish incidence relations between a line and a triangle. Historically they are attributed to Moritz Pasch, whose influential works have been one century ago in the center of attention of many authors interested in foundations of Geometry.

Theorem 1.1.9 *If a line l does not intersect two sides of a triangle $\triangle ABC$, then it cannot intersect the third one, either.*

Proof Without loss of generality, we can assume l does not intersect neither $[AC]$ nor $[BC]$. By contradiction, we assume l intersects $[AB]$, so l contains a point between A and B. Then the axiom $\mathbf{O}_4$ affirms that l must contain either a point between A and C, or a point between B and C, in contradiction with the hypothesis. □

Theorem 1.1.10 *If a line l intersects two sides of a triangle $\triangle ABC$, then it cannot intersect the third one.*

Proof Let us assume, by contradiction, that the line l intersects all sides $[BC]$, $[AC]$, and $[AB]$ of the triangle $\triangle ABC$ in respectively D, E, and F. We can assume the order $\overline{EFD}$ on the line l. We consider the triangle $\triangle CDE$ and the secant line AB, which intersects $[DE]$ in F. It follows that AB intersects either $[DC]$ or $[EC]$ (axiom $\mathbf{O}_4$). In both cases, it follows that AB intersects either $[AC]$ or $[BC]$, respectively, in two points, which means that either $AB = BC$ or $AB = AC$ (axiom $\mathbf{I}_1$, in contradiction with the assumption that $\triangle ABC$ is a triangle. □

In what follows, we introduce the notion of *half-line*. Let O be a fixed point on a line l and let A, $B \in l$ be two points such that we have the order $\overline{OAB}$. Then we call A and B to be on the same side of the point O. This defines a binary relation on the set of points of l.

Theorem 1.1.11 *The binary relation defined above is an equivalence relation on the set of points of a line l.*

Proof Reflexivity is obviously true, as for $A = B$, we have clearly the order $\overline{OAA}$. The symmetry follows from the fact that the order $\overline{OAB}$ is the same as the order $\overline{BAO}$ (axiom $\mathbf{O}_1$). For the transitivity, we observe: if we have $\overline{OAB}$ and $\overline{OBC}$, then it follows the order $\overline{OAC}$. □

In this context, we can define a half-line as follows.

Definition 1.1.12 The equivalence class of a point on a line l with respect to a fixed point $O \in l$ is called the *half-line* with vertex (origin) O.

An equivalent formulation would be as follows: given a pair of points A and B, the half-line starting at A and pointing in the direction of B consists of all points P so that we have either the order $\overline{ABP}$, or the order $\overline{APB}$. A half-line AB is often called a *ray* emanated from A towards B.

Theorem 1.1.13 *Let O and A be two points on a line l. The set of points $A' \in l$ such that we have the order $\overline{A'OA}$ forms a half-line with origin O.*

Proof Let A' be an arbitrary point such that $\overline{A'OA}$. Let B be a representative of the equivalence class defined by A with respect to O, i.e. A and B are on the same side of O. Thus we have the order $\overline{OAB}$. Let $B' \in l$ such that we have the order $\overline{B'OB}$. From the orders $\overline{BAO}$ and $\overline{B'OB}$ it follows the order $\overline{AOB'}$. But the orders $\overline{A'OA}$ and $\overline{B'OA}$ exclude the order $\overline{A'OB'}$ (try to prove this assertion). Therefore the points A' and B' are on the same side of O, which proves the conclusion of the theorem. □

The theorem above affirms that a point O on a line l divides the line in two half-lines. For any point $A \neq O$, we denote one half-line by $(OA$, and the other half-line by $(OA'$, also called the *complementary* half-line of $(OA$.

The set of points of a half-line is a total ordered set. Indeed, for two points A and B on a half-line, we have either A coincides with B, of we have one of the orders $\overline{OAB}$ or $\overline{OBA}$. If we have the order $\overline{OAB}$, we say A *precedes* B. Therefore, in view of this total ordering, for any two distinct points A and B on a half-line, either A precedes B or B precedes A.

In view of this remark, we can arrange any finite set of points on a line l in the order of their precedence. Moreover, if we denote the ordered points by $A_1, A_2, \ldots$, then for any $i < j < k$ we have the order $\overline{A_i A_j A_k}$. This proves the following:

Theorem 1.1.14 *There is an order preserving, one-to-one correspondence between any set of n points on a line l and the set of natural numbers $\{1, 2, \ldots, n\}$.*

Similarly as in the case of half-lines, one can introduce the following binary relation of the set of points in a plane.

Definition 1.1.15 If l is a line in a plane π and A, B are two points in π such that $[AB] \cap l = \emptyset$, then we say that the points A and B are *on the same side* of the plane π with respect to the line l.

This defines a binary relation on the set of points of the plane π.

As before, we prove the following:

Theorem 1.1.16 *The binary relation defined above is an equivalence relation.*

Proof Reflexivity and symmetry are obviously true. We have to prove the transitivity of this relation. Let A, B and B, C on the same side of the plane π with respect to the line l. If follows that the intersections of l with $[AB]$, respectively $[BC]$, are empty. From a previous theorem it follows that $l \cap [AC] = \emptyset$, so the points A, C are on the same side of the plane with respect to the line l.   □

In view of the theorem above, we give the following:

Definition 1.1.17 Let l be a fixed line in a plane π and a point $A \in \pi - l$. The equivalence class of A with respect to the line l is defined to be the half-plane determined by A and l. The line l is called the *border* of this half-plane.

Then we have the following.

Let l be a fixed line, and let $A \notin l$. Then the set of points A' with the property that the segment $[AA']$ intersects the line l forms a half-plane of border l.

Definition 1.1.18 This half-plane is called the *complementary* half-plane of the half-plane determined by l and A.

Note that every line l in a plane, divides the plane in two half-planes, both with border l.

Definition 1.1.19 An *angle* is defined to be a pair of two half-lines h and k with the same origin O, denoted by $\angle(hk)$. The point O is called the *vertex* of the angle, and the half-lines h and k are called the *sides* of the angle.

If $h = (OA$ and $k = (OB$ are two half-lines defined by three non-collinear points O, A and B (O is the vertex of the angle), then we will also denote the angle $\angle(hk)$ by $\angle AOB$.

Let us consider an angle $\angle(hk)$ in a plane π. Then, there are two distinguished half-planes: one is determined by the underlying line of the half-line h and the points of the half-line k, and, similarly, the other one is determined by the underlying line of the half-line k and the points of the half-line h.

Definition 1.1.20 We call the *interior* of the angle $\angle(hk)$, the intersection of the two half-planes above. The *exterior* of the angle $\angle(hk)$ consists of all the points in the plane which are neither in the interior, nor on the sides of the angle $\angle(hk)$.

In a similar fashion, one can define the interior of a triangle as follows.

Definition 1.1.21 The *interior* of the triangle $\triangle ABC$ is the intersection of the interiors of its angles.

Consider n half-lines with common vertex O and assume that there exists a line $l \not\ni O$ which intersects all of them. We can order all the intersection points ($\overline{A_1A_2A_3}$, etc.). This gives us the notion of a half-line being *between* two other half-lines, and implicitly an order on the set of half-lines.

The following theorem is usually known as the crossbar theorem, or, sometimes, as the transversal theorem. In the present approach, the proof relies on axiom $\mathbf{O_4}$, Pasch's axiom.

Theorem 1.1.22 (Crossbar Theorem) *Let $\angle(hk)$ be an angle of vertex O. Let $A \in h$ and $B \in k$ two points different than O, and T a point in the interior of the angle $\angle(hk)$. Then the half-line $(OT$ intersects the segments $[AB]$ (Fig. 1.1).*

Proof Denote by H_A the half-plane determined by OB and the point A. Consider a point A' on the complementary half-line of $(OA$, and $H_{A'}$ the half-plane determined by OB and the point A'. We apply Pasch's axiom $\mathbf{O_4}$ for the triangle $\triangle AA'B$ and the half-line $(OT$, which intersects $[AA']$ in O. Then $(OT$ should intersect either $[AB]$ or $[A'B]$. If $(OT$ doesn't intersect $[AB]$, it exists a point $L \in [A'B] \cap (OT$, in collision with the fact that all points of $(OT$ are in $H_{A'}$.   □

As a final remark, we can observe that the complementary half-line of $(OT$, say $(OT'$ is included in the interior of the opposite angle of $\angle AOB$, say $\angle A'OB'$, therefore it cannot intersect neither $[A'B]$ nor $[AB]$, because they have empty intersection with the interior of $\angle A'OB'$.

Angles as $\angle AOT$ and $\angle TOB$ are called *adjacent angles*.

We introduce below the *axioms of congruence* and we study their immediate consequences. The congruence notion we introduce below is actually an equality

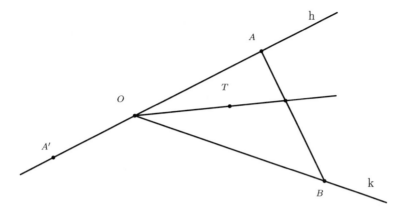

Fig. 1.1 Crossbar theorem

notion, but it is called different just to make distinction between equality of real numbers and equality of geometric objects. The relationship between the set of real numbers and Geometry is addressed later on.

The formulation of these axioms is after Arthur Rosenthal [10], which has considerably modified the original Hilbert's formulation of Axiom E_4, by omitting the symmetry and transitivity properties of the congruence of angles. These properties can be actually proven from the axioms below.

The following axioms introduce the concept of *congruence* (equality) of segments and angles. The notion of congruence is written using the special symbol $\equiv$, in order to eliminate any confusion between this geometric notion with the equality notion from set or number theories. We will reserve the equality symbol $=$ for when we define the *values* of segments and angles.

Axiom E_1 : If A and B are two points on a line l, and A′ is a point on a line l′, where l′ is not necessarily distinct from l, then there exists a point B′ on l′ such that $[AB] \equiv [A'B']$. *For every segment* $[AB] \equiv [BA]$.

As we can see from the previous axiom, the congruence $[AB] \equiv [A'B']$ is provided by the ability to construct the point B' on the line l' with the requested property.

Axiom E_2 : If $[A'B'] \equiv [AB]$ *and* $[A''B''] \equiv [AB]$, *then* $[A'B'] \equiv [A''B'']$.

Note that this axiom is not the transitivity property of congruence of segments. Transitivity will be proved in a theorem below. The next axiom establishes the additivity of the congruence of segments.

Axiom E_3 : Let $[AB]$ *and* $[BC]$ *be two segments of a line l, without common interior points, and let* $[A'B']$ *and* $[B'C']$ *be two segments without common interior points on a line l′, where l′ is not necessarily distinct from l. If* $[AB] \equiv [A'B']$ *and* $[BC] \equiv [B'C']$, *then* $[AC] \equiv [A'C']$.

The next axiom defines the congruence of angles in a plane.

Axiom E_4 : Let $\angle(hk)$ be an angle in a plane π, and let l' be a line in a plane π', where π' is not necessarily distinct from π. Let h' be a half-line of l', where h' is not necessarily distinct from h. Then in one of the half-planes determined by l', there uniquely exists a half-line k', such that $\angle(hk) \equiv \angle(h'k')$. For every angle, $\angle(hk) \equiv \angle(hk)$ (reflexivity), and $\angle(hk) \equiv \angle(kh)$ (symmetry).

As above, the congruence $\angle(hk) \equiv \angle(h'k')$ is provided by the ability to construct the angle $\angle(h'k')$ in one of the half-planes of π'.

Axiom E_5 : For any angles, if $\angle(h'k') \equiv \angle(hk)$ and $\angle(h"k") \equiv \angle(hk)$, then $\angle(h'k') \equiv \angle(h"k")$.

The next axiom is establishing conditions for congruences of angles of triangles. For an angle of a triangle $\triangle ABC$, say $\angle ABC$, we understand the angle determined by the half-lines $(BA$ and $(BC$.

Axiom E_6 : Let $\triangle ABC$ and $\triangle A'B'C'$ be two triangles. If $[AB] \equiv [A'B']$, $[AC] \equiv [A'C']$, and $\angle BAC \equiv \angle B'A'C'$, then:

$$\angle ABC \equiv \angle A'B'C' \qquad \angle ACB \equiv \angle A'C'B'.$$

The first two congruence axioms give the following result.

Theorem 1.1.23 *The congruence relation for segments is an equivalence relation.*

Proof We prove first the following statement: if we have two segments $[AB] \equiv [A'B']$, then $[AB] \equiv [B'A']$. Indeed, we have $[B'A'] \equiv [A'B']$ (axiom E_1). Therefore $[AB] \equiv [A'B']$ and $[B'A'] \equiv [A'B']$, so, using axiom E_2, it follows $[AB] \equiv [B'A']$.

Reflexivity now follows from axiom E_1 ($[AB] \equiv [BA]$) and, from the statement above it follows $[AB] \equiv [AB]$.

How to prove the symmetry? We have $[A'B'] \equiv [A'B']$, via the reflexivity proved above. Moreover, if $[AB] \equiv [A'B']$ it follows that $[A'B'] \equiv [AB]$, via Axiom E_2. It is very important to notice that only from this point on, we have the right to assert that $[AB] \equiv [CD]$ is the same as $[CD] \equiv [AB]$.

For transitivity, we consider $[AB] \equiv [A'B']$, and $[A'B'] \equiv [A"B"]$. But the congruence $[A'B'] \equiv [A"B"]$ implies the congruence $[A"B"] \equiv [A'B']$ (symmetry). Then, from $[AB] \equiv [A'B']$ and $[A"B"] \equiv [A'B']$, it follows the congruence $[AB] \equiv [A"B"]$ (axiom E_2). □

The congruence relation, being an equivalence relation, gives rise to a partition of the set of all segments in disjoint equivalence classes. This fact allows us to define all segments in an equivalence class to have the same *value*. We denote the value of a segment $[AB]$ by simply AB. Note that the same notation AB is also used for the line which passed through the points A and B. In general it is clear from the context if we refer to the line AB or to the value of the segment $[AB]$. Moreover, the congruence $[AB] \equiv [CD]$ can be also written as an equality of values, $AB = CD$, when there is no danger of confusion between equivalence classes and their representatives. In what follows, going back and forth between congruence of segments (or angles) and equality of their values, technically requires one to prove the independence of chosen representatives in a given equivalence class. For the simplicity of geometric arguments, we will omit these technical details.

Theorem 1.1.24 *Let $(OA$ be a half-line with origin O. If C and C' are two points on $(OA$ such that $[OC] \equiv [OC']$, then the points C and C' coincide.*

Proof Without loss of generality, we can assume the order $\overline{OCC'}$. Let I be a point which does not belong to the half line $(OA$ (Axiom $\mathbf{I}_3$). Then, in the triangles $\triangle OCI$ and $\triangle OC'I$, we have: $[OC] \equiv [OC']$, $[OI] \equiv [OI]$ and $\angle IOC \equiv \angle IOC'$. From Axiom $\mathbf{E}_6$ it follows $\angle OIC \equiv \angle OIC'$, therefore the half lines $(IC$ and $(IC'$ coincide as sets (Axiom $\mathbf{E}_4$). This implies $(IC \cap (OA = (IC' \cap (OA$, so C and C' coincides. □

Sometimes we write $C = C'$ whenever C and C' coincide. Notice that the equal sign which expresses the coincidence is not the same as the usual symbol $=$ of equality of numbers.

Note that Axiom $\mathbf{E}_3$ guarantees the additivity of the values of segments on same line. Indeed, if A, B, C and A', B', C' are points on the lines l and l', respectively, with orders $\overline{ABC}$ and $\overline{A'B'C'}$, respectively, such that $[AB] \equiv [A'B']$, $[BC] \equiv [B'C']$, then if follows directly from Axiom $\mathbf{E}_3$ that $[AC] \equiv [A'C']$. We can formally write the following equalities in terms of values of segments: $AC = AB + BC$, and $A'C' = A'B' + B'C'$.

Theorem 1.1.25 *The congruence relation for segments preserves the order relation.*

Proof Consider the points A, B, C on a line l, with the property that B is an interior point of the segment $[AC]$, i.e. we have the order $\overline{ABC}$. Moreover, let us consider the points A', B', C' on another line l', such that $[AB] \equiv [A'B']$, $[AC] \equiv [A'C']$, and B', C' are on the same half-line of vertex A'.

If we show that B' is interior to $[A'C']$, and $[B'C'] \equiv [BC]$, then it will follow the order $\overline{A'B'C'}$, which is the conclusion of our theorem. Indeed, assume the existence of another point $C'' \in l'$ with order $\overline{A'B'C''}$, such that $[B'C''] \equiv [BC]$. $[A'B'] \equiv [AB]$ and $[B'C''] \equiv [BC]$, so, by additivity, it follows $[A'C''] \equiv [AC]$. But $[A'C'] \equiv [AC]$, thus $[A'C''] \equiv [A'C']$, therefore it follows that $C' = C''$. Thus we have the desired order $\overline{A'B'C'}$. □

In view of the results above, one can define the *difference* operation among segments. Indeed, if $[AB]$ and $[AC]$ are two segments on a line l, such that the have order $\overline{ABC}$, then the difference of the values of $[AC]$ and $[AB]$ is the value of the segment $[BC]$, respecting the additivity property $AB + BC = AC$. Therefore we can write $AC - AB = BC$.

Definition 1.1.26 Two triangles $\triangle ABC$ and $\triangle A'B'C'$ are called *congruent*, and we denote by $\triangle ABC \equiv \triangle A'B'C'$, if they have congruent sides and congruent angles, respectively.

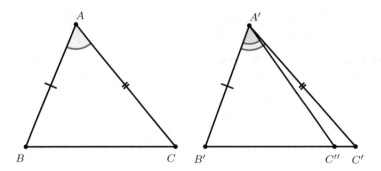

Fig. 1.2 Theorem SAS

Concretely, $\triangle ABC \equiv \triangle A'B'C'$ if the following six congruences are respected:

$$[AB] \equiv [A'B'], \quad [BC] \equiv [B'C'], \quad [CA] \equiv [C'A'],$$

$$\angle BAC \equiv \angle B'A'C', \quad \angle ABC \equiv \angle A'B'C', \quad \angle BCA \equiv \angle B'C'A'.$$

When there is no danger of confusion, we denote by $\angle A$ the angle $\angle BAC$. The first result about congruence of triangles is the following.

Theorem 1.1.27 *If a triangle $\triangle ABC$ has two congruent sides, then it has two congruent angles, too. In this case, we call the triangle $\triangle ABC$ to be isosceles.*

Proof Without loss of generality, we can assume $[AB] \equiv [AC]$. Then the triangles $\triangle BAC$ and $\triangle CAB$ are in the conditions of Axiom $\mathbf{E}_6$, thus $\angle ABC \equiv \angle ACB$. □

The next theorem is the first important congruence case of triangles.

Theorem 1.1.28 (SAS) *Let $\triangle ABC$ and $\triangle A'B'C'$ be two triangles, such that $[AB] \equiv [A'B'], [AC] \equiv [A'C'],$ and $\angle BAC \equiv \angle B'A'C'$. Then $\triangle ABC \equiv \triangle A'B'C'$ (Fig. 1.2). (This congruence case is called Side-Angle-Side (SAS).)*

Proof Using axiom $\mathbf{E}_6$, we have $\angle ABC \equiv \angle A'B'C'$ and $\angle ACB \equiv \angle A'C'B'$. The only congruence left to show is $[BC] \equiv [B'C']$. Consider a point C'' on the half-line $(B'C'$ such that $[BC] \equiv [B'C'']$ (Axiom $\mathbf{E}_1$). Consider now the triangles $\triangle ABC$ and $\triangle A'B'C''$. From $[AB] \equiv [A'B'], [BC] \equiv [B'C''],$ and $\angle ABC \equiv \angle AB'C''$, if follows from axiom $\mathbf{E}_6$ that $\angle BAC \equiv \angle B'A'C''$. From the hypothesis, we have $\angle BAC \equiv \angle B'A'C'$. Then we have C' and C'' such that the angles $\angle C'A'B'$ and $\angle C''A'B'$ are congruent. Since C' and C'' are in the same half-plane with respect to the line $A'B'$, it follows from axiom $\mathbf{E}_4$ that $(A'C'$ and $(A'C''$ coincide, thus $C' = C''$. □

The next theorem establishes the second case of triangle congruence.

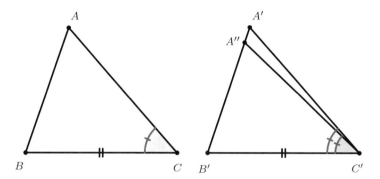

Fig. 1.3 Theorem ASA

Theorem 1.1.29 (ASA) *Let* $\triangle ABC$ *and* $\triangle A'B'C'$ *be two triangles, such that* $[BC] \equiv [B'C']$, $\angle ABC \equiv \angle A'B'C'$, *and* $\angle ACB \equiv \angle A'C'B'$. *Then* $\triangle ABC \equiv \triangle A'B'C'$ *(Fig. 1.3). (This congruence case is called Angle-Side-Angle (ASA).)*

Proof Let $A'' \in (B'A'$ such that $[BA] \equiv [B'A'']$. Consider the triangles $\triangle BAC$ and $\triangle B'A''C'$. Axiom $\mathbf{E}_6$ guarantees that $\angle BCA \equiv \angle B'C'A''$. Since A' and A'' are in the same half-plane with respect to $B'C'$, it follows that $(C'A'$ and $(C'A''$ coincide. Therefore, $A' = A''$. We apply Theorem SAS for the triangles $\triangle ABC$ and $\triangle A'B'C'$, where we now have $[AB] \equiv [A'B']$, $[BC] \equiv [B'C']$ and $\angle ABC \equiv \angle A'B'C'$. $\qquad\square$

Theorem 1.1.30 (Additivity of Angles) *If* $\angle(hl) \equiv \angle(h'l')$, *and* $\angle(lk) \equiv \angle(l'k')$, *where* l *and* l' *are half-lines interior to the angles* $\angle(hk)$ *and* $\angle(h'k')$, *then* $\angle(hk) \equiv \angle(h'k')$.

Proof Let H and K be two points such that $H \in h$ and $K \in k$. Using Crossbar Theorem, it follows that $l \cap [HK] \neq \emptyset$. Let $\{L\} = l \cap [HK]$. Now take $H' \in h'$ and $L' \in l'$ such that $[OH] \equiv [O'H']$ and $[OL] \equiv [O'L']$, and take K' on the half-line complement to $(L'H'$ such that $[L'K'] \equiv [LK]$. Notice that the congruence $\triangle OHL \equiv \triangle O'H'L'$ (case SAS) implies $[HL] \equiv [H'L']$ and $\angle OHL \equiv \angle O'H'L'$. But the segments $[HL]$, $[LK]$; $[H'L']$, $[L'K']$ satisfy the conditions of axiom $\mathbf{E}_3$, thus the triangles $\triangle OHK$ and $\triangle O'H'K'$ are congruent (case SAS). It follows that $\angle HOK \equiv \angle H'O'K'$, thus using axiom $\mathbf{E}_4$, it follows that the half-lines $(O'K'$ and k' coincide. $\qquad\square$

Suppose we are in the same hypothesis as in Theorem of Additivity of Angles;

Theorem 1.1.31 *If* $\angle(hk) \equiv \angle(h'k')$, *and* $\angle(hl) \equiv \angle(h'l')$, *then* $\angle(lk) \equiv \angle(l'k')$.

Proof Consider the triangles $\triangle ABC$ and $\triangle A'BC$ such that A and A' are in different half-planes with respect to the line BC. If $[AB] \equiv [A'B]$ and $[AC] \equiv [A'C]$, then triangles $\triangle ABC$ and $\triangle A'BC$ have congruent angles, respectively. Considering the segments $[AA']$ and $[BC]$, we distinguish two cases: $[AA'] \cap [BC] \neq \emptyset$ or $[AA'] \cap [BC] = \emptyset$. In each one of these cases, we apply the theorem for isosceles triangles

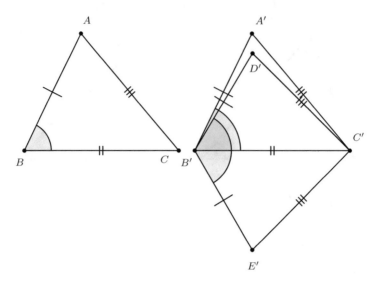

Fig. 1.4 Theorem SSS

in the case of triangles $\triangle ABA'$ and $\triangle ABA'$, respectively. The conclusion of the theorem follows then immediately. □

Now we are in the right context to prove the following side-side-side (SSS) congruence theorem of triangles. Note that in the proof we do not use neither the symmetry, nor the transitivity of the equality relation for angles! These properties are an immediate corollary to the following theorem.

Theorem 1.1.32 (SSS) *Let $\triangle ABC$ and $\triangle A'B'C'$ be two triangles, such that $[AB] \equiv [A'B']$, $[BC] \equiv [B'C']$, and $[CA] \equiv [C'A']$. Then $\triangle ABC \equiv \triangle A'B'C'$. This congruence case is called Side-Side-Side (SSS) (Fig. 1.4).*

Proof Consider the half-line $(B'D'$ such that $[B'D'] \equiv [AB]$ and $\angle D'B'C' \equiv \angle ABC$, D' in the half-plane determined by A' and $B'C'$.

Since $[BC] \equiv [B'C']$, $[BA] \equiv [B'D']$, and $\angle ABC \equiv \angle D'B'C'$, thus $\triangle ABC \equiv \triangle D'B'C'$ (case SAS). It follows that $[AC] \equiv [D'C']$. Let us construct a point E' in the complementary half-plane defined by the line $B'C'$ and the point A', such that $[B'E'] \equiv [B'D']$ and $\angle E'B'C' \equiv \angle C'B'D'$. It follows that $\triangle D'B'C' \equiv \triangle E'B'C'$ (case SAS), thus $[E'C'] \equiv [D'C'] \equiv [AC] \equiv [A'C']$. Similarly, $[E'B'] \equiv [B'D'] \equiv [AB] \equiv [A'B']$.

Then, using the fact that isosceles triangles have equal angles corresponding to equal sides, the triangles $\triangle A'B'C'$ and $\triangle E'B'C'$ are congruent (since we use the previous theorems with sum or difference of angles to prove that $\angle B'A'C' \equiv \angle B'E'C'$). Then $\angle A'B'C' \equiv \angle E'B'C'$, so in the half-plane determined by $B'C'$ and A' we have two half-lines $(B'D'$ and $(B'A'$, such that they determine $\angle A'B'C' \equiv \angle D'B'C'$.

Fig. 1.5 Right angle
existence

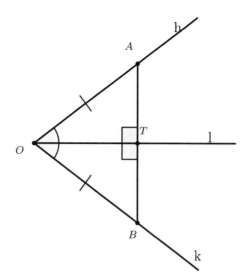

Therefore they are coincident and this means that the points A' and D' have to coincide.   □

Corollary 1.1.33 *The congruence relation for triangles is an equivalence relation.*

Corollary 1.1.34 *The congruence relation for angles is an equivalence relation.*

The details are left for the reader, and here it is used $\mathbf{E}_5$. We have to mention that this equivalence relation allows us to define a value for all representatives of a class, which can be denoted by $v(\angle(hk))$, with the same remarks we did in the case of segments.

Definition 1.1.35 Let $\angle hk$ be an angle. The angle formed by a ray of angle $\angle hk$ and the complement of the other ray is called the supplementary angle to the angle $\angle hk$.

Definition 1.1.36 Two angles which have the same vertex and complementary sides are called opposite (or complementary) angles.

We propose two problems to the reader.

Problem 1.1.37 *Supplementary angles of congruent angles are congruent.*

Problem 1.1.38 *Opposite angles are congruent.*

Hint for both problems: choose points on rays such that congruent triangles occurs (Fig. 1.5).

Definition 1.1.39 A right angle is an angle congruent to its supplementary angle. We denote by R the class of the right angles.

The following theorem establishes the existence of right angles in any Geometry respecting all axioms introduced so far.

Theorem 1.1.40 *There exist right angles.*

Proof Consider the congruent angles $\angle hl$ and $\angle lk$ such that all rays have the common point O and l belongs to the interior of $\angle hk$. Choose $A \in h$, $B \in k$ such that $[OA] \equiv [OB]$. Crossbar theorem tells us that it exists $\{T\} = l \cap [AB]$. It is easy to see using congruent triangles that $\angle ATO \equiv \angle BTO$, i.e. $\angle ATO$ and $\angle BTO$ are both right angles.   □

The supplementary angle of an right angle is a right angle itself. The angle $\angle ATB$ can be seen as the sum of the right angles $\angle ATO$ and $\angle OTB$, therefore its class is $R + R$, i.e. $2R$.

Definition 1.1.41 The points A and B are called symmetric with respect the line l. The line AT is called perpendicular to l, T is called the foot of the perpendicular line to l passing through A.

Theorem 1.1.42 *All right angles are congruent.*

Proof By contradiction, we assume that there exist two right angles $\angle BAD$ and $\angle B'A'D'$ which are not congruent. We consider the supplementary angles $\angle CAD$ and $\angle C'A'D'$, respectively. Consider the half-line $(AE$ such that $\angle BAE \equiv \angle B'A'D'$ and observe the equality of angles $\angle CAE \equiv C'A'D'$. Therefore we have $\angle CAE \equiv \angle C'A'D' \equiv \angle B'A'D' \equiv \angle BAE$. Let $(AF$ such that $\angle CAE \equiv \angle BAE$. It results $\angle CAE \equiv \angle CAF$, in collision with $\mathbf{E}_4$.   □

Theorem 1.1.43 *The perpendicular line from an exterior point to a given line is unique.*

Proof By contradiction, suppose AC and AC' are perpendicular lines to l, C, $C' \in l$. Consider the symmetric points B, B' of A with respect to l on each perpendicular line and choose $O \in l$ such that the order is $\overline{OC'C}$. It results $\triangle OCA \equiv \triangle OCB$ and $\triangle OC'A \equiv \triangle OC'B'$. We have $\angle BOC \equiv \angle AOC' \equiv \angle B'O'C'$ and $[OB] \equiv [OB'] \equiv [OA]$, i.e. B and B' coincide, therefore C coincides C'.   □

Definition 1.1.44 Let $[AB]$ and $[A'B']$ be two segments. If there exists a point C in the interior of the segment $[AB]$ such that $[AC] \equiv [A'B']$, we say that the segment $[A'B']$ is less than the segment $[AB]$, and we denote by $[A'B'] < [AB]$.

In the same time we may say that the segment $[AB]$ is greater than the segment $[A'B']$ and we denote by $[AB] > [A'B']$. Note that the order $\overline{ABC}$ on a line determines the inequalities $[AB] < [AC]$ and $[BC] > [AC]$. We may also define $[AB] \leq [A'B']$, etc. The inequality relation $\leq$ is a partial order relation on the set of segment and more, if $[AB] > [A'B']$ and $[CD] > [C'D']$, then $[AB] + [CD] > [A'B'] + [C'D']$. The inequality can be transferred to values with the notations established there.

Fig. 1.6 The exterior angle
theorem

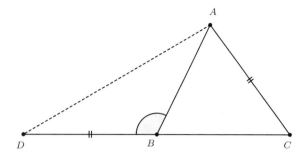

Definition 1.1.45 Let $\angle(h'k')$ and $\angle(hk)$ be two angles. If there is a line l in the interior of the angle $\angle(hk)$ such that $\angle(h'k') \equiv \angle(hl)$, then we can say that the $\angle(h'k')$ is less than the angle $\angle(hk)$, denoted by $\angle(h'k') \leq \angle(hk)$.

Or, we can say that the angle $\angle(hk)$ is greater than the angle $\angle(h'k')$, denoted by $\angle(hk) > \angle(h'k')$. We can easily define $\angle(hk) \geq \angle(h'k')$ or $\angle(h'k') \leq \angle(hk)$.

We do not insist and we left to the reader to prove that the inequality relations $\geq$ and $\leq$ are partial order relations on the set of angles.

Definition 1.1.46 Two lines which do not have any common point are called non-secant lines.

Definition 1.1.47 Consider the triangle $\triangle ABC$. The angle formed by the half-line $(BA$ and the complement half-line of $(BC$, say $(BD$, is called the exterior angle of the triangle $\triangle ABC$ with respect to the vertex B.

Consider the angle formed by the half-line $(BC$ and the complement half-line of $(BA$, say $(BF$. This angle is also the exterior angle of the triangle $\triangle ABC$ with respect to the vertex B, and of course $\angle ABD \equiv \angle CBF$ as opposite angles. Having in mind the previous definition we can prove

Theorem 1.1.48 (Exterior Angle Theorem) *The exterior angle of a triangle with respect to a given vertex is greater than both the angles of the triangle which are not adjacent to it (Fig. 1.6).*

Proof Let us fix the vertex to be B. We have to prove "the exterior angle of the triangle $\triangle ABC$ with respect the vertex B is greater than both the angles $\angle BAC$ and $\angle ACB$". Let D be a point on BC with the order $\overline{DBC}$ such that $[BD] \equiv [AC]$. We show that $\angle DBA > \angle BAC$. The other inequality results from $\angle ABD \equiv \angle CBF > \angle ACB$. We focus on the first inequality. By contradiction, let us suppose that $\angle ABD \equiv \angle BAC$. If we succeed to obtain a contradiction, the case $\angle ABD < \angle BAC$ is reduced to the previous case by considering $C_1 \in (BC)$ such that $\angle ABD \equiv \angle BAC_1$. Therefore, it remains to prove that $\angle ABD \equiv \angle BAC$ is impossible. In the given conditions it results $\triangle ABD \equiv \triangle CAB$, (SAS), i.e. $\angle DAB \equiv \angle ABC$. Since $\angle CAD \equiv \angle CAB + \angle BAD \equiv \angle ABD + \angle ABC \equiv \angle CBD = 2R$, equivalent to $A \in BC$, in collision to the fact that ABC is a triangle. $\qquad \square$

Corollary 1.1.49 *The sum of two among the three angles of triangle is less than the sum of two right angles.*

Proof To simplify the writing denote by B_e the exterior angle with respect to the vertex B. The exterior angle theorem asserts that $B_e > A$, $B_e > C$. It results $B_e + B > A + B$, therefore $A + B < 2R$.   □

Definition 1.1.50 An angle of a triangle which is greater than a right angle is called an obtuse angle. An angle of a triangle which is less than a right angle is called an acute angle.

Corollary 1.1.51 *A triangle can not have more than one obtuse angle.*

Proof Suppose there exists a triangle $\triangle ABC$ such that $A > R$ and $B > R$. Then $A + B > 2R$, in collision with the previous corollary.   □

We left for the reader the following very nice problems:

Problem 1.1.52 *Given a segment $[AB]$, there exists an unique point M, (called the midpoint of the segment $[AB]$), such that $[AM] \equiv [MB]$.*

Problem 1.1.53 *Given an angle $\angle hk$, there is an unique half-line l in its interior such that $\angle hl \equiv \angle lk$ (the half-line l is called the bisector of the $\angle hk$).*

The previous problems create an infinity of points in the interior of a given segment and an infinity of half-lines in the interior of an angle.

Definition 1.1.54 The perpendicular line from a vertex of a triangle on the line which contains the opposite side is called an altitude (or height) of the triangle.

Theorem 1.1.55 *At least one altitude among the three altitudes of a triangle lies in the interior of the triangle.*

Proof *(Hint)* Consider the altitude corresponding to the greatest angle of a triangle, say AD, $D \in BC$. B and C are mandatory acute angles. The order on BC has to be $\overline{BDC}$.   □

Theorem 1.1.56 *In the triangle $\triangle ABC$, $[AC] > [AB]$ if and only if $\angle B > \angle C$.*

Proof Consider $D \in [AC]$ such that $[AB] \equiv [AD]$. It results $\angle B > \angle ABD \equiv \angle ADB > \angle C$. Conversely, assume $\angle B > \angle C$ and by contradiction, $[AC] \leq [AB]$. If $[AC] \equiv [AB]$, then $\angle B \equiv \angle C$, contradiction. If $[AC] < [AB]$, then $\angle B < \angle C$, contradiction.   □

Theorem 1.1.57 (triangle inequality) *In every triangle $\triangle ABC$ the sum of two sides is bigger than the third side. For example $[BC] < [BA] + [AC]$.*

Proof Consider $D \in (BA$ such that the order is $\overline{BAD}$ and $[AD] \equiv [AC]$. It follows that $[BD] \equiv [BA] + [AD] \equiv [BA] + [AC]$. Since $\angle BDC \equiv \angle DCA < \angle DCB$ it follows that $[BD] > [BC]$, that is $[BA] + [AC] > [BC]$.   □

Fig. 1.7 The existence of a
non-intersecting line

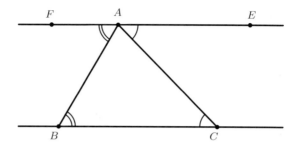

Theorem 1.1.58 $[A_1A_n] \leq [A_1A_2] + [A_2A_3] + \cdots + [A_{n-1}A_n]$

Proof (*Hint*) $[A_1A_n] \leq [A_1A_2] + [A_2A_n] \leq [A_1A_2] + [A_2A_3] + [A_3A_n] \leq \cdots$ □

Theorem 1.1.59 *Consider the triangles* $\triangle ABC$ *and* $\triangle A'B'C'$ *such that* $[AB] \equiv$
$[A'B'], \ [AC] \equiv [A'C']$. *If* $\angle A > \angle A'$, *then* $[BC] > [B'C']$.

Proof (*Hint*) Consider the half-line $(AD$ such that $\angle BAD \equiv \angle B'A'C'$ and $[AD] \equiv$
$[AC] \equiv [A'C']$. Observe that $(AD$ is included in the interior of the angle $\angle BAC$. We
have $\triangle ABD \equiv \triangle A'B'C'$. The triangle $\triangle ACD$ is isosceles, $[AC] \equiv [AD]$, therefore
$\angle DCB < \angle ADC < \angle BDC$, i.e. $[BC] > [BD] \equiv [B'C']$ (Fig. 1.7). □

Theorem 1.1.60 *From a point A exterior to a line d, one can construct at least one*
non-secant line to d.

Proof Consider the points B, C on d and a half-line $(AE$ in the half-plane determined by A and d such that B and E are in opposite half-planes with respect to the
line AC and $\angle EAC \equiv \angle BCA$. According to the exterior angle theorem we have
$(AE \cap (BC = \emptyset$. The complementary half-line $(AF$ has the property $\angle BAF \equiv$
$\angle ABC$. The same exterior angle theorem implies $(AF \cap (CB = \emptyset$. Therefore
$FE \cap d = \emptyset$. □

Definition 1.1.61 The angles $\angle EAC$ and $\angle ACB$ are called interior alternate angles.

The angles $\angle FAB$ and $\angle ABC$ are interior alternate angles, too.

The reader observes that until now there is no a parallelism axiom involved in the
construction we made. We are still in the absolute Geometry area mentioned at the
beginning of this chapter. The previous result is an important one. In the axiomatic
frame created before it exists at least one non-secant line through a point with respect
to a given line. There is only one or there are more? We left the answer for later.

The axioms before allow us to have infinitely many points on a line, but we
don't know if a line can be "filled" with points or if it is "unbounded." Until
now we can see that if we establish an origin O on a line and if we take a segment $[AB]$ we can construct on one half line the points $E_1, E_2, \ldots, E_n, \ldots$ such

that $[AB] \equiv [OE_1] \equiv [E_1E_2] \equiv [E_2E_3] \equiv \cdots \equiv [E_nE_{n+1}] \equiv \cdots$ and on the complementary half-line the points $E_{-1}, E_{-2}, \ldots, E_{-n}, \ldots$ such that $[AB] \equiv [OE_{-1}] \equiv [E_{-1}E_{-2}] \equiv [E_{-2}E_{-3}] \equiv \cdots \equiv [E_{-n}E_{-(n+1)}] \equiv \cdots$, therefore we can associate for any integer number a point on the line l. Combining with a result before related about the existence of the midpoint of every segment, we can see on the line l all rational points having the form $\dfrac{n}{2^m}$.

So, not all the real numbers can be "seen" on l. And still the problem of unboundedness persists. Why? Since even if $[OE_n] = [OE_1] + [E_1E_2] + \cdots + [E_{n-1}E_n] < [OE_{n+1}]$ the following example of segments bigger and bigger is bounded in the segment $[0, 1]$. It is about the sequence of intervals $(0, 1 - \dfrac{1}{n})$, $n \in \mathbb{N}$. Can we make any connection between the set of real numbers and the points of a line? We need to introduce the *axioms of continuity* at this point.

Axiom $\mathbf{C}_1$ (Axiom of Archimedes): Let $[AB]$ and $[CD]$ be two arbitrary segments such that $[CD] < [AB]$. Then, there exists a finite number of points A_1, A_2, $\ldots A_n$, $\ldots$ on the ray $(AB$, such that $[CD] \equiv [AA_1] \equiv [A_1A_2] \equiv [A_2A_3] \equiv \cdots \equiv [A_{n-1}A_n]$, the interiors of those segments have every two an empty intersection and finally, either $B = A_n$ or $B \in (A_{n-1}A_n)$.

In view of the additivity property of segments we can write that it exists $n \in \mathbb{N}$ such that

$$[AA_1] + [A_1A_2] + [A_2A_3] + \cdots + [A_{n-1}A_n] \equiv n[CD] \geq [AB]$$

and the inequality may refer to values. The Axiom of Archimedes multiplies values by natural numbers and we expect to understand the value of a segment as a real positive number describing the length of the segment. Considering $n = 1$ in the previous inequality, we have the old inequality between segments, therefore $\mathbf{C}_1$ offer us the chance to understand the appropriate nature of values attached to the segments and the unboundedness of the set of natural numbers.

The next axiom is attributed to Cantor and it will be involved in "completing" the line with points we don't know until now that they have to belong to a line.

Axiom $\mathbf{C}_2$ (Axiom of Cantor): Let $[A_1B_1]$, $[A_2B_2]$, $\ldots$ be a sequence of segments on a given line l, such that every segment is included in the interior of the precedent one, i.e. $[A_nB_n] \subset [A_{n-1}B_{n-1}]$ for all $n \geq 2$. If we assume that no segment is included in the interior of all segments $[A_nB_n]$, $n \in \mathbb{N}$, then there is an unique point M on the line l such that $\{M\} = [A_1B_1] \cap [A_2B_2] \cdots \cap [A_nB_n] \cap \cdots$

These two axioms of continuity allow us to use real positive numbers as the "values" of segments and angles. The results are a little bit more complicate and we try to suggest them without complete proofs.

Using the continuity axiom $\mathbf{C}_1$ we can assign the natural number 1 to the segment $[CD]$ and the number n to the value nCD. To every segment $[AB]$ we attach a system of coordinates on the line $d = AB$ such that A is the origin O and $1 = OA_1 = A_1A_2 = \cdots = A_{m-1}A_m = \cdots$.

According to C_1 it exists one integer $m \in \mathbb{N}$ such that $B \in [A_{n-1}A_n]$. If $B = A_{n-1}$ then $AB = n - 1$. If $B = A_n$ then $AB = n$. If $B = M_1$ is the midpoint of the segment $[A_{n-1}A_n]$ wee assign to the value AB is the 2-adic number $n - 1, 10000 \ldots 0 \ldots$. In fact at each step from now when B is the midpoint of a segment, we associate 1 and the other decimals after are 0. Suppose $B \in [A_{n-1}M_1]$. We consider the value of AB as $\overline{n - 1}, 0$ and we are looking after the next decimal observing where B is with respect to the midpoint M_2 of the segment $[A_{n-1}M_1]$. If $B \in [M_2M_1]$ the next decimal is 1, therefore the attached number is until now $\overline{n - 1}, 01$ and we continue looking at the position of B with respect to the midpoint M_3 of the segment $[M_2M_1]$. Imagine a little bit the position of B if the next three digits are 001 such that the value of AB is until now $\overline{n - 1}, 01001$. We can continue to discover digits until B is a midpoint of a segment when we stop with a 1 followed by 0 only, or we never stop because the point stays in the intersection of all segments which are like in axiom C_2. The real number

$$\overline{n - 1}, a_1 a_2 a_3 \ldots a_n \ldots$$

with $a_i = 1$ if B is in the "at the right" segment, or with $a_i = 0$ if B is in the "at the left" segment, is the 2-adic number attached to the value of the segment AB.

Then we can show that to every real number we can assign an unique point on l.

The theory can be extended to angles with the following two theorems.

Theorem 1.1.62 *Let $(a_1, b_1), (a_2, b_2), \ldots$ be a sequence of angles with common vertex O, with the property that the angle (a_{n+1}, b_{n+1}) is contained in the angle (a_n, b_n), for all $n \geq 1$. In the assumption that there is no angle contained in the interior of all angles in the sequence, then there is a unique half-line l in the intersection of the interior of all angles.*

Proof *(Hint)* Intersect all angles with a line l and denote the points of intersection with a_k by A_k and the points of intersection with b_k with B_k, etc.   □

Theorem 1.1.63 *Let $\angle(hk)$ and $\angle(h'k')$ two angles. There exists a natural number n such that $n\angle(hk) > \angle(h'k')$.*

Proof *(Hint)* Observe that the measure of the angle $\angle(h'k')$ is less then $2R$. If $\angle(hk) > R$ we take $n = 2$. If $R > \angle(hk) > \dfrac{R}{2}$ we take $n = 4$, etc.   □

We have two statements for angles analogue with the axioms C_1 and C_2. We can develop a similar theory for defining measure of angles, restricting all the proofs in the case of segments to the interval $(0, 2R)$. Then, there is a one-to-one correspondence between the set of angles and the interval $(0, 2R)$ of the real numbers.

We prove now in the Absolute Geometry frame the most important result regarding the sum of the angles of a triangle.

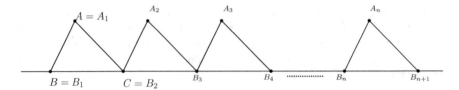

Fig. 1.8 Legendre's theorem

Theorem 1.1.64 (Legendre) *For any triangle, the sum of its angles is at most* $2R$ *(Fig. 1.8).*

Proof Consider the triangle $\triangle ABC$. We have to show $\angle A + \angle B + \angle C \leq 2R$. By contradiction, let us assume that $\angle A + \angle B + \angle C > 2R$. On the line BC we consider the points $B = B_1, C = B_2, B_3, \ldots, B_n, B_{n+1}$ in this order such that $B_1 B_2 = B_2 B_3 = \cdots = B_n B_{n+1}$ and in the same half-plane the points $A = A_1, A_2, A_3, \ldots, A_n$, such that $\triangle A_1 B_1 B_2 \equiv \triangle A_2 B_2 B_3 \equiv \cdots \triangle A_n B_n B_{n+1}$. It is easy to see that the following triangles are congruent, $\triangle A_1 B_2 A_2 \equiv \triangle A_2 B_3 A_3 \equiv \cdots \triangle A_{n-1} B_n A_n$, therefore $A_1 A_2 = A_2 A_3 = \cdots = A_{n-1} A_n$. It is easy to deduce that $\angle A > \angle A_1 B_2 A_2$ and then $BC = B_1 B_2 > A_1 A_2$. The polygonal line $B_1 A_1 A_2 \ldots A_n B_{n+1}$ is bigger than the segment $B_1 B_{n+1} = nBC$, that is

$$nBC < BA + (n-1)A_1 A_2 + AC.$$

This one can be written in the form

$$(n-1)(BC - A_1 A_2) < BA - BC + AC.$$

We know that $BC - A_1 A_2 > 0$, $BA - BC + AC > 0$, i.e. it exists the segments $[ST], [MK]$ such that $BC - A_1 A_2 = ST$, $BA - BC + AC = MK$ and $(n-1)ST < MK$. But in the last inequality the natural number n is arbitrary, in collision with C_1. Therefore $\angle A + \angle B + \angle C$ can not be greater than $2R$. It follows $\angle A + \angle B + \angle C \leq 2R$. $\qquad\square$

The next definition takes care that the values of angles are real numbers.

Definition 1.1.65 For any triangle $\triangle ABC$ we define the defect of it, denoted $\mathcal{D}(ABC)$, to be $\mathcal{D}(ABC) = 2R - \angle A - \angle B - \angle C$.

Legendre's theorem states that $\mathcal{D}(ABC) \geq 0$ for any triangle. Let us investigate what other properties the defect of triangles may have.

Theorem 1.1.66 *If $P \in (BC)$, where $[BC]$ is a side of the triangle $\triangle ABC$, then $\mathcal{D}(APB) + \mathcal{D}(APC) = \mathcal{D}(ABC)$.*

Proof (*Hint*) Denote by $\angle A_1 = \angle BAP$, $\angle A_2 = \angle CAP$, $\angle P_1 = \angle APB$, $\angle P_2 = \angle APC$ and observe that $\angle A_1 + \angle A_2 = \angle A$, $\angle P_1 + \angle P_2 = 2R$. Then

$$\mathcal{D}(APB) + \mathcal{D}(APC) = 2R - \angle A_1 - \angle P_1 - \angle B + 2R - \angle A_2 - \angle P_2 - \angle C =$$

$$= 2R - \angle A - \angle B - \angle C = \mathcal{D}(ABC)$$

.
$\square$

Theorem 1.1.67 *Consider a triangle* $\triangle ABC$ *and two points,* $B_1 \in (AB)$, $C_1 \in (AC)$. *Then* $\mathcal{D}(AB_1C_1) \leq \mathcal{D}(ABC)$.

Proof (*Hint*) Consider the triangles $\triangle AB_1C_1$, $\triangle BB_1C_1$, $\triangle BCC_1$ and apply the previous theorem as follows $\mathcal{D}(ABC) = \mathcal{D}(ABC_1) + \mathcal{D}(BCC_1) = \mathcal{D}(AB_1C_1) + \mathcal{D}(B_1C_1B) + \mathcal{D}(BCC_1) \geq \mathcal{D}(AB_1C_1)$. $\square$

If "the big triangle" $\triangle ABC$ has $\mathcal{D}(ABC) = 0$, then, mandatory "the small triangle" $\triangle AB_1C_1$ has to fulfill $\mathcal{D}(AB_1C_1) = 0$.

Pay attention to the following construction.

Consider a right-angle triangle $\triangle BAC$, $\angle A = R$ with $\mathcal{D}(BAC) = 0$. In this case observe that $\angle B + \angle C = R$ and construct D in the opposite half-plane with respect to BC and A such that $\triangle ABC \equiv \triangle DBC$. It results a quadrilateral $ABDC$ such that all angles are equal to R, and the opposite sides are equal, i.e. $[AB] \equiv [CD]$, $[AC] \equiv [BD]$. We may call this figure rectangle and it is easy to discover two more properties. The diagonals AD and BC are congruent and they cut in the middle of each one. Let us rename the rectangle $ABDC$ by $A_{00}A_{10}A_{11}A_{01}$.

We intend to pave the plane with tiles congruent to our created rectangle $A_{00}A_{10}A_{11}A_{01}$ for obtaining a so called grid.

On the half-lines $A_{00}A_{10}$ we consider the points $A_{20}, A_{30}, \ldots, A_{n0}, \ldots$ such that the segment $[AB]$ is seen repeatedly as $[A_{00}A_{10}] \equiv [A_{10}A_{20}] \equiv [A_{20}A_{30}] \equiv \cdots \equiv [A_{(n-1)0}A_{n0}] \equiv \cdots$ and on the half-line $A_{00}A_{01}$ we consider the points $A_{02}, A_{03}, \ldots, A_{0n}, \ldots$ such that the segment $[AC]$ is seen repeatedly as $[A_{00}A_{01}] \equiv [A_{01}A_{02}] \equiv [A_{02}A_{03}] \equiv \cdots \equiv [A_{0(n-1)}A_{0n}] \equiv \cdots$.

The tiles we create and put on the first row are consecutively

$$A_{00}A_{10}A_{11}A_{01}, \ A_{10}A_{20}A_{21}A_{11}, \ A_{20}A_{30}A_{31}A_{21},$$

$$A_{30}A_{40}A_{41}A_{31}, \ldots, \ A_{n0}A_{(n+1)0}A_{(n+1)1}A_{n1}, \ldots,$$

on the second row

$$A_{01}A_{11}A_{12}A_{02}, \ A_{11}A_{21}A_{22}A_{12},$$

$$A_{21}A_{31}A_{32}A_{22}, \ A_{31}A_{41}A_{42}A_{32}, \ldots, A_{n1}A_{(n+1)1}A_{(n+1)2}A_{n2}, \ldots, etc.$$

A "general" tile in this pavement is $A_{kp}A_{(k+1)p}A_{(k+1)(p+1)}A_{k(p+1)}$.

It is easy to see that the points A_{20}, A_{11}, A_{02} are collinear.

The same, the points A_{30}, A_{21}, A_{12}, A_{03} and in general

$$A_{n0}, \ A_{(n-1)1}, \ A_{(n-2)2}, \ldots, A_{2(n-2)}, \ A_{1(n-1)}, \ A_{0n}$$

are collinear points.

And it is also easy to observe that all triangles $A_{0n}A_{00}A_{n0}$ have the sum of angles equal to $2R$, i.e. $\mathcal{D}(A_{0n}A_{00}A_{n0}) = 0$. We are prepared to prove a very important theorem.

Theorem 1.1.68 *If there exists a right-angle triangle with defect* 0, *then all right-angles triangles have defect* 0, *i.e. the sum of their angles is* $2R$.

Proof Consider the right-angle triangle to be $\triangle BAC$, $\angle A = R$ and $\mathcal{D}(BAC) = 0$ and let a general right-angle triangle $\triangle EFG$, $\angle F = R$. According to Archimedes' axiom it exist $m \in \mathbb{N}, n \in \mathbb{N}$ such that $m \cdot AB > FE$, $n \cdot CA > FG$. Without of loosing the generality we suppose $n > m$. Then the triangle $\triangle EFG$ can be "arranged" such as $F = A_{00}$, $E \in (A_{00}A_{0n})$, $G \in (A_{00}A_{n0})$. According to the previous theory $0 \leq \mathcal{D}(EFG) \leq \mathcal{D}(A_{0n}A_{00}A_{n0}) = 0$, i.e. $\mathcal{D}(EFG) = 0$. $\qquad \square$

Theorem 1.1.69 *If there exists a triangle with defect* 0, *then all triangles have defect* 0, *i.e. the sum of their angles is* $2R$.

Proof Consider a triangle with defect 0, say $\triangle ABC$. It exists an altitude which intersects the opposite side in its interior, say AT, $T \in (BC)$. The altitude and the sides of the triangle determine two right-angle triangles, $\triangle ABT$ and $\triangle ATC$, both with 0 defect, because the initial triangle is with 0 defect. The previous theorem asserts that all right-angle triangles have 0 defect. Now, consider an arbitrary triangle $\triangle DEF$. Suppose that the altitude which intersects the opposite side is DP, $P \in (EF)$ The two right-angle triangles $\triangle DFP$ and $\triangle DEP$ are with 0 defect, therefore the defect of the triangle $\triangle DEF$ is 0. $\qquad \square$

There are only two situations that can happen. All the triangles have the sum of angles $2R$ or all the triangles have the sum of angles strictly less than $2R$. In the given context we cannot decide about the sum. The next axioms will clarify this aspect.

Definition 1.1.70 The collection of all properties and results deduced from all axioms of incidence, order, congruence and continuity above is called an *absolute Geometry.*

1.2 From Absolute Geometry to Euclidean Geometry Through Euclidean Parallelism Axiom

The question we asked before is how many non-secant lines can be constructed through an exterior point to a given line. We know that at least one can be constructed. Could there be two or more? The standard *Euclidean Parallelism Axiom* is stated as follows.

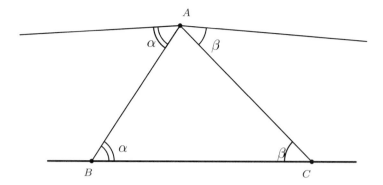

Fig. 1.9 Axiom of Euclidean parallelism

Axiom P: Given any line in a plane and given any point not incident to the given line, there exists at most one line that passes through the given and it is non-secant to the given line.

Definition 1.2.1 The collection of all properties and results deduced from all axioms of incidence, order, congruence, continuity and Euclidean Parallelism above is called a *Euclidean Geometry.*

A direct consequence: In Euclidean Geometry, i.e. in the axiomatic frame created by the axioms of incidence, order, congruence, continuity and the Euclidean Parallelism Axiom there exists an unique line that passes through a given point A and it is non-secant to the given line d. This unique non-secant line is called the *parallel line to the given line d* through the given point A.

Theorem 1.2.2 *In Euclidean Geometry, the sum of angles of any triangle is $2R$ (Fig. 1.9).*

Proof Consider a triangle $\triangle ABC$ and the unique parallel through A to BC. We have in mind the figure of the Theorem 23 where the parallel was FE with $A \in FE$. The interior alternate angles $\angle EAC$ and $\angle ACB$ are equal. The same for the interior alternate angles $\angle FAB$ and $\angle ABC$. Therefore, if we look at the angle $\angle FAE$ we observe that it is equal to the sum of the angles of the triangle $\triangle ABC$ and, in the same time it has the value $2R$. Since this particular triangle has the sum of angles equal to $2R$, all other triangles have the sum of angles equal to $2R$. ☐

In the case of the figure, $\angle A + \angle B + \angle C = \angle A + \alpha + \beta = 2R$.

Since $\angle A + \angle B + \angle C = 2R$, we deduce $2R - \angle A = \angle B + \angle C$. But $2R - \angle A$ is the value of the exterior angle A.

Corollary 1.2.3 (Exterior Angle Theorem in the Euclidean Geometry) *For every triangle, each exterior angle is the sum of the interior non-adjacent angles.*

Theorem 1.2.4 *If it exists a triangle with the sum of its angles equal to* $2R$ *the parallelism axiom is satisfied.*

Proof Assume that it exists a triangle $\triangle ABC$ with $\angle A + \angle B + \angle C = 2R$.

Therefore every triangle has the same property.

Let d be a line and M a point not on d. Let MN be the perpendicular line from M to d, $N \in d$. Let l be a perpendicular to MN in M. We know that l and d are non-secant lines. We have to prove that l is the only non-secant line through M to d.

Consider another line l' passing through M and denote by α the acute angle between MN and l'.

It makes sense to consider a triangle $M'N'P'$ such that

$$[MN] \equiv [M'N'], \ \angle M'N'P' = R, \ \angle N'M'P' = \alpha$$

and $\angle M'P'N' = R - \alpha$ (without to know that $R, \alpha, R - \alpha$ are the angles of a triangle, we do not know that only the angles R and α together with the side $M'N'$ determine a triangle).

Considering $P \in d$ with $[NP] \equiv [N'P']$, the triangles $\triangle MNP$ and $\triangle M'N'P'$ are congruent and one of the half-line of l' is coincident to MP. Therefore $l' \cap d = \{P\}$ i.e. l is the unique non-secant line through M to d. $\qquad\square$

The story of Euclidean Geometry may continue with many theorems which can be proven only in this axiomatic frame. But we are interested in introducing Non-Euclidean Parallelism and models of Non-Euclidean Geometry. Therefore we remain in the axiomatic frame corresponding to the axioms of incidence, order, congruence and continuity and, at this moment we add another axiom, more specific the denial of the Axiom of Euclidean Parallelism.

1.3 From Absolute Geometry to Non-Euclidean Geometry Through Non-Euclidean Parallelism Axiom

The Euclidean Parallelism Axiom, in the set theoretical language, can be written as:

$$\forall d, \ \forall A \notin d, \ \#\{a \mid A \in a, \ a \cap d = \emptyset\} \leq 1,$$

where $\#$ denotes the number of elements of a set. In what follows we assume the negation of the previous axiom, and we call this the axiom of Non-Euclidean Parallelism. In set theory language, this translates to:

$$\exists d_0, \ \exists A_0 \notin d_0, \ \#\{a \mid A_0 \in a, \ a \cap d_0 = \emptyset\} \geq 1.$$

Therefore the axiom of non-Euclidean parallelism is the following.

Axiom PN_0: There exist both a line d_0 and a point A_0 exterior to d_0 with the property: at least two non-secant lines to d_0 passing through A_0 exist.

Definition 1.3.1 The collection of all properties and results deduced from all axioms of incidence, order, congruence, continuity and Non-Euclidean Parallelism above is called a *Non-Euclidean Geometry.*

We study below some important results in the context of this Geometry.

Theorem 1.3.2 *Axiom PN_0 acts as a global property, i.e. it holds for any line and any exterior point.*

Proof By contradiction, we assume that there is a point A and a line d which do not satisfy the property "there are at least two non-secant lines to d, passing through A". Then, through A passes exactly one non-secant line to d. So, the Axiom P is satisfied for the pair (A, d). If we choose B and C on d, it is easy to see that the sum of angles of the triangle $\triangle ABC$ is $2R$, and this is equivalent as we saw before with Axiom P for all pairs (M, l), $M \notin l$, in collision with our assumption.   □

We can restate the axiom of Non-Euclidean Geometry as follows.
Axiom PN: Given a line and a point exterior to the line, there exists at least two non-secant lines to the given line.
It is easy to prove

Theorem 1.3.3 *Let l be a given line in a plane and A be an exterior point to l. Let a_1 and a_2 be two lines in the same plane which pass through A and are non-secant to l. Then every line a passing through A and included in the interior of the angle $\angle a_1 a_2$ is non-secant to l.*

Proof (*Hint*) If a intersects l, then it does intersect a_1 or a_2, in collision with the fact that a is included in the interior of the angle $\angle a_1 a_2$.   □

Corollary 1.3.4 *In Non-Euclidean Geometry there are an infinite number of non-secant lines to a given line through an exterior point.*

Theorem 1.3.5 *In a Geometry which satisfies the groups of axioms of incidence, order, congruence, continuity and the Axiom NP, the sum of angles of a triangle is less than $2R$.*

Proof (*Hint*) If it exists a triangle with the sum of angles equal to $2R$, then Axiom P is valid, contradiction.   □

We conclude that the Non-Euclidean Geometry established by the Absolute Geometry together with the Axiom NP is completely different than the Euclidean Geometry established by the Absolute Geometry together with Axiom P. More other interesting results may be found in both geometries, but in the following, we are interested in offering examples of models of Euclidean and Non-Euclidean geometries.

Chapter 2
Basic Facts in Euclidean and Minkowski Plane Geometry

Entia non sunt multiplicanda praeter necessitatem.

W. Ockham

In Chap. 1, we found out that there exist different geometries in a plane. It depends on the axioms one chooses if Euclidean Geometry or Non-Euclidean Geometry are described. But how these geometries look like? In this chapter we present an algebraic model for Euclidean Geometry discussing some important theorems. We obtain a visual representation for the Euclidean Geometry of the plane. Making small changes in the algebraic construction of the Euclidean Geometry, it is possible to construct a Minkowski Geometry. This Geometry is deeply involved both with Physics and with Non-Euclidean Geometry. Later, we see how the models of Non-Euclidean Geometry are connected between them and how a Minkowski one is among them. The geometric objects in Minkowski Geometry seem to have a non-intuitive look, but the main theorems have a similar look with their Euclidean counterparts. Generally, Non-Euclidean models are more sophisticated and we need more mathematical tools in order to built them. This happens in the following chapters. One more comment: this chapter is not as formal as the previous one where we used the language style of an axiomatic theory. We can relax a little bit the mathematical language structure. The definitions appear often written as part of a mathematical algebraic description of geometric objects and italic letters are used to indicate them. The following notation is used: $A := B$. It means that the object A from the left side of the equality is described by definition through the object-expression B from the right part of the equality. The word iff has the meaning of "if and only if."

© The Editor(s) (if applicable) and The Author(s), under exclusive license
to Springer Nature Switzerland AG 2020
W. Boskoff and S. Capozziello, *A Mathematical Journey to Relativity*,
UNITEXT for Physics, https://doi.org/10.1007/978-3-030-47894-0_2

2.1　Pythagoras Theorems in Euclidean Plane

The idea to consider a system of coordinates on a line was discussed in the previous chapter.

The coordinates are real numbers and their set, geometrically represented as a line, is denoted by $\mathbb{R}$. In the following we suppose known

- the set of natural numbers denoted by $\mathbb{N}$,
- the set of integers denoted by $\mathbb{Z}$,

- the set of rational numbers denoted by $\mathbb{Q}$ and
- the set of irrational numbers denoted by $\mathbb{R} - \mathbb{Q}$.

In the same time we have proved that the values of angles are real numbers.

Basic facts about matrix theory, groups, vector spaces, trigonometric, exponential and logarithms functions are suppose known by the reader interested in the topic of this book.

When we are talking about a model of Euclidean Geometry in a plane, we have to start from the vector space $\mathbb{R}^2$ over the field $\mathbb{R}$. $x := (x_1, x_2)$, $y := (y_1, y_2)$ are called *vectors*. The vector space operations are $x + y := (x_1 + y_1, x_2 + y_2)$ and $\lambda x := (\lambda x_1, \lambda x_2)$.

The *Euclidean inner product of the vectors x and y* is defined by

$$\langle x, y \rangle := x_1 y_1 + x_2 y_2$$

and the *norm of x*, by $|x| := \sqrt{\langle x, x \rangle} = \sqrt{x_1^2 + x_2^2}$.

Two vectors are called *Euclidean orthogonal (or Euclidean perpendicular)* if their inner product is null.

According to the operations, the vector (x_1, x_2) can be thought as $x_1(1, 0) + x_2(0, 1)$, that is $(x_1, x_2) = x_1(1, 0) + x_2(0, 1) = x_1 e_1 + x_2 e_2$ so, the pair (x_1, x_2) can be seen also as a pair of coordinates of a point A of the plane.

The line determined by $x e_1$, $x \in \mathbb{R}$ is called the *x-axis*, and the line determined by $y e_2$, $y \in \mathbb{R}$ is called the *y-axis*.

Therefore, in the system of coordinates generated by the orthogonal vectors $e_1 = (1, 0); e_2 = (0, 1)$, the geometric meaning of the vector $x = (x_1, x_2)$ is the oriented segment $\overrightarrow{OA}$, lying from the origin O with the coordinates $(0, 0)$ and the endpoint A with the coordinates (x_1, x_2).

Let us consider the 2×2 *rotation matrix* $A_\alpha = \begin{pmatrix} \cos\alpha & -\sin\alpha \\ \sin\alpha & \cos\alpha \end{pmatrix}$ in which the basic trigonometric function sine and cosine are involved as components.

We define $A_\alpha x := \begin{pmatrix} \cos\alpha & -\sin\alpha \\ \sin\alpha & \cos\alpha \end{pmatrix} \begin{pmatrix} x_1 \\ x_2 \end{pmatrix}$

$A_\alpha x$, $A_\alpha y$ are two matrices with two lines and one column, and it makes sense to consider the inner product $\langle A_\alpha x, A_\alpha y \rangle$ by adding, after multiplying, the corresponding first and second components, i.e.

$$\langle A_\alpha x, A_\alpha y \rangle =$$

$$= (x_1 \cos\alpha - x_2 \sin\alpha)(y_1 \cos\alpha - y_2 \sin\alpha) + (x_1 \sin\alpha + x_2 \cos\alpha)(y_1 \sin\alpha + y_2 \cos\alpha).$$

Exercise 2.1.1 $\langle A_\alpha x, A_\alpha y \rangle = \langle x, y \rangle$.

Hint. Use $\sin^2\alpha + \cos^2\alpha = 1$.

Exercise 2.1.2 $|A_\alpha x| = |x|$.

Exercise 2.1.3 *If* $|x| = 1$, *then* $\langle A_\alpha x, x \rangle = \cos\alpha$.

If $|x| = 1$, then $|A_\alpha x| = |x| = 1$. Denote by u the unitary vector $A_\alpha x$. The previous relation for the unitary vectors u, x can be written in the form $\langle u, x \rangle = \cos\alpha$.

We can see the vector u as the rotation of the vector x, so the angle between these two vectors is α. For two arbitrary vectors a, b, the vectors $\dfrac{a}{|a|}$, $\dfrac{b}{|b|}$ are unitary and the previous relation becomes

$$\left\langle \frac{a}{|a|}, \frac{b}{|b|} \right\rangle = \cos\alpha,$$

i.e.

$$\frac{\langle a, b \rangle}{|a||b|} = \cos\alpha.$$

This last formula is known as the *Generalized Pythagoras Theorem*. Let us discuss why (Fig. 2.1).

Since we have the vectors $a = (a_1, a_2)$, $b = (b_1, b_2)$, we can think about the triangle OAB as the triangle determined by the points $O(0, 0)$, $A(a_1, a_2)$, $B(b_1, b_2)$.

Before continuing, we point out the meaning of the Euclidean Parallelism in this coordinate frame;

Let us consider $M(m_1, m_2)$ and $N(n_1, n_2)$.

Definition 2.1.4 The *lines AB and MN are Euclidean parallel* and we denote this by $MN||AB$, if the vectors

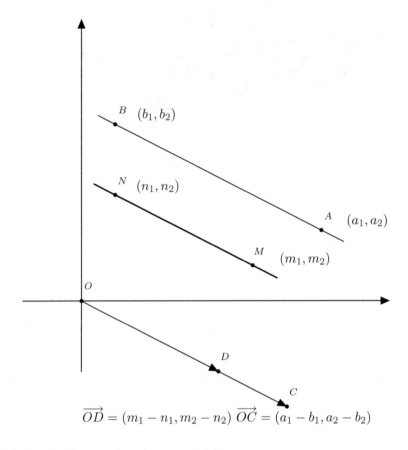

$$\overrightarrow{OD} = (m_1 - n_1, m_2 - n_2) \quad \overrightarrow{OC} = (a_1 - b_1, a_2 - b_2)$$

Fig. 2.1 Parallel lines seen through vector properties

$$\overrightarrow{OD} = (m_1 - n_1, m_2 - n_2)$$

and

$$\overrightarrow{OC} = (a_1 - b_1, a_2 - b_2)$$

are collinear, i.e. $\exists \beta \neq 0$ such that $(m_1 - n_1, m_2 - n_2) = \beta(a_1 - b_1, a_2 - b_2)$.

The *Generalized Pythagoras Theorem* in AOB asserts

$$|AB|^2 = |OA|^2 + |OB|^2 - 2|OA||OB|\cos\alpha,$$

where $|OA| = |\overrightarrow{OA}| = |a|$,

$$|AB| = |\overrightarrow{AB}| = |OC| = |a - b| = \sqrt{\langle a - b, a - b \rangle} = \sqrt{(a_1 - b_1)^2 + (a_2 - b_2)^2},$$

and $\angle AOB = \alpha$.

The formula explained and written above,

$$|AB| := \sqrt{(a_1 - b_1)^2 + (a_2 - b_2)^2}$$

is called *the Euclidean distance between the two points* $A(a_1, a_2)$, $B(b_1, b_2)$ of the plane.

Theorem 2.1.5 *In the previous notations, it is*

$$|AB|^2 = |OA|^2 + |OB|^2 - 2|OA||OB|\cos\alpha,$$

iff

$$\frac{\langle a, b \rangle}{|a||b|} = \cos\alpha.$$

Proof We observe that we need to prove only that

$$2|OA||OB|\cos\alpha = |OA|^2 + |OB|^2 - |AB|^2$$

is the same as

$$2|a||b| = 2\langle a, b \rangle.$$

Or, this means

$$|OA|^2 + |OB|^2 - |AB|^2 = 2\left(\vec{OA}, \vec{OB}\right),$$

and, in coordinates, this becomes a quick computation for the reader, that is

$$(a_1^2 + a_2^2) + (b_1^2 + b_2^2) - \left((a_1 - b_1)^2 + (a_2 - b_2)^2\right) = 2(a_1 b_1 + a_2 b_2)^2.$$

Corollary 2.1.6 If $\left(\vec{OA}, \vec{OB}\right) = 0$, i.e. the vectors a and b are orthogonal (Euclidean perpendicular), then we obtain the standard Pythagoras' Theorem.

The side AB is called a *hypotenuse*, and OA, OB are called *legs* of the triangle OAB.

Theorem 2.1.7 (Pythagoras' Theorem) *In the previous notations, it is*

$$|AB|^2 = |OA|^2 + |OB|^2.$$

The angle corresponding to orthogonal vectors is described by the condition $\cos\alpha = 0$, that is its measure is $\alpha = \dfrac{\pi}{2}$.

Therefore $\dfrac{\pi}{2}$ is the value of the right angle R. The sum of angles of a triangle in Euclidean Geometry becomes $\angle A + \angle B + \angle C = \pi$.

Theorem 2.1.8 (Thales Theorem) *Let us consider* $O(0,0)$, $A(x_1, x_2)$, $B(y_1, y_2)$, $A_1(\mu x_1, \mu x_2)$, $B_1(\lambda y_1, \lambda y_2)$. *Then,* $AB \| A_1 B_1$ *iff* $\lambda = \mu$.

Proof In coordinates $\overrightarrow{AB} = (y_1 - x_1, y_2 - x_2)$ and $\overrightarrow{A_1 B_1} = (\lambda y_1 - \mu x_1, \lambda y_2 - \mu x_2)$. The parallelism between AB and $A_1 B_1$ is equivalent to: $\exists \beta$ such that $(\lambda y_1 - \mu x_1, \lambda y_2 - \mu x_2) = \beta(y_1 - x_1, y_2 - x_2)$. Therefore

$$(\lambda - \beta)y_1 - (\mu - \beta)x_1 = 0,$$

$$(\lambda - \beta)y_2 - (\mu - \beta)x_2 = 0$$

for arbitrary x_1, x_2, y_1, y_2, that is $AB \| A_1 B_1$ iff $\lambda = \mu$.

Thales theorem can be written in the form:
Consider the triangle OAB, $A_1 \in OA$, $B_1 \in OB$.
Then $AB \| A_1 B_1$ *iff* $\dfrac{|OA|}{|OA_1|} = \dfrac{|OB|}{|OB_1|}$.

Problem 2.1.9 *Consider the triangle* OAB, $A_1 \in OA$, $B_1 \in OB$. *Then* $AB \| A_1 B_1$ *iff*
$$\frac{|OA|}{|OA_1|} = \frac{|OB|}{|OB_1|} = \frac{|AB|}{|A_1 B_1|}.$$

Hint. Construct a parallel from B to OA, denote by X the point of intersection between the parallel and $A_1 B_1$, apply Thales Theorem in the form $\dfrac{|B_1 X|}{|B_1 A_1|} = \dfrac{|B_1 B|}{|B_1 O|}$ and use the properties of proportions. $\square$

It is not very difficult to express line equations in the Euclidean plane.
If the line d passes through $A(a_1, a_2)$; $B(b_1, b_2)$ the *equation of d* is

$$y - a_2 = \frac{a_2 - b_2}{a_1 - b_1}(x - a_1).$$

The ratio denoted by m, $m := \dfrac{a_2 - b_2}{a_1 - b_1}$ is called a *slope* for the line d. The slope m has the value $m = \tan \alpha$, where α is the angle between the Ox and d in this order.

Exercise 2.1.10 *Show that two lines* d_1 *and* d_2 *are Euclidean perpendicular iff* $m_1 m_2 = -1$.

Hint. Use Euclidean exterior angle theorem and $\tan(\alpha + \beta)$ formula.

The *equation of a circle centered in* (a_1, a_2) *with radius r* is expressed with respect to the Euclidean distance between the center and a point (x, y) on the circle:

$$(x - a_1)^2 + (y - a_2)^2 = r^2.$$

The *interior of a circle C* is denoted by $int\,C$ and, between the two regions in which a circle divides the plane, it is the region containing its center. The Euclidean distance between the center and a point belonging to this region is less than the radius. The complementary region is called the *exterior of the circle*. The Euclidean distance between the center and a point belonging to this region is greater than the radius.

There are many properties related to circles and lines attached to triangles in the Euclidean Geometry. Some of them will be studied in the next chapter. The Euclidean plane is denoted by E^2.

2.2 Space-Like, Time-Like and Null Vectors in Minkowski Plane

When we are talking about a model of Minkowski Geometry in a plane, we have to start from the same vector space $\mathbb{R}^2$ over the field $\mathbb{R}$. Here, $x = (x_1, x_2)$, $y = (y_1, y_2)$ are called vectors, as in the Euclidean case.

The vector space operations are the same $x + y := (x_1 + y_1, x_2 + y_2)$ and $\lambda x := (\lambda x_1, \lambda x_2)$.

The *Minkowski product of the vectors x and y* is defined by

$$\langle x, y \rangle_M := x_1 y_1 - x_2 y_2$$

and the *Minkowski norm of x* by $|x|_M := \sqrt{|\langle x, x \rangle|_M} = \sqrt{|x_1^2 - x_2^2|}$.

Two vectors are called *Minkowski orthogonal* if their Minkowski product is null.

In a system of coordinates generated by the Minkowski orthogonal vectors $e_1 = (1, 0)$; $e_2 = (0, 1)$, the geometric meaning of the vector $x = (x_1, x_2)$ is the oriented segment $\overrightarrow{OA}$, lying from the origin O with the coordinates $(0, 0)$ and the endpoint A with the coordinates (x_1, x_2). This is exact as in the Euclidean case.

Even the parallelism is like in the Euclidean case; Consider $M(m_1, m_2)$ and $N(n_1, n_2)$.

Definition 2.2.1 The lines AB and MN are parallel and we denote this by $MN||AB$, if the vectors $\overrightarrow{OD} = (m_1 - n_1, m_2 - n_2)$ and $\overrightarrow{OC} = (a_1 - b_1, a_2 - b_2)$ are collinear, i.e. $\exists \beta \neq 0$ such that $(m_1 - n_1, m_2 - n_2) = \beta(a_1 - b_1, a_2 - b_2)$.

However, in a Minkowski space, we have three different kind of vectors $\overrightarrow{OA}$. Let us explain. There are space-like vectors, time-like vectors and null vectors.

A vector x is a *space-like vector* if $\langle x, x \rangle_M < 0$.

Examples are $b = (-1, 2)$, $e = (2, -3)$, or in general $a = (a_1, a_2)$ with $|a_1| < |a_2|$.

A vector x is a *time-like vector* if $\langle x, x \rangle_M > 0$.

Examples are $b = (3, 2)$, $e = (-4, -3)$, or in general $a = (a_1, a_2)$ with $|a_1| > |a_2|$.

A vector x is a *null vector* if $\langle x, x \rangle_M = 0$.

Examples are $b = (-1, 1), e = (2, 2)$, or in general $a = (a_1, a_2)$ with $|a_1| = |a_2|$.

The reader can observe that Minkowski orthogonal vectors have to be pairs, one space-like and one time-like. An example: $x = (x_1, x_2); v = (kx_2, kx_1)$.

Consider the 2×2 "hyperbolic rotation" matrix $A_\alpha = \begin{pmatrix} \cosh\alpha & \sinh\alpha \\ \sinh\alpha & \cosh\alpha \end{pmatrix}$ in which the basic hyperbolic trigonometric functions sine and cosine are involved as components.

$$\sinh\alpha = \frac{e^\alpha - e^{-\alpha}}{2}, \quad \cosh\alpha = \frac{e^\alpha + e^{-\alpha}}{2}.$$

This matrix is called a *hyperbolic rotation* and this name is legitimate by the next quick exercises.

As in the Euclidean case, $A_\alpha x$, $A_\alpha y$ are two matrices with two lines and one column, and it makes sense to consider the Minkowski product $\langle A_\alpha x, A_\alpha y \rangle_M$.

Exercise 2.2.2 $\langle A_\alpha x, A_\alpha y \rangle_M = \langle x, y \rangle_M$.

Hint. Use $\cosh^2\alpha - \sinh^2\alpha = 1$. Therefore

$$\langle A_\alpha x, A_\alpha y \rangle_M = (x_1 \cosh\alpha + x_2 \sinh\alpha)(y_1 \cosh\alpha + y_2 \sinh\alpha) -$$

$$-(x_1 \sinh\alpha + x_2 \cosh\alpha)(y_1 \sinh\alpha + y_2 \cosh\alpha) =$$

$$= x_1 y_1 - x_2 y_2 = \langle x, y \rangle_M$$

We leave the reader to prove that it does not exist α such that $A_\alpha e_1 = e_2$.

Or, more general, after rotating a time-like (space-like) vector we cannot obtain a space-like (time-like) vector. As we will see below, this property is related to causality in Relativity.

Exercise 2.2.3 $|A_\alpha x|_M = |x|_M$.

So, the matrices A_α preserves the Minkowski type and the Minkowski length of vectors.

Definition 2.2.4 If $u = A_\alpha v$ we say that α is the *oriented hyperbolic angle* between v and u. Obviously, $-\alpha$ is the oriented hyperbolic angle between u and v.

Next, we discuss about time-like vectors properties.

A *future-pointing time-like vector* v fulfills the property $\langle v, e_1 \rangle_M > 0$. An example is $v = (3, 2)$.

Otherwise the vector v is a *past-pointing space-like vector*. $v = (-3, -2)$ is an example, and the reader can observe that if we consider the lines $d_1 : x_2 = x_1$ and $d_2 : x_2 = -x_1$ which describe the *null cone*, a future-pointing time like vector is a vector $v = \overrightarrow{OA}$ with $A = (a_1, a_2)$ included in the interior of the angle $\angle d_1 d_2$ (i.e. $|a_1| > |a_2|$) such that $a_1 > 0$.

Exercise 2.2.5 *If v is a future-pointing time-like vector, then $A_\alpha v$ is a future-pointing time-like vector.*

Hint. Since we have proved that the time-like property is kept after a hyperbolic rotation, it remains to prove that the future-pointing property is preserved.

Or $\langle A_\alpha v, e_1 \rangle_M = a_1 \cosh \alpha + a_2 \sinh \alpha$.

We have $|a_1| > |a_2|$ and $|\sinh \alpha| < \cosh \alpha$, i.e. $|a_1 \cosh \alpha| > |a_2 \sinh \alpha|$.

It remains to observe that there are triangles in this Minkowski Geometry in which the meaning of angle does not exist. The triangles in which we can discuss about angles are called pure triangles, i.e. in such triangles all the sides are time vectors, all pointing towards the future (or, all pointing towards the past). How we can create such a triangle? We start with two, say, future-pointing time-like vectors, $x = (x_1, x_2)$, $y = (ky_1, ky_2)$ and we choose $k > 0$ such that $y - x = (ky_1 - x_1, ky_2 - x_2)$ is future-pointing time-like vector.

Exercise 2.2.6 *If x and y are future-pointing time-like vectors then*
1. $\langle x, y \rangle_M > 0$
2. $x + y$ is a future-pointing time-like vector
3. $\langle x, y \rangle_M \geq |x|_M |y|_M$, where the equality happens iff $y = kx$
4. $|x + y|_M \geq |x|_M + |y|_M$, the equality happens iff $y = kx$.

2.3 Minkowski-Pythagoras Theorems

Let us start with a simple exercise.

Exercise 2.3.1 *If x is a space-like vector such that $|x|_M = 1$ then $\langle A_\alpha x, x \rangle_M = -\cosh \alpha$.*

Hint. $\langle A_\alpha x, x \rangle_M = (x_1 \cosh \alpha + x_2 \sinh \alpha) x_1 - (x_1 \sinh \alpha + x_2 \cosh \alpha) x_2 = (x_1^2 - x_2^2) \cosh \alpha = -\cosh \alpha$.

Denote by u the unitary vector $A_\alpha x$. The previous relation for the unitary vectors u, x can be written in the form $\langle u, x \rangle_M = -\cosh \alpha$.

For two arbitrary future-pointing space-like vectors a, b, the vectors $\dfrac{a}{|a|_M}, \dfrac{b}{|b|_M}$ are unitary and the previous relation becomes

$$\left\langle \frac{a}{|a|_M}, \frac{b}{|b|_M} \right\rangle_M = -\cosh\alpha,$$

i.e.

$$\frac{\langle a, b\rangle_M}{|a|_M|b|_M} = -\cosh\alpha.$$

According to the Euclidean case, this last formula can be called the *Generalized Minkowski-Pythagoras theorem*.

Consider a *Minkowski right triangle* OAB, i.e. a triangle such that the vectors $\overrightarrow{OA}$ and $\overrightarrow{OB}$ are Minkowski orthogonal, that is $\left\langle \overrightarrow{OA}, \overrightarrow{OB} \right\rangle_M = 0$. The side AB is called a *Minkowski hypotenuse*, and OA, OB are called *Minkowski legs* of the triangle OAB.

An example is given by $\overrightarrow{OA} = (0, a)$, $a > 0$; $\overrightarrow{OB} = (b, 0)$, $b > 0$. When we consider the vector $\overrightarrow{AB} = (b, -a)$ it depends by the absolute values $|a|$, b if this vector is a time-like vector, a space-like vector or a null vector.

So, the Minkowski-Pythagoras Theorem asserts that "*in a Minkowski right triangle, the square of the Minkowski hypotenuse is the difference of the square of Minkowski legs.*"

The endpoints of unitary space-like vectors determine a Minkowski circle. The equation of this circle is $x^2 - y^2 = -1$. From the Euclidean point of view this is a hyperbola equation.

The endpoints of unitary time-like vectors determine a Minkowski circle, too. The equation of this circle is $x^2 - y^2 = 1$.

Exercise 2.3.2 *What kind of triangle is determined by three arbitrary points of the Minkowski circle $x^2 - y^2 = -1$?*

The answer is: a *pure time-like triangle*, i.e. a triangle in which each side is a time-like vector pointing the future (or all three pointing the past).

There are a lot of nice geometric properties for Minkowski circles, some of them similar to Euclidean properties. For our purpose the facts highlighted above are enough to continue. The Minkowski plane is denoted by M^2.

Chapter 3
Geometric Inversion, Cross Ratio, Projective Geometry and Poincaré Disk Model

Virtus unita fortior agit.

Abstract This chapter is devoted to a first model of Non-Euclidean Geometry. To construct this model, we need to deal with one of the most important transformations of the Euclidean plane, the geometric inversion. We still need some other acquirements, therefore we meet the Projective Geometry. An invariant described by a special projective map of a circle allows us to construct a non-Euclidean distance inside the disk. Elaborating the previous model we highlight the Poincaré disk model.

3.1 Geometric Inversion and Its Properties

The geometric inversion is a classical transformation of elementary Euclidean Plane Geometry. To describe it, let us consider a circle centered at O and radius R, denoted $C(O, R)$.

A *geometric inversion of center* $O \in E^2$ *and radius* R maps each point $M \in E^2$, $M \neq O$ to the point N on the radius OM such that $|OM| \cdot |ON| = R^2$, where $|OM|$ is the Euclidean length between the points O and M.

The circle $C(O, R)$ is called a *circle of inversion*.

The points *M and N are called homologous inverse points* with respect to the previous geometric inversion determined by the circle of inversion $C(O, R)$. R^2 is called *power of inversion*.

We prefer to use "inversion of center $O \in E^2$ and power R^2" instead of "inversion of center $O \in E^2$ and radius R".

Suppose we know the homologous inverse points M and N with respect to a geometric inversion having O as a center and R^2 as a power and the order of points on the radius OM is O, M, N, or O, N, M i.e. $O \notin (MN)$. This is a *direct inversion*, when the oriented segments OM and ON have the same direction.

W. Boskoff and S. Capozziello, *A Mathematical Journey to Relativity*, UNITEXT for Physics, https://doi.org/10.1007/978-3-030-47894-0_3

Now choose N' the symmetric of N with respect to O, i.e. $|ON'| = |ON|$. We have N' on the radius OM, $|OM| \cdot |ON'| = R^2$ and $O \in (MN')$.

Therefore, for the inverse N of a point M, with respect to a given inversion, we have two possibilities:

(1) O does not belong to the segment (MN),
(2) O belongs to the segment (MN).

To conclude, when we are talking about an inversion and the inverse N belongs to the radius OM, we have to specify if it is the *direct geometric inversion* i.e. we are talking about the map

$$T_{O;R^2} : E^2 - \{O\} \to E^2 - \{O\}, \ T_{O;R^2}(M) = N, \ O \notin (MN), \ |OM| \cdot |ON| = R^2,$$

or we are talking about the map

$$T^s_{O;R^2} : E^2 - \{O\} \to E^2 - \{O\}, \ T^s_{O;R^2}(M) = N, \ O \in (MN), \ |OM| \cdot |ON| = R^2,$$

which can be called a *symmetric geometric inversion*.

All next results are done for the direct geometric inversion, that is for the map $T_{O;R^2}$. All the properties obtained can be easily transferred by symmetry with respect O for the symmetric geometric inversion. When in a problem we use an inversion, the reader finds the information if it is a direct one or symmetric one looking at the map involved, i.e. $T_{O;R^2}$ or $T^s_{O;R^2}$.

The main *properties of the direct geometric inversion* are:
1. If $T_{O;R^2}(M) = N$, then $T_{O;R^2}(N) = M$.
In simple words, if N is the inverse of M, then M is the inverse of N.
That is,

$$T_{O;R^2}(T_{O;R^2}(M)) = M.$$

This property can be written in a simpler form as

$$T^2 = id_{E^2 - \{O\}}$$

and it highlights that the geometric inversion is an idempotent transformation.

2. $T_{O;R^2}(C(O, R)) = C(O, R)$, that is the circle of inversion is invariant under the inversion it generates.

3. A line d which passes through the pole of inversion is invariant under inversion, i.e.

$$T_{O;R^2}(d - \{O\}) = d - \{O\}.$$

Before to continue, some notions are needed. A *cyclic quadrilateral ABCD* is a quadrilateral which vertexes A, B, C, D belong to a circle Γ, called the *circumcircle* of the quadrilateral.

4. If $T_{O;R^2}(A) = B$ and $T_{O;R^2}(C) = D$, then $ABDC$ is a cyclic quadrilateral.

Fig. 3.1 Inversion main figure

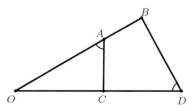

Fig. 3.2 The inverse point of an interior point M

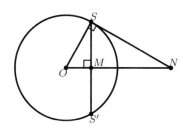

Fig. 3.3 The inverse point of an exterior point M

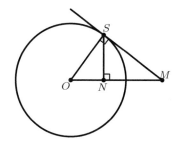

Hint: From $|OA| \cdot |OB| = |OC| \cdot |OD| = R^2$ it results $\dfrac{|OA|}{|OD|} = \dfrac{|OC|}{|OB|}$. Then, triangles $\triangle OAC$ and $\triangle ODB$ are similar, i.e. $\angle OAC = \angle CDB$, that is the quadrilateral $ABDC$ is a cyclic one (Fig. 3.1).

Why the circle of inversion is important? Because it allows us to construct the inverse of a point.

5. Construction of the inverse of a given point (Fig. 3.2).

Suppose M belongs to $int C(O, R)$.

We consider the radius OM and the perpendicular line to OM in M which intersects the circle at the points S and S'. Next, we refer to S. The tangent at S intersects the radius OM in N.

If we look at the right triangle $\triangle OSN$, $R^2 = |OS|^2 = |OM| \cdot |ON|$, i.e. N is the (direct) inverse of M.

Suppose M is outside the circle of inversion (Fig. 3.3).

We construct the radius OM and one of the tangent to the circle, MS, $S \in C(O, R)$. The perpendicular from S to OM intersects OM in N. In the right triangle $\triangle OSM$ we have $R^2 = |OS|^2 = |OM| \cdot |ON|$, i.e. N is the inverse of M.

Fig. 3.4 The inverse of a
line d such that $O \notin d$

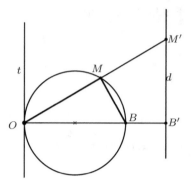

Let us observe that in the above two situations the circle pases through $SMS'N$ is orthogonal to the circle of inversion. If $M \in C(O, R)$, $N = M$, that is the inverse of M is M itself.

6. Consider a line $d \subset E^2$, $O \notin d$. Then, $T_{O;R^2}(d) = C - \{O\}$, i.e. the inverse of the line d is a circle $C - \{O\}$, such that the tangent line in O to the circle C is parallel to d (Fig. 3.4).

Proof Denote by B' the intersection between d and the perpendicular line from O to d. The inverse of B' is B. Consider a point $M' \in d$ and its inverse M. The quadrilateral $B'BMM'$ is cyclic, therefore $\angle OMB$ is a right one, i.e. when M' belongs to d, M belongs to the circle having $(OB]$ as a diameter. Since the diameter BO is perpendicular to the tangent denoted by t in O to the circle, it results $d \parallel t$. $\square$

7. The inverse of a circle C passing through O is line d, $O \notin d, d \parallel t$, where t is the tangent at O to the circle C.

Proof The inversion $T_{O;R^2}$ is an idempotent transformation. If we are looking backwards at the previous property of inversion the result is obvious. $\square$

8. The inverse of a circle C_1, $O \notin C_1$ is a circle C_2, $O \notin C_2$, i.e. $T_{O;R^2}(C_1) = C_2$ (Fig. 3.5).

Proof Consider the radius OO_1 where O_1 is the center of the circle C_1. Denote $\{A, B\} := OO_1 \cap C_1$ and suppose the order of points is O, A, O_1, B.

Consider A_1, B_1 the inverses of A, B respectively. Since $|OA| < |OB|$ and $|OA| \cdot |OA_1| = |OB| \cdot |OB_1| = R^2$ it results $|OA_1| > |OB_1|$. Without losing the generality, we can suppose the order of points on the radius OO_1 is O, A, O_1, B, B_1, A_1. Consider $M \in C_1$ and its inverse M_1.

Using the cyclic quadrilaterals AA_1M_1M and BB_1M_1M it results both $\angle OAM = \angle MM_1A_1$ and $\angle ABM = \angle MM_1B_1$.
Since $\angle OAM = \dfrac{\pi}{2} + \angle MBA = \angle MM_1A_1 = \angle MM_1B_1 + \angle B_1M_1A_1$, we have

$$\angle B_1M_1A_1 = \frac{\pi}{2},$$

Fig. 3.5 Inversion of circles

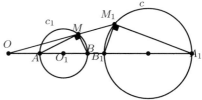

Fig. 3.6 The inverse of a
circle orthogonal to the
fundamental circle of
inversion

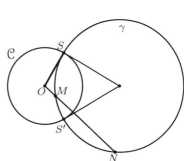

that is M_1 belongs to a circle of diameter $B_1 A_1$. □

9. Consider $T_{O,R^2}(A) = A_1$, $T_{O,R^2}(B) = B_1$. Then $|A_1 B_1| = R^2 \cdot \dfrac{|AB|}{|OA| \cdot |OB|}$.

Proof The triangles $\triangle OAB$ and $\triangle OB_1 A_1$ are similar, therefore $\dfrac{|A_1 B_1|}{|AB|} = \dfrac{|OA_1|}{|OB|}$.

It results

$$\frac{|A_1 B_1|}{|AB|} = \frac{|OA_1|}{|OB|} \cdot \frac{|OA|}{|OA|},$$

that is $|A_1 B_1| = R^2 \cdot \dfrac{|AB|}{|OA| \cdot |OB|}$. □

10. Orthogonal circles to the circle of inversion are preserved by inversion (Fig. 3.6).

Proof Denote by S, S' the intersection points between the circle of inversion $C(O, R)$ and the orthogonal circle γ. Consider $M, N \in \gamma$ such that O, M, N are collinear points. Since $|OM| \cdot |ON| = |OS|^2 = R^2$, it results $T_{O,R^2}(M) = N$, i.e. $T_{O,R^2}(\gamma) = \gamma$. □

11. The inversion is a *conformal map*, i.e. it preserves the angles between curves.

Proof The angle between two curves at their point of intersection, S, is the angle between the tangent lines at S to the curves. Let $\Gamma_1, \Gamma_2; \Gamma_1^1, \Gamma_2^2$ be four curves such that $T_{O,R^2}(\Gamma_1) = \Gamma_1^1, T_{O,R^2}(\Gamma_2) = \Gamma_2^2$, $O \notin \Gamma_1 \cup \Gamma_2 \cup \Gamma_1^1 \cup \Gamma_2^2$, $S \in \Gamma_1 \cap \Gamma_2$, $T_{O,R^2}(S) = S'$, $T_{O,R^2}(M_1) = M_1^1$, $T_{O,R^2}(N_1) = N_1^1$. The quadrilateral $SM_1 M_1^1 S'$ and $SM_2 M_2^2 S'$ are cyclic therefore $\angle M_1 S M_2 = \angle M_1^1 S' M_2^2$. When

Fig. 3.7 Ptolemy's Theorem

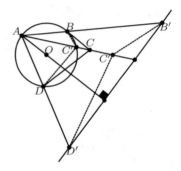

radius OM approaches OS the previous angles are still equal. The limit position highlights the previous angles as angles between tangent lines to the curves.   □

Examples of problems solved by inversion.

Problem 3.1.1 (Ptolemy's Theorem) The products of the lengths of two diagonals of a quadrilateral is less than or equal to the sum of the products of opposite sides and the equality holds if and only if the quadrilateral is a cyclic one (Fig. 3.7).

Solution. (Hint) Consider the inversion of center A and arbitrary power $k > 0$ and denote by B', C', D' the inverses of the points B, C, D. We have

$$|B'D'| = k \cdot \frac{|BD|}{|AB| \cdot |AD|}, \quad |B'C'| = k \cdot \frac{|BC|}{|AB| \cdot |AC|}, \quad |C'D'| = k \cdot \frac{|CD|}{|AC| \cdot |AD|}.$$

Replacing in $|B'D| \le |B'C'| + |C'D'|$, and taking into account that the equality happens when B', C', D' are collinear it results the statement.   □

Problem 3.1.2 Consider two pairs of circles, γ_x, Γ_x; γ_y, Γ_y which pass through the same point O having the centers on perpendicular axes Ox, Oy. Then the four points of mutual intersection are cyclic.

Solution. (Hint) Consider an inversion of center O and power $k > 0$, $T_{O,k}$. The circles γ_x, Γ_x; γ_y, Γ_y which pass through O are mapped into a rectangle $A'B'C'D'$ whose vertexes comes from A, B, C, D respectively. Since a rectangle allows a circumcircle, by inversion, this circumcircle comes from the circle containing the initial points A, B, C, D.   □

Problem 3.1.3 Two circles intersect at A and B. The tangent lines at A to the circles intersect the circles at M and N. Let B_1 be the symmetric of A with respect to B. Prove that the quadrilateral AMB_1N is a cyclic one.

Solution. Consider an inversion of center A and power $k > 0$, $T_{A,k}$. The three lines AM, AN and AB are transformed after the rule: $T_{A,k}(AM) = AM$, $T_{A,k}(AN) = AN$, $T_{A,k}(AB) = AB$, $T_{A,k}(M) = M_1$, $T_{A,k}(N) = N_1$, $T_{A,k}(B) = B'$. Since the circles passing through A are transformed into lines parallel to the tangents AM and AN it is easy to deduce that the quadrilateral $AM_1B'N_1$ is a parallelogram. The point B_1 is mapped by inversion into B_1' such that $|AB_1| \cdot |AB_1'| = k = 2|AB| \cdot \frac{1}{2}|AB'|$, i.e. B_1' is the center of the previous parallelogram. Therefore the diagonal $M_1 N_1$ which contains B_1' comes from the inversion of a circle containing the points A, M, B_1, N. $\qquad\square$

For the next problem the reader has to know what is an inscribed circle for a given triangle, and the fact that "the lines which connect the vertexes to the opposite tangent points (of the circle with the sides) are concurrent lines". The point of concurrence is called *Gergonne's point*.

Problem 3.1.4 Denote by $C(O, R)$ the circumcircle of the triangle $\triangle ABC$, A_1, B_1, C_1 the midpoints of the sides $[BC]$, $[CA]$, $[AB]$ respectively.
Prove that the circles Γ_{AOA_1}, Γ_{BOB_1}, Γ_{COC_1} have a common point E, $E \neq O$.

Solution. An inversion of center O and power R^2 preserves A, B, C and $C(O, R)$.
The circles Γ_{AOA_1}, Γ_{BOB_1}, Γ_{COC_1} are mapped into lines passing through A, B, C respectively.
$T_{O, R^2}(A_1) = A_2$ such that $|OA_1| \cdot |OA_2| = |OB|^2 = |OC|^2 = R^2$, i.e. A_2 is the intersection between the tangents at B and C to $C(O, R)$. In the same way we obtain the points B_2 and C_2. If we look at the triangle $\triangle A_1 B_1 C_1$ which has as inscribed circle $C(O, R)$, the lines $A_1 A$, $B_1 B$, $C_1 C$ intersects at Gergonne's point. The inverse of Gergonne's point is E. $\qquad\square$

3.2 Cross Ratio and Projective Geometry

Consider four distinct collinear points A, B, C, D on the line d. Attach them the coordinates x_A, x_B, x_C, x_D, respectively. Choose two possible ordered pairs, say (A,B); (C,D), that is, consider the ordered pairs of coordinates (x_A, x_B); (x_C, x_D).

Definition 3.2.1 The cross ratio of four ordered points is the real number

$$[AB; CD] := \frac{x_C - x_A}{x_C - x_B} : \frac{x_D - x_A}{x_D - x_B}.$$

One can see that the definition can be written in the form

$$[AB; CD] := \frac{CA}{CB} : \frac{DA}{DB},$$

but in this case we have to point that if the order on d for the points A and C is given by $x_A < x_C$, then the meaning of CA is $|CA|$, and if $x_A > x_C$ we have $CA = -|CA|$.

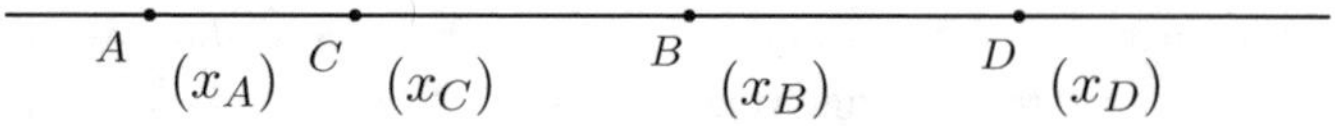

Fig. 3.8 Cross ratio

Exercise 3.2.2 If the order of points A, B, C, D on d is given by $x_A < x_B < x_C < x_D$, then $[AB; CD] > 0$.

Exercise 3.2.3 If the order of points A, B, C, D on d is given by $x_A < x_C < x_B < x_D$, then $[AB; CD] < 0$.

Exercise 3.2.4 If the order of points A, B, C, D on d is given by $x_A < x_B < x_C < x_D$, then

$$[AD; CB] = \frac{x_C - x_A}{x_C - x_D} : \frac{x_B - x_A}{x_B - x_D} > 0.$$

Observe that in this last case the ordered pairs are (A, D); (C, B) and the cross ratio can be written in the equivalent form $[AD; CB] := \dfrac{CA}{CD} : \dfrac{BA}{BD}$ with the meaning explained above (Fig. 3.8).

Exercise 3.2.5 $[AD; BC] + [AB; DC] = 1$ if and only if the order of points on the line d is A, B, C, D.

Hint.

$$[AD; BC] + [AB; DC] = \frac{BA \cdot CD}{BD \cdot CA} + \frac{DA \cdot CB}{DB \cdot CA} = \frac{BA \cdot CD + DA \cdot CB}{BD \cdot CA} = 1.$$

If $A(x_A)$, $B(x_B)$, ... etc.,

$$(x_B - x_A)(x_D - x_C) + (x_D - x_A)(x_C - x_B) = (x_C - x_A)(x_D - x_B)$$

iff the order is as in the statement before.

Exercise 3.2.6 $[AD; BC] = [DA; CB]$.

Exercise 3.2.7 Consider $A(-1)$, $B(0)$, $C(1)$, $D(x)$. Determine x such that

$$[AC; BD] = -1.$$

Hint. If we write the given condition, it results $x + 1 = x - 1$. There is no real x.

To maintain the possibility to have four distinct points with a given cross ratio, as well as for a given cross ratio and three distinct points to exist a fourth point such that the cross ratio is a given one, we have to accept that for each line d it exists an abstract point, denoted ∞, such that for $A \neq B$, $\dfrac{\infty A}{\infty B} = 1$. This point is called *point at infinity for the line d*.

Fig. 3.9 Pappus' Theorem

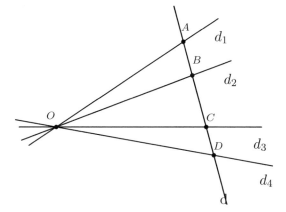

The cross ratio of collinear points can be extended to *pencils of lines*. Consider the lines d_1, d_2, d_3, d_4 and $\{O\} = d_1 \cap d_2 \cap d_3 \cap d_4$. Let d be an arbitrary line and $\{A\} = d \cap d_1$; $\{B\} = d \cap d_2$; $\{C\} = d \cap d_3$; $\{D\} = d \cap d_4$. Choose two ordered pairs of lines, say (d_1, d_2); (d_3, d_4).

By definition $[d_1 d_2; d_3, d_4] := [AB; CD]$.

If we look at this definition it seems that it depends on the line d we choose. Therefore, we have to prove that if we choose another line d' and $\{A'\} = d' \cap d_1$; $\{B'\} = d' \cap d_2$; $\{C'\} = d' \cap d_3$; $\{D'\} = d' \cap d_4$, then $[AB; CD] = [A'B'; C'D']$.

Theorem 3.2.8 (Pappus' Theorem) *The cross ratio of four lines in a pencil depends only by the angles of the pencil (Fig. 3.9).*

Proof We are in the case: the pencil of lines d_1, d_2, d_3, d_4 with $\{O\} := d_1 \cap d_2 \cap d_3 \cap d_4$, the arbitrary line d and $\{A\} = d \cap d_1$; $\{B\} = d \cap d_2$; $\{C\} = d \cap d_3$; $\{D\} = d \cap d_4$; suppose the order on d being A, B, C, D and use four times sine theorem:

$$\frac{CA}{\sin \angle COA} = \frac{OC}{\sin \angle OAC};$$

$$\frac{CB}{\sin \angle COB} = \frac{OC}{\sin \angle OBD};$$

$$\frac{DA}{\sin \angle DOA} = \frac{OD}{\sin \angle OAD};$$

$$\frac{DB}{\sin \angle DOB} = \frac{OD}{\sin \angle OBD};$$

therefore $[AB; CD] := \dfrac{CA}{CB} : \dfrac{DA}{DB} = \dfrac{\sin \angle COA}{\sin \angle COB} : \dfrac{\sin \angle BOA}{\sin \angle BOD}.$ □

Fig. 3.10 Existence of poits at infinity

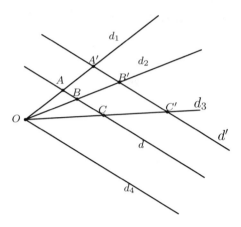

Observe that in fact the cross ratio depends on the sine of angles.

Another approach can be

Theorem 3.2.9 *If O is the origin and the lines of the pencil are $d_k : y = m_k \cdot x$, $k \in \{1, 2, 3, 4\}$, then*

$$[d_1 d_2; d_3 d_4] = \frac{m_3 - m_1}{m_3 - m_2} : \frac{m_4 - m_1}{m_4 - m_2}.$$

Proof Consider d having the equation $x = 1$. The points A, B, C, D on d have the coordinates $(1, m_1), (1, m_2), (1, m_3), (1, m_4)$.   □

The pencils of lines allow us to better understand the points at infinity of lines. As above, consider the lines d_1, d_2, d_3, d_4 having the property $\{O\} = d_1 \cap d_2 \cap d_3 \cap d_4$. Let d be parallel to d_4 and $\{A\} = d \cap d_1$; $\{B\} = d \cap d_2$; $\{C\} = d \cap d_3$ (Fig. 3.10).

In this case we have to consider the point at infinity to define $[d_1 d_2; d_3 d_4]$;

$[d_1 d_2; d_3 d_4] = [AB; C\infty] = \dfrac{|CA|}{|CB|}$. If we consider another line, say d', such that $d' \parallel d$ and if we denote by $\{A'\} = d' \cap d_1$, $\{B'\} = d' \cap d_2$, $\{C'\} = d \cap d_3$, then $[d_1 d_2; d_3, d_4] := [A'B'; C\infty]$. The lines d, d', d_4 have empty intersection in E^2. This abstract point who doesn't belong to the Euclidean plane, say ∞, can be taught as the intersection of parallel lines d, d' with d_4.

We define for all parallel lines the same abstract point ∞.

If, in a system of coordinates, all the parallel lines have the slope m, we may think that this point at infinity is attached to this slope. We can even denote this point by ∞_m. An interesting question can be asked: which geometrical structure will be assigned to $\{\infty_m, \ m \in \mathbb{R}\}$? It can be taught as an abstract line? Or it is more intuitive to be taught as an abstract circle? Or it is something else? We see the answer a little bit later.

The cross ratio can be extended to four points distinct points A, B, C, D on a circle Γ. Choose $M \in \Gamma$ and the pencil determined by the rays $d_1 = MA, d_2 = MB, d_3 = MC, d_4 = MD$.

By definition, $[AB; CD]_\Gamma := [d_1d_2; d_3d_4]$.

Pappus' theorem shows that this definition is independent of the choice of M. Here, it is important our observation related to the fact that the cross ratio of pencils depends on sine of angles. Since $\sin\alpha = \sin(\pi - \alpha)$, the point M can be chosen even between two consecutive points, that is we can have, for a given sense on our circle, even the order A, B, M, C, D. The cross ratio is the same as for the order, say M, A, B, C, D.

Next theorem shows that the previous cross ratio $[AC; BD]_\Gamma$ can be transferred to the segment lines BA, BC, DA, DC determined by the four distinct points on the circle. We keep our notation generated by the order of points, now on the circle. To have a clear statement, for a chosen sense on our circle, let us consider the points M, A, B, C, D in this order. Denote the angles of the pencil created by $\angle AMB = \alpha, \angle BMC = \beta, \angle CMD = \gamma$.

Theorem 3.2.10 $[AC; BD]_\Gamma = \dfrac{BA}{BC} : \dfrac{DA}{DC}$

Proof Consider the segment line $[AD]$ and its intersection with MB, MC denoted by B_1, C_1 respectively. The order on the segment line $[AD]$ is then A, B_1, C_1, D. If we denote by R the radius of Γ we have $|AB| = 2R\sin\alpha, |BC| = 2R\sin\beta, |AD| = 2R\sin(\alpha + \beta + \gamma), |CD| = 2R\sin\gamma$. Taking the order into consideration, we can write

$$[AC; BD]_\Gamma = [d_1d_3; d_2d_4] = \frac{\sin\alpha}{\sin\beta} : \frac{\sin(\alpha + \beta + \gamma)}{\sin\gamma} = \frac{BA}{BC} : \frac{DA}{DC}$$

$\square$

Exercise 3.2.11 $[AD; BC]_\Gamma + [AB; DC]_\Gamma = 1$ if and only if the order of points on the circle Γ is A, B, C, D.

Solution. (Hint) We use the previous theorem, i.e. we express the

$$[AD; BC]_\Gamma + [AB; DC]_\Gamma = \frac{BA \cdot CD}{BD \cdot CA} + \frac{DA \cdot CB}{DB \cdot CA} = \frac{BA \cdot CD + DA \cdot CB}{BD \cdot CA} = 1$$

iff $BA \cdot CD + DA \cdot CB = BD \cdot CA$, that is Ptolomy's equality must happen. $\square$

Theorem 3.2.12 *Let I be an interior point of $C(O, r)$. Consider the chords AA', BB', CC', DD' such that $\{I\} = AA' \cap BB' \cap CC' \cap DD'$ and the order is $A, B, C, D, A', B', C', D'$.*

Then $[A'C'; B'D']_\Gamma = [AC; BD]_\Gamma$.

Proof Let us observe that the symmetric inversion $T^s_{I, R^2 - OI^2}$ maps the circle $C(O, R)$ in itself, since $R^2 - OI^2$ is the power of I with respect to the circle, and $|IA| \cdot |IA'| = |IB| \cdot |IB'| = |IC| \cdot |IC'| = |ID| \cdot |ID'| = R^2 - OI^2$, that is

$T^s_{I,R^2-OI^2}(A) = A'$, $T^s_{I,R^2-OI^2}(B) = B'$, $T^s_{I,R^2-OI^2}(C) = C'$, $T^s_{I,R^2-OI^2}(D) = D'$. It results

$$|B'A'| = (R^2 - OI^2) \cdot \frac{|BA|}{|IB| \cdot |IA|}, \quad |B'C'| = (R^2 - OI^2) \cdot \frac{|BC|}{|IB| \cdot |IC|},$$

$$|D'A'| = (R^2 - OI^2) \cdot \frac{|DA|}{|ID| \cdot |IA|}, \quad |D'C'| = (R^2 - OI^2) \cdot \frac{|DC|}{|ID| \cdot |IC|}.$$

Taking into consideration the established order and the theorem which transfers the cross ratio from circle to segment lines, we obtain $[A'C'; B'D']_\Gamma = [AC; BD]_\Gamma$. $\square$

We obtain a similar result for a point J outside the circle using a direct inversion T_{J,OJ^2-R^2} and $T_{J,OJ^2-R^2}(A) = A'$, etc. $[A'C'; B'D']_\Gamma = [AC; BD]_\Gamma$

More general, *an inversion $T_{O,k}$ leaves unchanged the cross ratio of four collinear points or the cross ratio of four cyclic points*. It doesn't matter if the four collinear points are mapped into cyclic points, or the cyclic points are mapped into cyclic (or collinear) points. This result is a fundamental one.

Definition 3.2.13 A projective map of a circle $C(O, R)$ is a one to one function $f : C(O, R) \to C(O, R)$ such that for any four points A_i, $i \in \{1, 2, 3, 4\}$ and their images $B_i = f(A_i)$, it happens

$$[A_1 A_2; A_3 A_4]_{C(O,R)} = [B_1 B_2; B_3 B_4]_{C(O,R)}.$$

Definition 3.2.14 The points which correspond in a projective map f, eg. A_i and $f(A_i)$, are called homologous points.

According to a previous result, let observe that the symmetric inversion T^s_{I,R^2-OI^2}, $I \in int C(O, R)$ is a projective map of the circle $C(O, R)$.

More, $T^s_{R^2-OI^2}$ can be identified with the simpler map determined by the point I denoted

$$I : C(O, R) \to C(O, R), \ I(A) = A',$$

where $A' \neq A$ is the other intersection between AI and the circle $C(O, R)$.

The same for the direct inversion T_{I,OJ^2-R^2}, $J \in ext C(O, R)$. This is a projective map and can be identified with $J : C(O, R) \to C(O, R)$, $J(A) = A'$, where $A' \neq A$ is the other intersection between JA and the circle $C(O, R)$.

If we consider $I_1, I_2 \in int C(O, R)$ it can be obtained that $T^s_{I_1,R^2-OI_1^2} \circ T^s_{I_2,R^2-OI_2^2} : C(O, R) \to C(O, R)$ is a projective map.

Definition 3.2.15 A projective map between two lines d_1 and d_2 is a one to one function $f : d_1 \to d_2$ such that for any four points $A_i \in d_1$, $i \in \{1, 2, 3, 4\}$ and their images $B_i = f(A_i) \in d_2$, it happens

$$[A_1 A_2; A_3 A_4] = [B_1 B_2; B_3 B_4].$$

Since the previous definition has also sense for $f : d \to d$, we may talk about *projective maps on d*.

Theorem 3.2.16 *A projective map between two lines is determined by three pairs of homologous points.*

Proof Denote the homologous points in the form $A \to B$ instead of $B = f(A)$, because this notation will help us later.

Then, we know the three pairs of homologous points $A_0 \to B_0$, $A_1 \to B_1$, $A_2 \to B_2$.

We have to show that for any four arbitrary points A_i, A_j, A_k, A_l and their homologous B_i, B_j, B_k, B_l the relation $[A_i A_j; A_k A_l] = [B_i B_j; B_k B_l]$ is deduced from $[A_0 A_1; A_2 A_i] = [B_0 B_1; B_2 B_i]$ using successively the indexes i, j, k and l.

The idea is to find somehow a procedure of replacement of the homologous points initially given.

We also have $[A_1 A_2; A_0 A_j] = [B_1 B_2; B_0 B_j]$ and $[A_1 A_2; A_i A_0] = [B_1 B_2; B_i B_0]$.
It results

$$[A_1 A_2; A_0 A_j] \cdot [A_1 A_2; A_j A_0] = [B_1 B_2; B_0 B_j] \cdot [B_1 B_2; B_i B_0],$$

that is

$$[A_1 A_2; A_i A_j] = [B_1 B_2; B_i B_j].$$

We succeeded to replace the pair of homologous points $A_0 \to B_0$.

Then, the from previous relation and $[A_1 A_k; A_j A_i] = [B_1 B_k; B_i B_j]$ we have

$$[A_1 A_2; A_i A_j] \cdot [A_1 A_k; A_j A_i] = [B_1 B_2; B_i B_j] \cdot [B_1 B_k; B_i B_j],$$

i.e.

$$[A_2 A_k; A_j A_i] = [B_2 B_k; B_j B_i].$$

Finally, taking into consideration the previous result and $[A_l A_2; A_j A_i] = [B_l B_2; B_j B_i]$ it results

$$[A_2 A_k; A_i A_j] \cdot [A_l A_2; A_j A_i] = [B_2 B_k; B_j B_i] \cdot [B_l B_2; B_j B_i],$$

that is

$$[A_l A_k; A_j A_i] = [B_l B_k; B_j B_i].$$

$\square$

Consequences: The way the previous theorem was proved makes it to hold for projective maps: we can imagine between two circles, between a line and a circle, on the same line or on the same circle.

This theorem can be easily extended to projective pencils: they are determined by three pairs of homologous rays. Generally speaking, a projective map is determined by the knowledge of three pairs of homologous points.

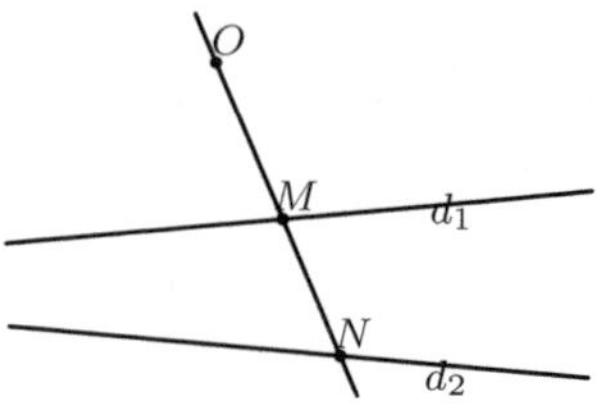

Fig. 3.11 Projectivity
between two lines

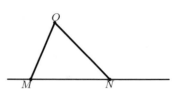

Fig. 3.12 Projectivity
determined by an angle on a
line

Fig. 3.13 Projectivity
determined by an angle
between two lines

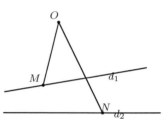

*If two projective maps f and f_1 has the same three pairs of homologous points,
then $f = f_1$.*

Other examples of projective maps:

1. Consider two distinct lines d_1, d_2 and a point O in E^2 who doesn't belong to
   $d_1 \cup d_2$. A moving ray through O intersects d_1 in M and d_2 in N. Then, using
   Pappus' theorem, $M \to N$ is a projective map between d_1 and d_2 (Fig. 3.11).
2. Two points moving with the same speed on two distinct lines, or on a same line,
   determine a projective map (Fig. 3.12).
3. Consider a point $O \notin d$ and a constant angle given angle with its vertex in O
   rotating around O. The first side of the angle intersect d in M and the second side
   in N. Again, using Pappus' theorem, $M \to N$ is a projective map on d (Fig. 3.13).
4. The example above may be extended. Consider two lines d_1 and d_2. The given
   constant angle intersects the lines such that M is on d_1 and N is on d_2. Using
   Pappus' theorem, $M \to N$ is a projective map between d_1 and d_2.

For a projective map on d, denote the coordinate of M by x and the coordinate of
N by y, where $M \to N$ are homologous points.

Theorem 3.2.17 *A projective map on d determines a function* $h(x) = \dfrac{Ax + B}{Cx + d}$,
A, B, C, D being real constants such that $AD - BC \neq 0$.

Fig. 3.14 Interior involution
of a circle

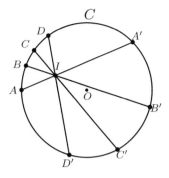

Proof (Hint) Suppose the three given homologous points are $0 \to y_0$, $1 \to y_1$, $x_2 \to y_2$. The condition $[xx_2; 01] = [yy_2; y_0y_1]$ becomes

$$\frac{x}{x_2} : \frac{x-1}{x_2-1} = \frac{y - y_0}{y_2 - y_0} : \frac{y - y_1}{y_2 - y_1}.$$

After computations it results the desired formula. After another computation the coefficients verify $AD - BC \neq 0$. □

Theorem 3.2.18 *The function $h(x) = \dfrac{Ax + B}{Cx + d}$, A, B, C, D being constants such that $AD - BC \neq 0$ describes a projective map on d.*

Proof (Hint) Replace $y_k = h(x_k)$ by $\dfrac{Ax_k + B}{Cx_k + D}$ in $[y_1y_2; y_3y_4]$ and use $AD - BC \neq 0$ to simplify. A straightforward computation shows that $[x_1x_2; x_3x_4] = [y_1y_2; y_3y_4]$.

□

Definition 3.2.19 A projective map on d which interchanges a pair of homologous points is called geometric involution, or simple, involution.

Theorem 3.2.20 *All pairs of homologous points interchange in an involution.*

Proof Consider $A \to B$, $B \to A$, $M \to N$ and $N \to X$. We wish to prove that $X = M$. Since $[AB; MN] = [BA; NX]$ it results $[AB; MN] = [AB; XN]$ i.e. $X = M$. □

The same considerations hold for involutions of a circle. They have a pair of homologous points which can be interchanged. In fact all pairs of homologous points can be interchanged. Therefore $M \to N$ implies $N \to M$, too (Fig. 3.14).

We saw above two examples of involutions of a circle: T^s_{I,R^2-OI^2} is an interior involution, i.e. it is described by the point $I \in int C(O, R)$. T_{J,OJ^2-R^2} is an exterior involution, i.e. it is described by the point $J \in ext C(O, R)$.

Consider a projective map between two lines, $f : d_1 \rightarrow d_2$. Denote $\{O\} := d_1 \cap d_2$ and suppose that $f(O) = O$. Such a point is called a *self-homologous point*. A projective map between two lines as above with a self-homologous point is called a *perspective map*.

Theorem 3.2.21 *For a perspective map between d_1 and d_2, the lines which connect homologous points have a common point.*

Proof The perspective map is determined by $O \rightarrow O$, $A_1 \rightarrow A_2$, $B_1 \rightarrow B_2$. Denote $\{I\} = A_1 A_2 \cap B_1 B_2$ and consider the pencil of lines IO, IA_1, IB_1, IM where $A_1, B_1, M \in d_1$. Suppose that $f(M) = N$, $N \in d_2$ and denote by $\{N'\} = IM \cap d_2$. The perspective map f implies $[OA_1; B_1 M] = [OA_2; B_2 N]$, and Pappus' theorem implies $[OA_1; B_1 M] = [OA_2; B_2 N']$. Therefore $N = N'$ and the arbitrary line MN which connects homologous points contains I. $\square$

Consider a set of arbitrary indexes denoted by $\mathbb{I}$, $O_1, O_2 \in E^2$. Also consider both the lines passing through O_1, denoted by α_i, $i \in \mathbb{I}$ and the lines passing through O_2, denoted by β_i, $i \in \mathbb{I}$. Let denote by $O_1(\alpha)$, $O_2(\beta)$ the two pencils of lines. The next definition makes sense even if $O_1 = O_2$.

Definition 3.2.22 Two pencils of lines are projective if there exists an one to one map $f : O_1(\alpha) \rightarrow O_2(\beta)$ such that for any four rays of the first pencil, say $\alpha_1, \alpha_2, \alpha_3, \alpha_4$ and their images $\beta_1, \beta_2, \beta_3, \beta_4$, we have $[\alpha_1 \alpha_2; \alpha_3 \alpha_4] = [\beta_1 \beta_2; \beta_3 \beta_4]$. α_1 and β_1 are called homologous rays.

Example 3.2.23 Consider a line d and a projective map $f : d \rightarrow d$. Choose four arbitrary points A_1, A_2, A_3, A_4 on d and their images via f, B_1, B_2, B_3, B_4. We have $[A_1 A_2; A_3 A_4] = [B_1 B_2; B_3 B_4]$. Therefore $[OA_1 OA_2; OA_3 OA_4] = [OB_1 OB_2; OB_3 OB_4]$. It results that $O(OA)$ and $O(Of(A))$ are projective pencils of lines.

Example 3.2.24 Consider a line d and a projective map $f : d \rightarrow d$. Choose four arbitrary points A_1, A_2, A_3, A_4 on d and their images via f, B_1, B_2, B_3, B_4. We have $[A_1 A_2; A_3 A_4] = [B_1 B_2; B_3 B_4]$. Therefore $[O_1 A_1 O_1 A_2; O_1 A_3 O_1 A_4] = [O_2 B_1 O_2 B_2; O_2 B_3 O_2 B_4]$. It results that $O_1(O_1 A)$ and $O_2(O_2 f(A))$ are projective pencils of lines.

Example 3.2.25 Consider the lines d, d' and a projective map $f : d \rightarrow d'$. Choose four arbitrary points A_1, A_2, A_3, A_4 on d and their images via f, B_1, B_2, B_3, B_4 on d'. We have $[A_1 A_2; A_3 A_4]_d = [B_1 B_2; B_3 B_4]_{d'}$. Therefore $[O_1 A_1 O_1 A_2; O_1 A_3 O_1 A_4] = [O_2 B_1 O_2 B_2; O_2 B_3 O_2 B_4]$. It results that $O_1(O_1 A)$ and $O_2(O_2 f(A))$ are projective pencils of lines.

We left to the reader to prove: "A projective map between two pencils of lines in determined by three pairs of homologous rays."

Theorem 3.2.26 (Steiner) *Consider two projective pencils of lines. Their homologous rays intersect on a conic.*

Proof To simplify the proof consider a projective map f on the line of equation $y = 1$,

$O_1(0, 0)$, $O_2(1, 0)$. If $M(x, 1)$ and $N\left(\dfrac{Ax + B}{Cx + D}, 1\right)$ are the homologous points, the equations of the lines O_1M and O_2N are $Y = \dfrac{1}{x}X$ and $Y = \dfrac{1}{\dfrac{Ax + B}{Cx + D} - 1}(X - 1)$.

It is not necessary to compute the coordinates of the intersection $O_1M \cap O_2N$. We can substitute x from the first equation, i.e. $x = \dfrac{X}{Y}$, and replace in the second. It results

$$Y = \frac{1}{\dfrac{A\dfrac{X}{Y} + B}{C\dfrac{X}{Y} + D} - 1}(X - 1),$$

therefore, the coordinates of the intersection point lie on the conic of equation

$$-CX^2 + (A - C - D)XY + (B - D)Y^2 + CX + DY = 0.$$

The reader has to observe that Steiner's conic contains O_1 and O_2. □

Problem 3.2.27 Consider M and N on the hypotenuse BC of an isosceles rectangle triangle ABC such that $MN^2 = BM^2 + CN^2$. Prove that $\angle MAN = \dfrac{\pi}{4}$.

Solution. Consider a system of coordinates such that $A(0, a)$, $B(-a, 0)$, $C(a, 0)$, $M(x, 0)$, $N(y, 0)$. $MN^2 = BM^2 + CN^2$ can be written in the form

$$(x + a)^2 + (a - y)^2 = (y - x)^2,$$

that is $y = \dfrac{ax + a^2}{-x + a}$. This is a projective map on BC.

Since $x = -a$ implies $y = 0$, it results $B \to O$. $x = 0$ implies $y = a$, therefore $O \to C$. For $x = a$ it results $y = \infty$, i.e. $C \to \infty$.

So, this projective map on BC is determined by $B \to O$, $O \to C$, $C \to \infty$.

If we consider a rotating angle $\angle MAN = \dfrac{\pi}{4}$ we observe three important positions

$$M = B, N = O; \quad M = O, N = C; \quad M = C, N = \infty$$

such that $\angle BAO = \angle OAC = \angle CO\infty = \dfrac{\pi}{4}$. Therefore the rotating angle leads to a projective map on BC determined by $B \to O$, $O \to C$, $C \to \infty$. The two projective map are coincident, therefore always $\angle MAN = \dfrac{\pi}{4}$. □

Fig. 3.15 Karya's Point

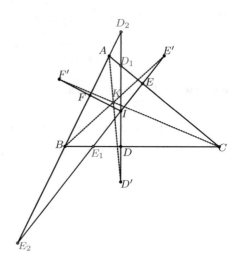

Problem 3.2.28 (Karya's point) Let I be the incenter of the triangle $\triangle ABC$ and D', E', F' be the symmetric of I with respect the sides BC, CA, AB.
Then, $AD' \cap BE' \cap CE' \neq \emptyset$.

Solution. (D. Barbilian) Denote by D, E, F the contacts of the incircle with the sides BC, CA, AB respectively. Denote also by D_1, D_2 the intersection points of ID with AC and AB, respectively. The same, $\{E_1\} = IE \cap BC$; $\{E_2\} = IE \cap BA$. Consider first two moving points $M \in ID$, $N \in IE$ who start to move with the same speed from I in the direction of D, respectively E (Fig. 3.15).

We know that $M \to N$ is a projective map between the lines ID and IE. It results a projective map between the pencils $A(AM)$ and $B(BN)$. The intersection point between AM and BN lies on a conic. A conic is determined by the knowledge of five distinct points of it. The initial moment $M = N = I$ implies that I belongs to the conic. When $M = D$ it results $N = E$, therefore Gergonne's point of the triangle ABC, G_e, belongs to the conic. Now consider M, N moving from I to D_1, E_1 respectively. Since the triangles IED_1 and IED_1 are congruent, when $M = D_1$ it results $N = E_1$, therefore $AM \cap BN = \{C\}$. According to Steiner's theorem the conic contains A and B. But it is easy to observe for this projective map why. When $M = D_2$, $AD_2 \cap BN = \{B\}$, and when $N = E_2$, $AM \cap BE_2 = \{A\}$. Therefore Steiner conic for the projective pencils $A(AM)$, $B(BN)$ is determined by A, B, C, I, G_e. The same for the Steiner conics determined by the projective pencils $A(AM), C(CP)$ and $B(BN), C(CP)$. Therefore the three Steiner conics are coincident, i.e. $AM \cap BN \cap CP \neq \emptyset$ when $|IM| = |IN| = |IP|$. In the particular case of the statement, a particular point belongs to Steiner's conic. We are talking about Karya's point. If we choose the points at infinity of AM, BN, CP we can see that the orthocenter H belongs to the Steiner conic. As you can see, this projective solution allows us to highlight many other points of intersection among AM, BN and CP described by the condition $|IM| = |IN| = |IP|$.   □

There is a special case when Steiner's conic is a line only.

Definition 3.2.29 Two projective pencils of lines, $O_1(\alpha)$ and $O_2(\beta)$, $O_1 \neq O_2$, are called perspective pencils if the ray $O_1 O_2$ is self-homologous, i.e. $O_1 O_2 \to O_1 O_2$.

Theorem 3.2.30 *If $O_1(\alpha)$ and $O_2(\beta)$, $O_1 \neq O_2$ are perspective pencils, then the homologous rays intersection lies on a line.*

Proof Denote $p = O_1 O_2$. The perspective map is determined by $p \to p$, $\alpha_1 \to \beta_1$, $\alpha_2 \to \beta_2$. Denote $\{T_1\} = \alpha_1 \cap \beta_1$, , $\{T_2\} = \alpha_2 \cap \beta_2$, . If $\alpha \to \beta$, and $\{T\} = \alpha \cap T_1 T_2$ we observe that $\alpha \to O_2 T$ belongs to the previous projective map. Therefore, $O_2 T = \beta$, and $\alpha \cap \beta$ always belongs to $T_1 T_2$. This line is called the perspective axis of the perspective pencils of lines $O_1(\alpha)$ and $O_2(\beta)$. □

Let us answer to the question: which is the geometrical structure of $\{\infty_m, \ m \in \mathbb{R}\}$?

First, it is easy to see that for a given line $d \subset E^2$ it exists two perspective pencils of lines, $O_1(\alpha)$ and $O_2(\beta)$, such that d is the perspective axis of the previous perspective pencils. If we have two perspective pencils there is one case in which the perspective axis doesn't exist: when the homologous rays of the perspective pencils are parallel. Exactly as in the case of the abstract infinity point of a line added to preserve a geometric rule, we do the same thing. In the case of parallel perspective pencils the perspective axis is an abstract line, called the *line at infinity of E^2*. Therefore we may denote $d_\infty := \{\infty_m, \ m \in \mathbb{R}\}$.

Perspective pencils allow us to construct a special line assigned to any projective map f on a circle: the axis of f. This line plays a crucial role in the construction of Poincaré disk model.

Consider for a projective map f on a circle Γ the homologous points M, M', $M \to M'$ which describe the projective map f. If we choose two particular pairs of homologous points, say $A \to A'$, $B \to B'$, the point $\{P\} = AB' \cap A'B$ allows us to create a function $g : \Gamma \to \Gamma$, $g(N) = N'$, $\{N'\} := NP \cap \Gamma$. It is obvious to observe that g is an involution of Γ.

Theorem 3.2.31 *(i) The map $f \circ g_P : \Gamma \to \Gamma$ is an involution of Γ.*
(ii) The locus of points $I_f \in E^2$ such that $f \circ g_{I_f}$ is an involution of Γ is a line.

Proof (i) f and g_P are projective maps on Γ, then $f \circ g_P$ is a projective map on Γ. It remains to prove that $f \circ g_P$ has a pair of homologous points which interchanges. We show that $f \circ g_P(A') = B'$ and $f \circ g_P(B') = A'$. Let's compute $f \circ g_P(A') = f(g_P(A')) = f(B) = B'$. In the same way $f \circ g_P(B') = f(g_P(B')) = f(A) = A'$, therefore $f \circ g_P$ is an involution of Γ.

(ii) Consider $A \to A'$ a given pair of homologous point of f and $M \to M'$ the general pair of homologous points of f. Therefore $A \to A'$ is a particular pair obtained from the general pair by replacing M by A. The pencils $A(AM')$ and $A'(A'M)$ are perspective, the self-homologous ray being AA'. The homologous rays intersection, i.e. $\{I_f\} = AM' \cap A'M$, lies on the perspective axis, therefore the locus is a line. □

This line is called the *axis of the projective map* f. The previous theorem shows that this line is $\{I_f \mid \{I_f\} = AM' \cap MA', \; M \in \Gamma\}$. A direct consequence appears.

Theorem 3.2.32 (i) If $A \to A'$, $B \to B'$, $C \to C'$ are homologous points of f on Γ, then the points $\{U\} = AB' \cap BA'$, $\{V\} = AC' \cap CA'$, $\{W\} = BC' \cap CB'$ are collinear.

  (ii) All projective maps of a circle can be written as a product of involutions (in a non unique way).

  (iii) Two interior involutions of Γ, I, J determine in an unique way the projective map of Γ, $f = I \circ J$ such that the axis of f is IJ.

  (iv) Denote $\{s, S\} = IJ \cap \Gamma$ such that the order is s, I, J, S. Then $f(s) = s$, $f(S) = S$ and $[IJ; Ss] > 1$.

  (v) If $M \in \Gamma$, $J(M) = M'$, $M' \in \Gamma$, $I(M') = N$, $N \in \Gamma$, for an arbitrary $X \in IJ$ there is an unique $Y \in IJ$ such that $XN \cap MY \in \Gamma$.

  (vi) $X \to Y$ is a projective map on the line IJ. (This map is called the axial decomposition of f).

  (vii) $[SsIJ] = [SsXY]$.

Proof (i) U, V, W are three particular points of $\{I_f \mid \{I_f\} = AM' \cap MA', M \in \Gamma\}$.

  (ii) If we choose the point, say $\{U\} = AB' \cap BA'$, then $f \circ U = L$ where $\{L\}$ is the intersection between the axis of f and $A'B'$. Therefore $f = L \circ I$.

  (iii) (iv) and (v) are obvious.

  (vi) Consider the projective map on IJ determined by particular positions of X and Y, $s \to s$, $S \to S$, $I \to J$. If remains to prove $[sSIX] = [sSJY]$. Consider the pencils $M(Ms, MS, MJ, MY)$ and $N(Ns, NS, NI, NX)$. Since the angles involved are equal it results

$$[MsMS; MJMY] = [NsNS; NINX],$$

   i.e. $[sSIX] = [sSJY]$.

  (vii) From $[sSIX] = [sSJY]$ it results $[SsIX] = [SsJY]$. If you write the last one equality it results

$$\frac{IS}{Is} : \frac{XS}{Xs} = \frac{JS}{js} : \frac{YS}{Ys}.$$

This one can be thought as

$$\frac{IS}{Is} : \frac{JS}{ss} = \frac{XS}{Xs} : \frac{YS}{Ys},$$

which means $[SsIJ] = [SsXY]$, or equivalently $[sSIJ] = [sSXY]$.   □

Before continuing, let us conclude in the following way.

For I and J belonging to the interior of our circle, we construct the projectivity of the circle $f := I \circ J$ determined by the product of the given interior involutions. Suppose $M \to N$ in this projectivity. For each $X \in IJ$, we construct

$Y \in IJ$ as in the previous theorem. The projectivity $X \to Y$ on IJ is determined by f. It is called the axial decomposition of the projectivity f (of the circle) on the line IJ. This was the most important step towards the construction of a non-Euclidean distance in the interior of the circle.

The next steps are the Theorems 3.3.1, 3.3.2 and 3.3.3 in the next section.

Let us observe: Consider $S, S' \in \Gamma$ such that the chord SS' is not a diameter; Then, the tangents at S, S' meet at the center of an orthogonal circle to Γ.

If $A \in int\Gamma$ and $S \in \Gamma$, we know that the orthogonal circle to Γ passing through A and S is constructed in the following way: the tangent at S meet the perpendicular bisector of the segment AS at the center of the orthogonal circle.

If $A, B \in int\Gamma$ the orthogonal circle to Γ passing through A and B is constructed in the following way: we construct A', the inverse of A in the direct inversion T_{O,R^2}, where O, R are the center, respectively the radius of Γ. The perpendicular bisectors of the triangle ABA' meet at the center of the orthogonal circle we are looking for. Observe that B', the inverse of B in the same inversion belongs to this circle.

Another more important observation is:

Proposition 3.2.33 *If $A, B, C, D \in \Gamma$ such that $\{L\} = AB \cap CD$, $L \in int\Gamma$, then the orthogonal circles determined by the chords AB, CD denoted by γ_{AB}, γ_{CD} respectively, meet in X, X' such that O, X, L, X' are collinear.*

Proof O and L have equal powers with respect γ_{AB}, γ_{CD}.
The powers are R^2 for O, $u := |LA| \cdot |LB| = |LC| \cdot |LD|$ for L respectively, therefore they belong to the radical axis of the two circles. But the radical axis passes through the points of intersection of the two orthogonal circles, i.e. O, X, L, X' are collinear. Extra, $|OX| \cdot |OX'| = R^2$, that is X and X' are inverse in T_{O,R^2}. □

3.3 Poincaré Disk Model

We underline some results proved above, results which are necessary to introduce the Poincaré disk model. If $I, J \in int\Gamma$, $f := I \circ J$ is a projective map on Γ such that IJ is its axis.

If $X \to Y$ are the homologous points in the axial decomposition of f on $a := IJ$, and $\{s, S\} = a \cap \Gamma$ such that the order is s, I, J, S; s, X, Y, S respectively, then $[IJSs] = [XYSs] = k > 1$. Therefore $[SsIJ] = k$ is an invariant of the axial decomposition of f. In fact k depends on I and J, that is $k = k_{IJ}$ is an invariant attached to the involutions I and J on the axis of the projective map $f = I \circ J$.

If we consider the orthogonal circle to Γ through s and S, denoted g, on the arc $g := g_{sS}$ from the $int\Gamma$ we can consider two special points I', J', $\{I'\} := OI \cap g_{sS}$, $\{J'\} := OJ \cap g_{sS}$.

A very important result will be proved:

$$[IJSs] = [I'J'Ss]_g^2.$$

Fig. 3.16 $[IJSs] =$
$[I'J'Ss]_g^2$

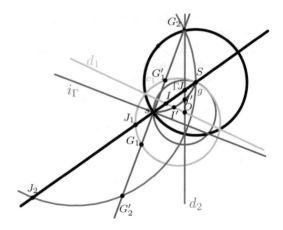

Let us describe again the context. Consider the circle Γ centered in O and $int\,\Gamma$ the disk enclosed by Γ. Let I and J be in $int\,\Gamma$ and denote by s and S the intersections of the line IJ with Γ. Suppose the order is s, I, J, S. Denote by g the orthogonal arc to Γ passing through s and S, and let I' and J' be the intersections of g with OI and OJ, respectively (Fig. 3.16).

We have to consider the direct inversion of pole S and power $\mu = (sS)^2$.

The point s is fixed by this transformation. The circle Γ, which passes through the pole of inversion, is transformed into the line $i(\Gamma)$ which passes through s. The arc g is transformed into the line $i(g)$, and $i(g) \perp i(\Gamma)$. Let $d_1 := OI$ and $d_2 := OJ$. The line d_1, which doesn't pass through the pole of inversion, is transformed into the circle c_1 passing through S. Furthermore, c_1 contains the images of I' and I, denoted by G_1 and J_1, respectively. In fact, since $d_1 \perp \Gamma$, then c_1 and $i(\Gamma)$ must also be orthogonal, which means that c_1 has the line $i(\Gamma)$ as a diameter. A similar reasoning can be done for the line d_2.

We introduce the following notations: $\{G_1, G_1'\} = c_1 \cap i(g)$, and $\{G_2, G_2'\} = c_2 \cap i(g)$.

Finally, we remark that S is mapped by this inversion into ∞.

Then we have the following:

$$[IJSs] = [J_1 J_2 \infty s] = \frac{|sJ_2|}{|sJ_1|},$$

$$[I'J'Ss]_g = [G_1 G_2 \infty S] = \frac{|sG_2|}{|sG_1|}.$$

The power of the point s with respect to c_1 yields:

$$|Ss| \cdot |sJ_1| = |sG_1| \cdot |sG_1'| = |sG_1|^2.$$

Similarly, the power of s with respect to c_2 yields:

$$|Ss| \cdot |sJ_2| = |sG_2| \cdot |sG_2'| = |sG_2|^2.$$

Therefore, we have

$$\frac{|sJ_1|}{|sJ_2|} = \left(\frac{|sG_1|}{|sG_2|}\right)^2.$$

This result actually means we proved the following

Theorem 3.3.1 $[IJSs] = [I'J'Ss]_g^2$.

It results that the points $I, J \in a$ generate $I', J' \in g_{sS}$ such that the invariant $k_{IJ} = [IJSs]$ generates the invariant $K_{I'J'} = [I'J'Ss]_g^2$ and $[IJSs] = [I'J'Ss]_g^2$.

Therefore we move points and invariants from the axis of a projective map of a circle Γ to an orthogonal arc g to Γ.

On the initial configuration, we apply a symmetric inversion of pole J' and power μ', where μ' is the power of J' with respect to the circle Γ.

Consequently, the circle Γ is mapped into Γ itself by this transformation.

The arc g becomes the line $i(g)$, which is a diameter in Γ.

The point I' is transformed into F_1', which lies on $i(g)$, such that $|J'I'| \cdot ||J'F_1'| = \mu$.

The pole J' is mapped into ∞.

The point $P \in \Gamma$ is transformed into $P' \in \Gamma$, such that $|J'P| \cdot |J'P'| = \mu'$ and P' is the second intersection of Γ with the line $J'P$.

Denote $i(s)$ and $i(S)$ the images of s and S through the previously described inversion. We have

$$[I'J'Ss]_g = [F_1'\infty i(S)i(s)] = \frac{|i(S)F_1'|}{|i(s)F_1'|} = \frac{\max_{P'\in\Gamma}|P'F_1'|}{\min_{P'\in\Gamma}|P'F_1'|}.$$

Furthermore,

$$|P'F_1'| = \mu' \cdot \frac{|PI'|}{|J'P| \cdot |I'J'|} = \frac{\mu'}{|I'J'|} \cdot \frac{|PI'|}{|PJ'|}.$$

This shows us that $|P'F_1'|$ reaches its maximum and minimum in the same time as the ratio $\dfrac{|PI'|}{|PJ'|}$ (Fig. 3.17).

Therefore, we have proved

Theorem 3.3.2

$$[I'J'Ss]_g = \frac{\max_{P\in\Gamma}\frac{|PI'|}{|PJ'|}}{\min_{P\in\Gamma}\frac{|PI'|}{|PJ'|}}.$$

Let see how these algebraic invariants generate distances in the interior of Γ. Denote

$$d_a(I, J) := \frac{1}{2}ln[IJSs],$$

Fig. 3.17 Poincaré Modified
Distance

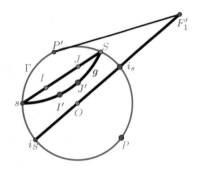

$$d_g(I', J') := ln[I'J'Ss]_g,$$

$$d(I', J') := \ln \frac{\max_{P \in \Gamma} \frac{|PI'|}{|PJ'|}}{\min_{P \in \Gamma} \frac{|PI'|}{|PJ'|}}.$$

The previous proved facts allow us to assert

Theorem 3.3.3 $d_a(I, J) = d_g(I', J') = d(I', J')$.

Consider two arbitrary sets K and U.

Definition 3.3.4 The function $f : K \times U \to \mathbb{R}_+^*$ is called an *influence of the set K over U* if for any $A, B \in U$ the ratio $g_{AB}(P) = \dfrac{f(P, A)}{f(P, B)}$ has a maximum $M_{AB} \in \mathbb{R}$ when $P \in K$.

Note that $g_{AB} : K \to \mathbb{R}_+^*$. If we assume the existence of max $g_{AB}(P)$, when $P \in K$, then there also exists $m_{AB} = \min_{P \in K} g_{AB}(P) = \dfrac{1}{M_{BA}}$.

Consider $d : U \times U \to \mathbb{R}_+$ given by

$$d(A, B) = \ln \frac{\max_{P \in K} g_{AB}(P)}{\min_{P \in K} g_{AB}(P)}.$$

It is easy to prove that the previous formula leads to a semi-distance, i.e.:
(1) if $A = B$ then $d(A, B) = 0$; (2) d is symmetric; (3) d satisfies triangle inequality.

(1) and (2) are obvious. For (3) let A, B, C distinct points in J and the pair of points $S_0, s_0 \in K$, $S_1, s_1 \in K$, $S_2, s_2 \in K$ such that

$$\max_{P \in K} g_{AB}(P) = \frac{f(S_0, A)}{f(S_0, B)}, \quad \min_{P \in K} g_{AB}(P) = \frac{f(s_0, A)}{f(s_0, B)}$$

$$\max_{P \in K} g_{AC}(P) = \frac{f(S_1, A)}{f(S_1, C)}, \quad \min_{P \in K} g_{AC}(P) = \frac{f(s_1, A)}{f(s_1, C)},$$

$$\max_{P \in K} g_{BC}(P) = \frac{f(S_2, B)}{f(S_2, C)}, \quad \min_{P \in K} g_{BC}(P) = \frac{f(s_2, B)}{f(s_2, C)}.$$

If S_0, S_2 are replaced by S_1 and s_0, s_2 are replaced by s_1 we obtain

$$d(A, B) + d(B, C) = \ln\left[\left(\frac{f(S_0, A)}{f(S_0, B)} : \frac{f(s_0, A)}{f(s_0, B)}\right) \cdot \left(\frac{f(S_2, B)}{f(S_2, C)} : \frac{f(s_2, B)}{f(s_2, C)}\right)\right] \geq$$

$$\geq \ln\left(\frac{f(S_1, A)}{f(S_1, C)} : \frac{f(s_1, A)}{f(s_1, C)}\right) = d(A, C).$$

In particular for $f(P, A) = |PA|$, $K = \Gamma$ is a circle and $U := int\,\Gamma$ its interior, we obtain that our last formula among the previous three is a semi-distance on $int\,\Gamma$. But there is no pair $(A, B) \in U \times U$, $A \neq B$, such that the ratio $g_{AB}(P) = \dfrac{f(P, A)}{f(P, B)}$ is constant for all $P \in K$ (in the case when $K = \Gamma$ is a circle and $U := int\,\Gamma$), that is if $d(A, B) = 0$ it results $A = B$, i.e. all three equal formulas $d_a(I, J) = d_g(I', J') = d(I', J')$ are distances.

Definition 3.3.5

$$d(I', J') = \ln \frac{\max_{P \in \Gamma} \frac{|PI'|}{|PJ'|}}{\min_{P \in \Gamma} \frac{|PI'|}{|PJ'|}}.$$

is called a *Poincaré distance between the points* I' *and* J' *of the disk.*

We prefer to consider this general form of the distance d, because if we change K and U, we can obtain available distances on U which come only from the existence of the asked maximum. The reader will see this in the cases of the "semi-plane" and "exterior of the disk" models for non-euclidean Geometry.

All these beautiful geometric facts were possible because of the axial projective map derived from a projective map of a circle.

Problem 3.3.6 Show that for three points A, B, C in this order on the orthogonal arc g to the circle Γ, A, B, $C \in int\,\Gamma$, we have $d(A, C) = d(A, B) + d(B, C)$.

Solution. Suppose the order is s, A, B, C, S where s, S are the "ends" of the arc g belonging to the center. In fact, the ratios $\dfrac{|PA|}{|PC|}$, $\dfrac{|PA|}{|PB|}$, $\dfrac{|PB|}{|PC|}$ have their maximum when $P = S$ and the minimum when $P = s$. And now just add. $\square$

When the orthogonal arc is a diameter and $s(-1)$, $I(0)$, $J(x)$, $S(1)$, $x > 0$, then

$$d(I, J) = \ln\frac{1 + x}{1 - x}.$$

We can observe that when $J \to S$, i.e. $x \to 1$ then $d(I, J) \to \infty$. The disk becomes unbounded with respect this distance.

What kind of Geometry do we have inside the disk? Next, we prove that it is a non-Euclidean one. A point $I \in int\,\Gamma$ is called an *n-point* in our Geometry. The points of the circle Γ are called ∞-*points*.

An orthogonal arc of circle to Γ is called an *n-line*. Such an n-line is uniquely determined by two n-points, by two ∞-points, or by an ∞-point and an n-point. Two n-lines intersect at most at an n-point. Three n-points are called *n-collinear* if they belong to an n-line.

It is easy to show that there exist non-intersecting n-lines. If two chords Ss and $S's'$ do not intersect in the interior of the disk, then the orthogonal to Γ arcs of circles having the same endpoints are n-lines with empty intersection, that is non-intersecting n-lines. Through an n-point which doesn't belong to a given n-line we can construct at least two non-intersecting n-lines with respect to the given n-line.

In fact, if the given n-line is the orthogonal arc γ_{sS} and $I \notin \gamma_{sS}$, among the infinitely many non-intersecting n-lines there exist two special ones, γ_{sI}, γ_{SI} which are called n-parallels to γ_{sS}.

The *angle between two n-lines* is, by definition, the Euclidean angle between the tangents to the arcs at the common point.

An *n-triangle* is determined by three non-n-collinear points. The sides of an n-triangle are n-lines.

What about the sum of the angles in an n-triangle?

According to the theory described in the previous chapter, it is enough to study what happens in the case of one given triangle. We can choose a triangle with one vertex at the center O of Γ and two other n-points, A and B. Consider the Euclidean triangle AOB. The sum of the angles of the Euclidean triangle is π. The angle at O, i.e. AOB is common to both triangles, but each other n-angle is less than the corresponding Euclidean angle. Therefore, the sum of the angles of the n-triangle is less than π.

More about this model of Non-Euclidean Geometry and some other models connected to this one can be understood only after we study Differential Geometry.

Chapter 4
Surfaces in 3D-Spaces

Ab initio res.

This chapter is devoted to the Differential Geometry of a surface in a 3-space. We need to know basic calculus. All functions which appear from now are smooth, i.e. they are indefinitely differentiable functions in one or several variables at each point of their domain of definition. First, we see surfaces in an Euclidean 3-dimensional space and we understand how the Euclidean inner product induces, via the first fundamental form, a way to measure lengths and angles for vectors belonging to tangent planes to the surface. We can also measure lengths of curves who belong to surfaces, areas of regions and the Gaussian curvature of a surface at each point. If at beginning, the curvature seems to be dependent on the embedding in the ambient Euclidean space, after we prove Gauss' formulas, we step into the intrinsic theory of surfaces where Gauss' equations and the Theorema Egregium offer another perspective the surfaces can be seen as pieces of a plane endowed with a metric, and this metric only determines the curvature. In Minkowski 3-spaces we have the same picture, the Minkowski product determines a non-Euclidean metric of a surface which allows us to conclude about the intrinsic Geometry of it. Therefore, in both cases the surface becomes irrelevant for our study. In fact we study the Geometry of a metric and we obtain relevant geometric aspects about the piece of plane endowed with that metric. This point of view will be continued in the next chapter when we better understand the nature of geometric objects which appear in Differential Geometry. Both chapters regarding Differential Geometry were adapted using ideas from [27–30].

4.1 Geometry of Surfaces in a 3D-Euclidean Space

Before developing our considerations, it is worth stressing that a standard notation in Differential Geometry is the *Einstein summation convention*, or simply *Einstein notation*, $a_i b^i := \sum_{i=1}^{n} a_i b^i$.

© The Editor(s) (if applicable) and The Author(s), under exclusive license to Springer Nature Switzerland AG 2020
W. Boskoff and S. Capozziello, *A Mathematical Journey to Relativity*, UNITEXT for Physics, https://doi.org/10.1007/978-3-030-47894-0_4

It can be taught for double or triple sums, that is $a_{ij}x^i y^j = \sum\limits_{i,j=1}^{n} a_{ij}x^i y^j$ or

$a_{ijk}x^i y^j z^k = \sum\limits_{i,j,k=1}^{n} a_{ijk}x^i y^j z^k$. One can adopt this convention for multiple sums, the sums being thought before the indexes up and down or down and up denoted by the same letter.

If below one reads something like $\Gamma^i_{sj}\Gamma^s_{kl}$, this means $\sum\limits_{s=1}^{n} \Gamma^i_{sj}\Gamma^s_{kl}$.

The index s from the previous formula is called a dummy index because we can replace the letter s by r and the meaning of the formula $\Gamma^i_{rj}\Gamma^r_{kl}$ is the same, i.e.

$\sum\limits_{r=1}^{n} \Gamma^i_{rj}\Gamma^r_{kl}$.

The number n is related to the dimension of the set endowed with a coordinate system, set in which we develop Differential Geometry concepts. In the case of surfaces, $n = 2$.

The Euclidean three dimensional space, denoted by E^3, can be thought as the vector space $\mathbb{R}^3$ over the field $\mathbb{R}$ endowed with the *Euclidean inner product*

$$\langle a, b \rangle := a_0 b_0 + a_1 b_1 + a_2 b_2,$$

where $a = (a_0, a_1, a_2)$, $b = (b_0, b_1, b_2)$.

If we consider a frame generated by the vectors $\overrightarrow{i} = (1, 0, 0)$, $\overrightarrow{j} = (0, 1, 0)$ and $\overrightarrow{k} = (0, 0, 1)$, the components of a vector a with respect to this basis become coordinates in the new frame, that is, we can assign them to a point A. We can write $A(a_0, a_1, a_2)$ and this point can be seen as the endpoint of the vector a whose origin is in the point $(0, 0, 0)$.

Euclidean perpendicular vectors correspond to null inner product, i.e. a and b are perpendicular (or orthogonal) if $\langle a, b \rangle = 0$. With respect to the Euclidean inner product the previous basis is an orthogonal one.

The *length of the vector a* is, by definition, $||a|| := \sqrt{\langle a, a \rangle} = \sqrt{a_0^2 + a_1^2 + a_2^2}$.

The Cauchy–Schwartz inequality for the triples (a_0, a_1, a_2), (b_0, b_1, b_2) is

$$(a_0 b_0 + a_1 b_1 + a_2 b_2)^2 \leq (a_0^2 + a_1^2 + a_2^2)(b_0^2 + b_1^2 + b_2^2),$$

that is, for vectors the inequality can be written in terms of inner product and norm in the form $\langle a, b \rangle^2 \leq ||a||^2 \cdot ||b||^2$. The equality happens when the triples are proportional: this fact corresponds to collinear vectors.

If the two vectors a and b are not collinear, they determine a plane.

In this plane it makes sense to define the *angle α between the nonzero vectors a and b* by the formula

$$\cos \alpha := \frac{\langle a, b \rangle}{||a|| \cdot ||b||}.$$

The length of a vector a becomes the distance between the origin $O(0, 0, 0)$ at the point $A(a_0, a_1, a_2)$. The *Euclidean distance* between two points $A(a_0, a_1, a_2)$, $B(b_0, b_1, b_2)$ is given by the formula

$$d(A, B) := ||a - b|| = \sqrt{\langle a - b, a - b \rangle} = \sqrt{(a_0 - b_0)^2 + (a_1 - b_1)^2 + (a_2 - b_2)^2}.$$

We can denote the Euclidean distance $||OA||$ by our previous notation $|OA|$. We prefer this last notation and we keep in our mind that $||AB|| = |AB|$.

The *crossproduct of two vectors* is the vector given by the formula

$$a \times b = (a_1 b_2 - a_2 b_1, -a_0 b_2 + a_2 b_0, a_0 b_1 - a_1 b_0).$$

It is easier to remember it from the formal developing of the following determinant,

$$\begin{vmatrix} \vec{i} & \vec{j} & \vec{k} \\ a_0 & a_1 & a_2 \\ b_0 & b_1 & b_2 \end{vmatrix}.$$

Since $\langle a \times b, a \rangle = 0$ and $\langle a \times b, b \rangle = 0$ the vector $a \times b$ is orthogonal to the plane determined by the vectors a and b.

Problem 4.1.1 $||a \times b|| = ||a|| \cdot ||b|| \cdot \sin \alpha$

Solution. (Hint)
$$||a \times b||^2 = (a_1 b_2 - a_2 b_1)^2 + (a_0 b_2 - a_2 b_0)^2 + (a_0 b_1 - a_1 b_0)^2 =$$

$$(a_0^2 + a_1^2 + a_2^2)(b_0^2 + b_1^2 + b_2^2) - (a_0 b_0 + a_1 b_1 + a_2 b_2)^2 = ||a||^2 \cdot ||b||^2 \cdot (1 - \cos^2 \alpha) =$$

$$= ||a||^2 \cdot ||b||^2 \cdot \sin^2 \alpha$$

□

From the square of the Generalized Pythagoras Theorem relation

$$\langle a, b \rangle^2 = ||a||^2 \cdot ||b||^2 \cos^2 \alpha$$

and the previous square of the cross product formula, adding it results

$$\langle a, b \rangle^2 + ||a \times b||^2 = ||a||^2 \cdot ||b||^2 .$$

This formula will be used below to obtain Kepler's first law.

Definition 4.1.2 A surface in the Euclidean three dimensional space E^3 is a smooth mapping of an open set $U \subset \mathbb{R}^2$ into E^3 with an extra property: at each point $f(x)$ of the surface, there exists a tangent plane.

Let us explain this definition.

$f : U \longrightarrow \mathbb{R}^3$ is written as $f(x) = (f^1(x), f^2(x), f^3(x))$, where $x = (x^1, x^2)$. We consider the vectors

$$\frac{\partial f}{\partial x^1}(x) = \left(\frac{\partial f^1}{\partial x^1}(x), \frac{\partial f^2}{\partial x^1}(x), \frac{\partial f^3}{\partial x^1}(x)\right),$$

$$\frac{\partial f}{\partial x^2}(x) = \left(\frac{\partial f^1}{\partial x^2}(x), \frac{\partial f^2}{\partial x^2}(x), \frac{\partial f^3}{\partial x^2}(x)\right).$$

If the matrix

$$\begin{pmatrix} \dfrac{\partial f^1}{\partial x^1}(x) & \dfrac{\partial f^1}{\partial x^2}(x) \\ \dfrac{\partial f^2}{\partial x^1}(x) & \dfrac{\partial f^2}{\partial x^2}(x) \\ \dfrac{\partial f^3}{\partial x^1}(x) & \dfrac{\partial f^3}{\partial x^2}(x) \end{pmatrix},$$

$x = (x^1, x^2) \in U \subset \mathbb{R}^2$, has rank 2, then $\dfrac{\partial f}{\partial x^i}(x)$, $i \in \{1, 2\}$ are linear independent vectors and the tangent plane at $f(x)$ exists and it has the equation

$$\begin{vmatrix} X - f^1(x) & Y - f^2(x) & Z - f^3(x) \\ \dfrac{\partial f^1}{\partial x^1}(x) & \dfrac{\partial f^2}{\partial x^1}(x) & \dfrac{\partial f^3}{\partial x^1}(x) \\ \dfrac{\partial f^1}{\partial x^2}(x) & \dfrac{\partial f^2}{\partial x^2}(x) & \dfrac{\partial f^3}{\partial x^2}(x) \end{vmatrix} = 0.$$

The *tangent plane* is denoted by $T_{f(x)}f$; the linear independent vectors $\left\{\dfrac{\partial f}{\partial x^1}(x), \dfrac{\partial f}{\partial x^2}(x)\right\}$ determine a *basis for the tangent plane* $T_{f(x)}f$.

Any vector $X(x)$ which belongs to $T_{f(x)}f$ can be written in the form

$$X(x) = X^1(x)\frac{\partial f}{\partial x^1}(x) + X^2(x)\frac{\partial f}{\partial x^2}(x)$$

where the coefficient $X^1, X^2 : U \longrightarrow \mathbb{R}$ are smooth maps. The line which is perpendicular to the tangent plane at the point $f(x)$ is called a normal line to the surface and has the equation

$$\frac{X - f^1(x)}{\begin{vmatrix} \dfrac{\partial f^2}{\partial x^1}(x) & \dfrac{\partial f^3}{\partial x^1}(x) \\ \dfrac{\partial f^2}{\partial x^2}(x) & \dfrac{\partial f^3}{\partial x^2}(x) \end{vmatrix}} = \frac{Y - f^2(x)}{\begin{vmatrix} \dfrac{\partial f^3}{\partial x^1}(x) & \dfrac{\partial f^1}{\partial x^1}(x) \\ \dfrac{\partial f^3}{\partial x^2}(x) & \dfrac{\partial f^1}{\partial x^2}(x) \end{vmatrix}} = \frac{Z - f^3(x)}{\begin{vmatrix} \dfrac{\partial f^1}{\partial x^1}(x) & \dfrac{\partial f^2}{\partial x^1}(x) \\ \dfrac{\partial f^1}{\partial x^2}(x) & \dfrac{\partial f^2}{\partial x^2}(x) \end{vmatrix}}.$$

The reader understands that the equation of this line is generated by the vector $\dfrac{\partial f}{\partial x^1} \times \dfrac{\partial f}{\partial x^2}$. This vector generates the normal unitary vector $N(x)$,

Definition 4.1.3 $N(x) := \dfrac{\dfrac{\partial f}{\partial x^1}(x) \times \dfrac{\partial f}{\partial x^2}(x)}{\left\| \dfrac{\partial f}{\partial x^1}(x) \times \dfrac{\partial f}{\partial x^2}(x) \right\|}$ is called the Gauss map of the surface f at the point $f(x)$.

Definition 4.1.4 The frame $\left\{ \dfrac{\partial f}{\partial x^1}(x), \dfrac{\partial f}{\partial x^2}(x), N(x) \right\}$ is called a Gauss frame attached to the surface f at $f(x)$.

At each point of a surface this frame is a vector basis in E^3.

The partial derivatives of the vectors of this frame can be written with respect to the Gauss frame using some coefficients. In some sense, the Differential Geometry deals with the geometric meaning of these coefficients.

One more comment about the tangent vectors $\left\{ \dfrac{\partial f}{\partial x^1}(x), \dfrac{\partial f}{\partial x^2}(x) \right\}$. Denoted by $T_x U$, it is the 2-dimensional vector space having the origin at $x \in U$. Since the surface f is the map $f : U \longrightarrow \mathbb{R}^3$, it has sense to consider the linear map $df_x : T_x U \longrightarrow \mathbb{R}^3$,

$$df_x = \begin{pmatrix} \dfrac{\partial f^1}{\partial x^1}(x) & \dfrac{\partial f^1}{\partial x^2}(x) \\ \dfrac{\partial f^2}{\partial x^1}(x) & \dfrac{\partial f^2}{\partial x^2}(x) \\ \dfrac{\partial f^3}{\partial x^1}(x) & \dfrac{\partial f^3}{\partial x^2}(x) \end{pmatrix}.$$

If $e_1 := (1, 0)$, $e_2 = (0, 1)$ are the vector of the basis in $T_x U$, then it is easy to see that $df_x(e_1) = \dfrac{\partial f}{\partial x^1}(x)$, $df_x(e_1) = \dfrac{\partial f}{\partial x^2}(x)$. Let us observe that, according to the meaning of the objects involved, we can write $df_x e_i$ instead of $df_x(e_i)$.

Therefore a vector $X(x) = X^1(x)e_1 + X^2(x)e_2 \in T_x U$ is mapped into the vector

$$df_x(X) = df_x[X^1(x)e_1 + X^2(x)e_2] = X^1(x)\dfrac{\partial f}{\partial x^1}(x) + X^2(x)\dfrac{\partial f}{\partial x^2}(x) \in T_{f(x)}f.$$

We can now define the first fundamental form of a surface. Let the map $f : U \longrightarrow \mathbb{R}^3$ be our surface and let $df_x : T_x U \subseteq \mathbb{R}^2 \longrightarrow \mathbb{R}^3$ be the map previously described. Remember that we denote by $\langle \cdot, \cdot \rangle : \mathbb{R}^3 \times \mathbb{R}^3 \longrightarrow \mathbb{R}$ the Euclidean inner product in E^3 described by the formula

$$\langle a, b \rangle = a_0 b_0 + a_1 b_1 + a_2 b_2,$$

where $a = (a_0, a_1, a_2)$, $b = (b_0, b_1, b_2)$.

Definition 4.1.5 The map $I_x(\cdot,\cdot) : T_x\mathbb{R}^2 \times T_x\mathbb{R}^2 \longrightarrow \mathbb{R}$, defined by $I_x(X, Y) := \langle df_x X, df_x Y \rangle$, is called the first fundamental form of the surface f at the point $f(x)$.

The matrix of this bilinear map with respect to the basis $\{e_1, e_2\}$ in $T_x U$ has the coefficients $(g_{ij}(x))_{i,j=1,2}$,

$$g_{11}(x) = I_x(e_1, e_1) = \langle df_x e_1, df_x e_1 \rangle = \left\langle \frac{\partial f}{\partial x^1}(x), \frac{\partial f}{\partial x^1}(x) \right\rangle$$

$$g_{12}(x) = I_x(e_1, e_2) = I_x(e_2, e_1) = g_{21}(x) = \langle df_x e_1, df_x e_2 \rangle = \left\langle \frac{\partial f}{\partial x^1}(x), \frac{\partial f}{\partial x^2}(x) \right\rangle$$

$$g_{22}(x) = I_x(e_2, e_2) = \langle df_x e_2, df_x e_2 \rangle = \left\langle \frac{\partial f}{\partial x^2}(x), \frac{\partial f}{\partial x^2}(x) \right\rangle.$$

The way these coefficients are described by the Euclidean inner product of E^3 encapsulates the way in which the ambient space endows with its Geometry each tangent plane of the surface.

Therefore, if $X(x) = X^1(x)e_1 + X^2(x)e_2, \ Y(x) = Y^1(x)e_1 + Y^2(x)e_2 \in T_x U$ then

$$I_x(X, Y) := \langle df_x X, df_x Y \rangle$$

leads to $I_x(X, Y) =$

$$= g_{11}(x)X^1(x)Y^1(x) + g_{12}(x)X^1(x)Y^2(x) + g_{21}(x)X^2(x)Y^1(x) + g_{22}(x)X^2(x)Y^2(x).$$

The Einstein summation convention highlights a simplified formula

$$I_x(X, Y) := \langle df_x X, df_x Y \rangle = g_{ij}(x)X^i(x)Y^j(x).$$

A direct consequence of the definition is

$$I_x(X, X) = \langle df_x X, df_x X \rangle \geqslant 0.$$

Proposition 4.1.6 *The coefficients of the first fundamental form satisfy*

$$det(g_{ij}(x)) = g_{11}(x)g_{22}(x) - g_{12}(x)g_{21}(x) > 0$$

and

$$\left\| \frac{\partial f}{\partial x^1}(x) \times \frac{\partial f}{\partial x^2}(x) \right\| = \sqrt{det(g_{ij}(x))}.$$

Proof Cauchy–Bunyakowsky–Schwartz inequality implies for the non-collinear vectors $\frac{\partial f}{\partial x^1}(x)$ and $\frac{\partial f}{\partial x^2}(x)$ that

$$det\, g_{ij}(x) =$$

$$\left\langle \frac{\partial f}{\partial x^1}(x), \frac{\partial f}{\partial x^1}(x) \right\rangle \left\langle \frac{\partial f}{\partial x^2}(x), \frac{\partial f}{\partial x^2}(x) \right\rangle - \left\langle \frac{\partial f}{\partial x^1}(x), \frac{\partial f}{\partial x^2}(x) \right\rangle \left\langle \frac{\partial f}{\partial x^2}(x), \frac{\partial f}{\partial x^1}(x) \right\rangle$$

$$= \left\| \frac{\partial f}{\partial x^1}(x) \right\|^2 \left\| \frac{\partial f}{\partial x^2}(x) \right\|^2 - \left\langle \frac{\partial f}{\partial x^1}(x), \frac{\partial f}{\partial x^2}(x) \right\rangle^2 > 0.$$

The second equality is provided by the previous formula written as $||a \times b||^2 = ||a||^2 \cdot ||b||^2 \cdot (1 - \cos^2 \alpha)$ for $a = \dfrac{\partial f}{\partial x^1}(x)$ and $b = \dfrac{\partial f}{\partial x^2}(x)$. $\square$

If we consider a change of coordinates $\varphi : \overline{U} \longrightarrow U$, then our surface in the new coordinates is $\bar{f} = f \circ \varphi : \overline{U} \longrightarrow \mathbb{R}^3$.

Theorem 4.1.7 *The first fundamental form is preserved by a change of coordinates.*

Proof If $\bar{x} \in \overline{U}$, $x = \varphi(\bar{x})$, $\overline{X}, \overline{Y} \in T_{\bar{x}}\overline{U}$ and $X = d\varphi_{\bar{x}}\overline{X}$, $Y = d\varphi_{\bar{x}}\overline{Y} \in T_x U$ we have

$$\overline{I}_{\bar{x}}\left(\overline{X}, \overline{Y}\right) = \left\langle d\,\bar{f}_{\bar{x}}\overline{X}, d\,\bar{f}_{\bar{x}}\overline{Y} \right\rangle = \left\langle d\,(f \circ \varphi)_{\bar{x}}\,\overline{X}, d\,(f \circ \varphi)_{\bar{x}}\,\overline{Y} \right\rangle =$$

$$= \left\langle df_x\left(d\varphi_{\bar{x}}\overline{X}\right), df_x\left(d\varphi_{\bar{x}}\overline{Y}\right) \right\rangle = \left\langle df_x(X), df_x(Y) \right\rangle = I_x(X, Y).$$

$\square$

Definition 4.1.8 An isometry of the Euclidean 3-dimensional space E^3 is a map $B : \mathbb{R}^3 \longrightarrow \mathbb{R}^3$ which preserves distances.

Our surface $f : U \longrightarrow \mathbb{R}^3$ is transformed by an isometry into another surface $\tilde{f} = B \circ f : U \longrightarrow \mathbb{R}^3$. A vector is transformed by an isometry into another vector with the same length. Two vectors with the same application point M_0 are transformed into two vectors having as application point $B(M_0)$. The transformed vectors have their lengths preserved. It is easy to observe that their initial angle between them is also preserved. Taking into consideration these observations, the initial first fundamental form is preserved by isometries of the Euclidean space.

Definition 4.1.9 A smooth function $a : I \subset \mathbb{R} \longrightarrow \mathbb{R}^3$, $a(t) = (a_1(t), a_2(t), a_3(t))$ is called a curve of the Euclidean 3-dimensional space E^3.

If $x = x(t) = (x^1(t), x^2(t))$ in U, we obtain $f(x(t)) = (f^1(x(t)), f^2(x(t)),$ $f^3(x(t)))$, that is an one parameter function with the image contained in the image of our surface. It makes sense to define for $x : I \subset \mathbb{R} \longrightarrow U \subset \mathbb{R}^2$, the map $c :=$ $f \circ x : I \subset \mathbb{R} \longrightarrow \mathbb{R}^3$.

The map c is called a *curve on the surface* $f : U \longrightarrow \mathbb{R}^3$.

Two properties of a curve on a surface are stated by the following theorem. The tangent vector $\dot{c}(t)$ belongs to the tangent plane to the surface $T_{c(t)}f$ and, the length of this tangent vector depends on the coefficients g_{ij} of the first fundamental form.

Theorem 4.1.10 *If c is a curve in the surface $f : U \longrightarrow \mathbb{R}^3$ then*

(i) $\dot{c}(t) = \dot{x}^1(t) \dfrac{\partial f}{\partial x^1}(x(t)) + \dot{x}^2(t) \dfrac{\partial f}{\partial x^2}(x(t)) \in T_{f(x(t))}f, \quad \forall t \in I.$

(ii) $\|\dot{c}(t)\|^2 = g_{11}(x(t)) \cdot (\dot{x}^1(t))^2 + 2g_{12}(x(t)) \cdot \dot{x}^1(t) \cdot \dot{x}^2(t) + g_{22}(x(t)) \cdot$ $(\dot{x}^2(t))^2.$

Proof Chain rule implies

(i) $\dot{c}(t) = \dfrac{d}{dt}(f \circ x)(t) = \dfrac{\partial f}{\partial x^1}(x(t)) \cdot \dot{x}^1(t) + \dfrac{\partial f}{\partial x^2}(x(t)) \cdot \dot{x}^2(t) \in T_{f(x(t))}f.$

(ii) Since

$$\|\dot{c}(t)\|^2 = \langle \dot{c}(t), \dot{c}(t) \rangle =$$

$$= \left\langle \frac{\partial f}{\partial x^1}(x(t)) \cdot \dot{x}^1(t) + \frac{\partial f}{\partial x^2}(x(t)) \cdot \dot{x}^2(t), \frac{\partial f}{\partial x^1}(x(t)) \cdot \dot{x}^1(t) + \frac{\partial f}{\partial x^2}(x(t)) \cdot \dot{x}^2(t) \right\rangle$$

it results the statement.   □

Formula (ii) highlights a quadratic form denoted by ds^2 which acts after the rule

$$ds^2(v, v) = g_{11} \cdot (v^1)^2 + 2g_{12} \cdot v^1 v^2 + g_{22} \cdot (v^2)^2,$$

if the vector v is $v = (v^1, v^2)$.

Taking into account that $dx^i(v) = v^i$, the previous formula of the quadratic form can be written as

$$ds^2 = g_{11} \cdot (dx^1)^2 + g_{12} \cdot dx^1 dx^2 + g_{21} \cdot dx^2 dx^1 + g_{22} \cdot (dx^2)^2$$

or, using the Einstein notation,

$$ds^2 = g_{ij}(x)dx^i dx^j.$$

This quadratic form induced by the first fundamental form is called *a metric for the surface f.*

Exercise 4.1.11 If $x = (x^1, x^2)$, prove the following equality:

$$
\begin{pmatrix}
\dfrac{\partial f^1}{\partial x^1}(x) & \dfrac{\partial f^2}{\partial x^1}(x) & \dfrac{\partial f^3}{\partial x^1}(x) \\[2mm]
\dfrac{\partial f^1}{\partial x^2}(x) & \dfrac{\partial f^2}{\partial x^2}(x) & \dfrac{\partial f^3}{\partial x^2}(x)
\end{pmatrix}
\begin{pmatrix}
1 & 0 & 0 \\
0 & 1 & 0 \\
0 & 0 & 1
\end{pmatrix}
\begin{pmatrix}
\dfrac{\partial f^1}{\partial x^1}(x) & \dfrac{\partial f^1}{\partial x^2}(x) \\[2mm]
\dfrac{\partial f^2}{\partial x^1}(x) & \dfrac{\partial f^2}{\partial x^2}(x) \\[2mm]
\dfrac{\partial f^3}{\partial x^1}(x) & \dfrac{\partial f^3}{\partial x^2}(x)
\end{pmatrix} =
$$

$$
= \begin{pmatrix} g_{11}(x) & g_{12}(x) \\ g_{21}(x) & g_{22}(x) \end{pmatrix}.
$$

The above formula is another way to explain how the inner product coefficients of the Euclidean space, restricted to a point of the tangent plane of the surface, produces the coefficients of the first fundamental form.

In fact this formula,

$$
df_x^T \cdot \begin{pmatrix} 1 & 0 & 0 \\ 0 & 1 & 0 \\ 0 & 0 & 1 \end{pmatrix} \cdot df_x = \begin{pmatrix} g_{11}(x) & g_{12}(x) \\ g_{21}(x) & g_{22}(x) \end{pmatrix},
$$

is a metric change at the level of each tangent plane; the Euclidean metric

$$
ds^2 = dX_1^2 + dY_1^2 + dZ_1^2
$$

endows the surface with a metric induced by the new coordinates

$$
X_1 = f^1(x^1, x^2); \quad Y_1 = f^2(x^1, x^2); \quad Z_1 = f^3(x^1, x^2),
$$

that is with the metric

$$
ds^2 = g_{ij}(x)dx^i dx^j.
$$

The *length of a curve* $c = f \circ x : I \longrightarrow \mathbb{R}^3$ on the surface $f : U \longrightarrow \mathbb{R}^3$ between the points $c(a)$ and $c(b)$ where $a, b \in I, a < b$ is given by

$$
L_c = \int_a^b \| \dot{c}(t) \| \, dt.
$$

It follows that it can be expressed in terms of the first fundamental form by

$$
L_c = \int_a^b \sqrt{g_{11}(x(t)) \cdot (\dot{x}^1(t))^2 + 2g_{12}(x(t)) \cdot \dot{x}^1(t)\dot{x}^2(t) + g_{22}(x(t)) \cdot (\dot{x}^2(t))^2} \, dt,
$$

or, using the Einstein notation

$$L_c = \int_a^b \sqrt{g_{ij}\left(x\left(t\right)\right) \cdot \dot{x}^i\left(t\right) \cdot \dot{x}^j\left(t\right)}dt.$$

Definition 4.1.12 For two curves $c = f \circ x : I \longrightarrow \mathbb{R}^3$ and $\bar{c} : f \circ \bar{x} : \bar{I} \longrightarrow \mathbb{R}^3$ on the surface $f : U \longrightarrow \mathbb{R}^3$, the angle between them at the common point $\bar{x}\left(\bar{t}_0\right) = x\left(t_0\right)$, is the acute angle between the two tangents to the curves at the common point.

The angle α of the curves c and $\bar{c}$ at their common point $c\left(t_0\right) = \bar{c}\left(\bar{t}_0\right)$ can be computed by the formula

$$\cos\alpha = \frac{\left\langle \dot{c}\left(t_0\right), \dot{\bar{c}}\left(\bar{t}_0\right)\right\rangle}{\left\|\dot{c}\left(t_0\right)\right\| \cdot \left\|\dot{\bar{c}}\left(\bar{t}_0\right)\right\|}$$

that is, it can be expressed in terms of the first fundamental form by the formula

$$\cos\alpha = \frac{g_{ij}\left(x\left(t_0\right)\right) \cdot \dot{x}^i\left(t_0\right) \cdot \dot{\bar{x}}^j\left(\bar{t}_0\right)}{\sqrt{g_{rs}\left(x\left(t_0\right)\right) \cdot \dot{x}^r\left(t_0\right) \cdot \dot{x}^s\left(t_0\right)} \cdot \sqrt{g_{pq}\left(\bar{x}\left(\bar{t}_0\right)\right) \cdot \dot{\bar{x}}^p\left(\bar{t}_0\right) \cdot \dot{\bar{x}}^q\left(\bar{t}_0\right)}}.$$

We observe that the lengths of tangent vectors to curves in surfaces depend on the coefficients of the first fundamental form; the length of curves in surfaces depends on the coefficients of the first fundamental form; the angle between two tangent vectors and, as a consequence, the angle between two curves depend on the coefficients of the first fundamental form.

Even if we do not prove here, the area of a region on the surface depends on the coefficients of the first fundamental form. The formula for the *area of a region* $f(D), D \subset U$ is

$$\sigma_{f(D)} = \iint_D \sqrt{\det\left(g_{ij}\left(x\right)\right)}dx^1 dx^2.$$

The words "depends on the coefficients of the first fundamental form" can be replaced by "depends on the metric of the surface". Let us conclude:

Definition 4.1.13 All the geometric properties depending on the coefficients of the first fundamental form, that is, depending on the metric of the surface, are called intrinsic geometric properties of a surface.

Therefore, we may say that

- the length of a curve,
- the angle between two curves,
- the area of a region,

all these are quantities belonging to the intrinsic Geometry of the surface. The change of coordinates and the isometries preserve the intrinsic nature of geometric properties.

We may ask if, for a given surface, geometric properties exist which do not belong to the intrinsic Geometry of the surface. For this purpose, we need to study geometric properties depending on the vector N.

Consider the Gauss frame $\left\{\dfrac{\partial f}{\partial x^1}(x), \dfrac{\partial f}{\partial x^2}(x), N(x)\right\}$ at each point $f(x)$ on the surface $f : U \subset \mathbb{R}^2 \longrightarrow \mathbb{R}^3$. Since the length of Gauss map is 1, if one considers the derivative of $\langle N(x), N(x)\rangle = 1$, it results both $\left\langle \dfrac{\partial N}{\partial x^1}(x), N(x)\right\rangle = 0$ and $\left\langle \dfrac{\partial N}{\partial x^2}(x), N(x)\right\rangle = 0$.

Therefore the vectors $\dfrac{\partial N}{\partial x^1}(x)$ and $\dfrac{\partial N}{\partial x^2}(x)$ are orthogonal to the Gauss vector $N(x)$ at each point on the surface, i.e. $\left\{\dfrac{\partial N}{\partial x^1}(x), \dfrac{\partial N}{\partial x^2}(x)\right\} \subset T_{f(x)}f$.

Definition 4.1.14 The map $II_x(\cdot, \cdot) : T_x\mathbb{R}^2 \times T_x\mathbb{R}^2 \longrightarrow \mathbb{R}$, defined by

$$II_x(X, Y) := -\langle dN_x X, df_x Y\rangle ,$$

is called the second fundamental form of the surface f at the point $f(x)$.

The matrix of this bilinear map with respect to the basis $\{e_1, e_2\}$ in $T_x U$ is described by its coefficients $\left(h_{ij}(x)\right)_{i,j=1,2}$,

$$h_{11}(x) = II_x(e_1, e_1) = -\langle dN_x e_1, df_x e_1\rangle = -\left\langle \frac{\partial N}{\partial x^1}(x), \frac{\partial f}{\partial x^1}(x)\right\rangle$$

$$h_{12}(x) = II_x(e_1, e_2) = -\langle dN_x e_1, df_x e_2\rangle = -\left\langle \frac{\partial N}{\partial x^1}(x), \frac{\partial f}{\partial x^2}(x)\right\rangle$$

$$h_{21}(x) = II_x(e_2, e_1) = -\langle dN_x e_2, df_x e_1\rangle = -\left\langle \frac{\partial N}{\partial x^2}(x), \frac{\partial f}{\partial x^1}(x)\right\rangle$$

$$h_{22}(x) = II_x(e_2, e_2) = -\langle dN_x e_2, df_x e_2\rangle = -\left\langle \frac{\partial N}{\partial x^2}(x), \frac{\partial f}{\partial x^2}(x)\right\rangle .$$

Exactly as in the case of the first fundamental form, the way these coefficients are described by the Euclidean inner product of $\mathbb{R}^3$ encapsulates the way in which the Euclidean ambient space allows its Geometry to produce these coefficients.

Therefore, if $X(x) = X^1(x)e_1 + X^2(x)e_2$, $Y(x) = Y^1(x)e_1 + Y^2(x)e_2 \in T_x U$ then

$$II_x (X, Y) := - \langle dN_x X, df_x Y \rangle,$$

that is

$$II_x (X, Y) =$$

$$= h_{11}(x)X^1(x)Y^1(x) + h_{12}(x)X^1(x)Y^2(x) + h_{21}(x)X^2(x)Y^1(x) + h_{22}(x)X^2(x)Y^2(x).$$

The Einstein summation convention leads to a simplified formula

$$II_x (X, Y) := - \langle dN_x X, df_x Y \rangle = h_{ij}(x)X^i(x)Y^j(x).$$

Theorem 4.1.15 $h_{12}(x) = h_{21}(x)$.

Proof Starting from the relations

$$\left\langle N(x), \frac{\partial f}{\partial x^1}(x) \right\rangle = 0 \text{ and } \left\langle N(x), \frac{\partial f}{\partial x^2}(x) \right\rangle = 0,$$

it results

$$\left\langle \frac{\partial N}{\partial x^2}(x), \frac{\partial f}{\partial x^1}(x) \right\rangle + \left\langle N(x), \frac{\partial^2 f}{\partial x^2 \partial x^1}(x) \right\rangle = 0$$

and

$$\left\langle \frac{\partial N}{\partial x^1}(x), \frac{\partial f}{\partial x^2}(x) \right\rangle + \left\langle N(x), \frac{\partial^2 f}{\partial x^1 \partial x^2}(x) \right\rangle = 0,$$

that is, using the definitions of the coefficients h_{12} and h_{21}, we finally obtain

$$h_{12}(x) = \left\langle N(x), \frac{\partial^2 f}{\partial x^1 \partial x^2}(x) \right\rangle = \left\langle N(x), \frac{\partial^2 f}{\partial x^2 \partial x^1}(x) \right\rangle = h_{21}(x).$$

$\square$

Using the same arguments as in the case of the first fundamental form, we deduce that the second fundamental form is preserved by changes of coordinates and isometries of the Euclidean 3-dimensional space.

From the previous theorem (regarding the symmetry of the first fundamental form), we know that $\left\langle N, \frac{\partial N}{\partial x^i} \right\rangle = 0$, i.e. $\frac{\partial N}{\partial x^i}(x) \in T_{f(x)} f$.

It implies the existence of the coefficients $h^i_j(x)$ $i, j \in 1, 2$, such that

$$-\frac{\partial N}{\partial x^i}(x) = h^s_i(x)\frac{\partial f}{\partial x^s}(x).$$

Fig. 4.1 Sphere

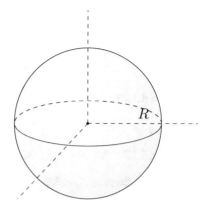

In the previous formula we used the Einstein notation, the dummy index being s. These formulas are called *Weingarten's formulas* and they can be written in their matrix form (Fig. 4.1).

Theorem 4.1.16 (Weingarten's formulas)

$$h_{ij}(x) = h_i^s(x) g_{sj}(x).$$

Proof The inner product of both members of the equality

$$-\frac{\partial N}{\partial x^i}(x) = h_i^s(x) \frac{\partial f}{\partial x^s}(x)$$

by $\dfrac{\partial f}{\partial x^j}(x)$ leads to

$$h_{ij}(x) = \left\langle -\frac{\partial N}{\partial x^i}(x), \frac{\partial f}{\partial x^j}(x) \right\rangle = h_i^s(x) g_{sj}(x).$$

$\square$

Corollary 4.1.17 $det(h_i^j(x)) = \dfrac{det(h_{ij}(x))}{det(g_{ij}(x))}$

Hint. Use $det(AB) = det A \cdot det B.$ $\square$

Definition 4.1.18 The Gaussian curvature of the surface f at a point $f(x)$ is denoted by $K(x)$ and is given by the formula $K(x) := det(h_j^i(x))$.

A direct consequence is the formula $K(x) = \dfrac{det(h_{ij}(x))}{det(g_{ij}(x))}$

Problem 4.1.19 Compute the Gaussian curvature at a point of a plane.

Hint. You may use $f(x^1, x^2) = (x^1, x^2, 0)$; $K(x) = 0$;

Problem 4.1.20 Compute the Gaussian curvature at a point of a circular cylinder.

Hint: $f(x^1, x^2) = (R \cos x^1, R \sin x^1, x^2)$; $K(x) = 0$;

Problem 4.1.21 Compute the Gaussian curvature at a point of a circular cone.

Hint. $f(x^1, x^2) = (x^2 \cos x^1, x^2 \sin x^1, x^2)$; $K(x) = 0$;

Problem 4.1.22 Compute the Gaussian curvature at a point of a sphere.

Hint. The parameterization is

$$f(x^1, x^2) = (R \sin x^2 \cos x^1, R \sin x^2 \sin x^1, R \cos x^2), \ x^1 \in (0, \pi), \ x^2 \in (0, 2\pi).$$

If $x^1 = x, x^2 = y$, the metric of the sphere is

$$ds^2 = R^2 dx^2 + R^2 \sin^2 x dy^2.$$

It's curvature is a positive constant at each point, that is $K(x) = \dfrac{1}{R^2}$.

Problem 4.1.23 Compute the Gaussian curvature of the surface

$$f(x^1, x^2) = (\sqrt{2}x^2 \cos x^1, \sqrt{2}x^2 \sin x^1, (x^2)^2 \sin 2x^1)$$

and observe that it is strictly negative.

Theorem 4.1.24 (The Geometric Interpretation of the Gaussian Curvature) *The absolute value of the Gaussian curvature of a surface f at the point $f(x)$ is given by the ratio of the areas determined by the vectors* $\left\{ -\dfrac{\partial N}{\partial x^1}(x), \ -\dfrac{\partial N}{\partial x^2}(x) \right\}$ *respectively* $\left\{ \dfrac{\partial f}{\partial x^1}(x), \ \dfrac{\partial f}{\partial x^2}(x) \right\}$,

$$|K(x)| = \frac{\left\| \dfrac{\partial N}{\partial x^1}(x) \times \dfrac{\partial N}{\partial x^2}(x) \right\|}{\left\| \dfrac{\partial f}{\partial x^1}(x) \times \dfrac{\partial f}{\partial x^2}(x) \right\|}.$$

Proof The vectors $\left\{ \dfrac{\partial f}{\partial x^1}(x), \ \dfrac{\partial f}{\partial x^2}(x), \ -\dfrac{\partial N}{\partial x^1}(x), \ -\dfrac{\partial N}{\partial x^2}(x) \right\}$, belonging to the tangent plane $T_{f(x)}f$, are related by the Weingarten's formulas

$$-\frac{\partial N}{\partial x^i}(x) = h_i^s(x) \frac{\partial f}{\partial x^s}(x), \ i \in \{1, 2\}.$$

Denote by $e_{ij}(x) = \left\langle \dfrac{\partial N}{\partial x^i}(x), \dfrac{\partial N}{\partial x^j}(x) \right\rangle$. We have both

$$e_{ij}(x) = \left\langle \frac{\partial N}{\partial x^1}(x), \frac{\partial N}{\partial x^2}(x) \right\rangle = \left\langle h_i^s(x) \frac{\partial f}{\partial x^s}(x), h_j^r(x) \frac{\partial f}{\partial x^r}(x) \right\rangle = h_i^s(x) h_j^r(x) g_{rs}(x)$$

and

$$\left\| \frac{\partial N}{\partial x^1}(x) \times \frac{\partial N}{\partial x^2}(x) \right\| = \sqrt{\det(e_{ij}(x))}.$$

It results

$$\left\| \frac{\partial N}{\partial x^1}(x) \times \frac{\partial N}{\partial x^2}(x) \right\|^2 = \det(e_{ij}(x)) = \det h_i^s(x) \cdot \det h_j^r(x) \cdot \det(g_{rs}(x)),$$

therefore

$$\left\| \frac{\partial N}{\partial x^1}(x) \times \frac{\partial N}{\partial x^2}(x) \right\|^2 = \det(h_i^j(x))^2 \cdot \left\| \frac{\partial f}{\partial x^1}(x) \times \frac{\partial f}{\partial x^2}(x) \right\|^2.$$

$\square$

There are some comments related to the Gaussian curvature:

- The Gaussian curvature of a surface at a point remains invariant under a change of coordinates. This happens because the first fundamental form and the second fundamental form are preserved (at the corresponding points) by a change of coordinates as we showed above.
- The same happens for the Gaussian curvature if we consider an isometry of the Euclidean space E^3.
- The Gaussian curvature is related to the fact that the surface "lives" in the ambient 3-dimensional Euclidean space E^3. So, the curvature is an "extrinsic" property of a surface.
- Isometries are maps which preserve distances in the Euclidean 3-space E^3. If we wrap around a flat plane into a cylinder or a cone the distances are preserved. The Gaussian curvature at corresponding points is the same, i.e. $K(x) = K(\bar{x}) = 0$.

Gauss showed that regions of two surfaces can be wrapped around one to another if, at corresponding points, we have the same Gaussian curvature. To obtain this result Gauss proved a special theorem, called *Theorema Egregium*, which shows that the Gaussian curvature belongs to the intrinsic Geometry of the surface. The Latin word Egregium means "remarkable". The theory of surfaces was developed by Gauss in 1827 in a paper entitled "General Investigations of Curved Surfaces", the original title being in Latin *Disquisitiones Generales circa Superficies Curvas* [31].

Next, we intend to obtain this result related to the Gaussian curvature. It shows us that the Differential Geometry of surfaces depends only on the first fundamental form, i.e. on the metric of the surface. A new language can be developed and Differential Geometry makes sense in a more abstract and general "environment" as we can see in the next chapter.

To continue, we define the *Christoffel symbols of first kind*, as[1]

$$\Gamma_{ij,k}(x) := \frac{1}{2}\left(\frac{\partial g_{ik}}{\partial x^j}(x) + \frac{\partial g_{jk}}{\partial x^i}(x) - \frac{\partial g_{ij}}{\partial x^k}(x)\right)$$

and the *Christoffel symbols of second kind*, as

$$\Gamma^i_{jk}(x) := g^{is}(x)\Gamma_{jk,s}(x) = \frac{1}{2}\, g^{is}(x)\left(\frac{\partial g_{js}}{\partial x^k}(x) + \frac{\partial g_{ks}}{\partial x^j}(x) - \frac{\partial g_{jk}}{\partial x^s}(x)\right),$$

where $g^{ij}(x)$ is the inverse of the matrix of the first fundamental form, $g_{ij}(x)$.

An important observation is the fact that these matrices are inverse each other and they can be written using the Einstein notation in the form

$$g^{sj}(x)g_{is}(x) = g_{is}(x)g^{sj}(x) = \delta^j_i.$$

Of course, one can calculate each g^{ij} exactly as one did it when studying the inverse of a matrix, that is $g^{11} = \dfrac{g_{22}}{det\,(g_{ij})}$, etc.

4.2 Intrinsic Geometry of Surfaces

The Weingarten formulas

$$-\frac{\partial N}{\partial x^i}(x) = h^s_i(x)\frac{\partial f}{\partial x^s}(x)\,,\ i \in \{1,2\}$$

involve the partial derivative of N, a vector of Gauss frame, written with respect to the Gauss frame. The previous formulas lead to the Gaussian curvature at a point of a surface. As we said, the Differential Geometry of surfaces is related to the coefficients appearing when one considers the partial derivatives of vectors in the Gauss frame. It remains to see what happens considering the partial derivatives of $\dfrac{\partial f}{\partial x^1}$ and $\dfrac{\partial f}{\partial x^2}$.

---

[1]Sometimes, such symbols are defined also as $\{ij, k\}$.

Theorem 4.2.1 (Gauss' Formulas) *Consider a given surface* $f : U \longrightarrow \mathbb{R}^3$ *and let* $\left\{ \dfrac{\partial f}{\partial x^1}(x), \dfrac{\partial f}{\partial x^2}(x), N(x) \right\}$ *be the Gauss frame at an arbitrary point* $f(x)$ *of the surface. If* $h_{ij}(x)$ *are the coefficients of the second fundamental form, then:*

$$\frac{\partial^2 f}{\partial x^i \partial x^k}(x) = \Gamma^s_{ik}(x) \cdot \frac{\partial f}{\partial x^s}(x) + N(x) \cdot h_{ik}(x)$$

Proof The vector $\dfrac{\partial^2 f}{\partial x^i \partial x^k}(x)$ can be expressed as a linear combination of the Gauss frame vectors, that is

$$\frac{\partial^2 f}{\partial x^i \partial x^k}(x) = A^s_{ik}(x) \cdot \frac{\partial f}{\partial x^s}(x) + a_{ik} \cdot N(x).$$

The inner product of both members of N leads to $a_{ik} = h_{ik}$, therefore we have

$$\frac{\partial^2 f}{\partial x^i \partial x^k}(x) = A^s_{ik}(x) \cdot \frac{\partial f}{\partial x^s} + h_{ik}(x) \cdot N(x).$$

Now, the inner product of both members of $\dfrac{\partial f}{\partial x^j}(x)$ leads to

$$\left\langle \frac{\partial^2 f}{\partial x^i \partial x^k}(x), \frac{\partial f}{\partial x^j}(x) \right\rangle = A^s_{ik}(x) \cdot g_{sj}(x),$$

which implies $A^s_{ik}(x) = A^s_{ki}(x)$. On the other hand, if we apply the partial derivative with respect to x^k to the equality $g_{ij}(x) = \left\langle \dfrac{\partial f}{\partial x^i}(x), \dfrac{\partial f}{\partial x^j}(x) \right\rangle$ we have

$$\frac{\partial g_{ij}}{\partial x^k}(x) = \left\langle \frac{\partial^2 f}{\partial x^k \partial x^i}(x), \frac{\partial f}{\partial x^j}(x) \right\rangle + \left\langle \frac{\partial f}{\partial x^i}(x), \frac{\partial^2 f}{\partial x^k \partial x^j}(x) \right\rangle,$$

and this can be written as

$$\frac{\partial g_{ij}}{\partial x^k}(x) = A^s_{ik}(x) \cdot g_{sj}(x) + A^s_{jk}(x) \cdot g_{si}(x).$$

If $i \to j \to k \to i$ we obtain two more relations

$$\frac{\partial g_{jk}}{\partial x^i}(x) = A^s_{ji}(x) \cdot g_{sk}(x) + A^s_{ki}(x) \cdot g_{sj}(x),$$

$$\frac{\partial g_{ki}}{\partial x^j}(x) = A^s_{kj}(x) \cdot g_{si}(x) + A^s_{ij}(x) \cdot g_{sk}(x).$$

If we add the first two and we extract the last one, we finally obtain

$$\frac{\partial g_{jk}}{\partial x^i}(x) + \frac{\partial g_{ik}}{\partial x^j}(x) - \frac{\partial g_{ij}}{\partial x^k}(x) = 2A_{ij}^s(x) \cdot g_{sk}(x),$$

and then

$$A_{ij}^r(x) = \Gamma_{ij}^r(x).$$

$\square$

The formulas

$$\frac{\partial g_{ij}}{\partial x^k}(x) = A_{ik}^s(x) \cdot g_{sj}(x) + A_{jk}^s(x) \cdot g_{si}(x)$$

are called *Ricci's equations* and, according to the above notations considering the Christoffel symbols, they are

$$\frac{\partial g_{ij}}{\partial x^k}(x) = \Gamma_{ik}^s(x) \cdot g_{sj}(x) + \Gamma_{jk}^s(x) \cdot g_{si}(x).$$

In order to simplify the notation we cancel x in all the formulas below.

We define the *Riemann symbols of second type* by

$$R_{ijk}^h = \frac{\partial \Gamma_{ik}^h}{\partial x^j} - \frac{\partial \Gamma_{ij}^h}{\partial x^k} + \Gamma_{mj}^h \Gamma_{ik}^m - \Gamma_{mk}^h \Gamma_{ij}^m,$$

the *Riemann symbols of first type* by $R_{ijkl} = g_{is} R_{jkl}^s$, and the *Ricci symbols* by: $R_{ij} = R_{isj}^s$. All these symbols depend only on g_{ij}, i.e. they belong to the intrinsic Geometry of surfaces.

Let us first observe that the metric coefficients g_{ij} allow us to lower indexes as in the formula

$$R_{ijkl} = g_{is} R_{jkl}^s.$$

The components of the inverse matrix of the metric coefficients allow us to rise indexes, that is

$$R_{jkl}^i = g^{is} R_{sjkl}.$$

The last formula can be derived if we multiply the last formula by g_{mi} and we consider the sum after the dummy index i. It results $g_{mi} R_{jkl}^i = g_{mi} g^{is} R_{sjkl}$, that is the true equality $R_{mjkl} = R_{mjkl}$.

For now, if we have a multi-index quantity, say T_{lmn}^{ij}, we can derive $T_{\alpha lmn}^j$ by the rule

$$T_{\alpha lmn}^j := g_{\alpha i} T_{lmn}^{ij}$$

and

$$T^{ij}_{lmn} := g^{i\alpha} T^{j}_{\alpha lmn},$$

etc.

Theorem 4.2.2 (Gauss' (1) and Codazzi–Mainardi's (2) equations)
The following assertions are equivalent :

(i) $\dfrac{\partial^3 f}{\partial x^i \partial x^j \partial x^k} = \dfrac{\partial^3 f}{\partial x^j \partial x^k \partial x^i}$;

(ii) $R_{ijkl} = h_{ik} \cdot h_{jl} - h_{il} \cdot h_{jk}$;   (1)

$\dfrac{\partial h_{ij}}{\partial x^k} + \Gamma^s_{ij} \cdot h_{sk} = \dfrac{\partial h_{ik}}{\partial x^j} + \Gamma^s_{ik} \cdot h_{sj}.$ (2)

Proof $(i) \Rightarrow (ii)$

We consider the partial derivative with respect to x^i of the Gauss formulas

$$\frac{\partial^2 f}{\partial x^j \partial x^k} = \Gamma^s_{jk} \cdot \frac{\partial f}{\partial x^s} + N \cdot h_{jk}.$$

We obtain

$$\frac{\partial^3 f}{\partial x^i \partial x^j \partial x^k} = \frac{\partial \Gamma^s_{jk}}{\partial x^i} \cdot \frac{\partial f}{\partial x^s} + \Gamma^s_{jk} \cdot \frac{\partial^2 f}{\partial x^i \partial x^s} + \frac{\partial N}{\partial x^i} \cdot h_{jk} + N \cdot \frac{\partial h_{jk}}{\partial x^i}.$$

Let us take into consideration Gauss' and Weingarten's formulas, the last as $\dfrac{\partial N}{\partial x^i} = -h^r_i \dfrac{\partial f}{\partial x^r}$; It results

$$\frac{\partial^3 f}{\partial x^i \partial x^j \partial x^k} = \left(\frac{\partial \Gamma^r_{jk}}{\partial x^i} + \Gamma^r_{is} \cdot \Gamma^s_{jk} - h_{jk} \cdot h^r_i \right) \cdot \frac{\partial f}{\partial x^r} + \left(\frac{\partial h_{jk}}{\partial x^i} + \Gamma^s_{jk} \cdot h_{si} \right) \cdot N,$$

and using $i \to j \to k \to i$ we obtain the formula

$$\frac{\partial^3 f}{\partial x^j \partial x^k \partial x^i} = \left(\frac{\partial \Gamma^r_{ki}}{\partial x^j} + \Gamma^r_{js} \cdot \Gamma^s_{ki} - h_{ki} \cdot h^r_j \right) \cdot \frac{\partial f}{\partial x^r} + \left(\frac{\partial h_{ki}}{\partial x^j} + \Gamma^s_{ki} \cdot h_{sj} \right) \cdot N.$$

Comparing the coefficients of $\dfrac{\partial f}{\partial x^r}$ and N the following equality holds :

$$\frac{\partial h_{jk}}{\partial x^i} + \Gamma^s_{jk} \cdot h_{si} = \frac{\partial h_{ki}}{\partial x^j} + \Gamma^s_{ki} \cdot h_{sj}$$

for the coefficients of N, and

$$\frac{\partial \Gamma^r_{jk}}{\partial x^i} + \Gamma^r_{is} \cdot \Gamma^s_{jk} - h_{jk} \cdot h^r_i = \frac{\partial \Gamma^r_{ki}}{\partial x^j} + \Gamma^r_{js} \cdot \Gamma^s_{ki} - h_{ki} \cdot h^r_j$$

for the coefficients of $\dfrac{\partial f}{\partial x^r}$. The first equality means the *Codazzi–Mainardi equations*. The second one can be rearranged in the form

$$\frac{\partial \Gamma^r_{kj}}{\partial x^i} - \frac{\partial \Gamma^r_{ki}}{\partial x^j} + \Gamma^r_{is} \cdot \Gamma^s_{kj} - \Gamma^r_{js} \cdot \Gamma^s_{ki} = h_{jk} \cdot h^r_i - h_{ki} \cdot h^r_j,$$

i.e.

$$R^r_{kij} = h_{jk} \cdot h^r_i - h_{ki} \cdot h^r_j.$$

Multiplying by g_{lr} we obtain the *Gauss equations*

$$R_{lkij} = g_{lr} \cdot R^r_{kij} = g_{lr} \cdot h^r_i \cdot h_{jk} - g_{lr} \cdot h^r_j \cdot h_{ki} = h_{li} \cdot h_{jk} - h_{lj} \cdot h_{ki}.$$

Gauss' equations can be written in the form

$$R_{ijkl} = h_{ik} \cdot h_{jl} - h_{il} \cdot h_{jk}.$$

$(ii) \Rightarrow (i)$

Starting from Gauss' and Codazzi–Mainardi's equations, if we separate and multiply in a convenient way by $\dfrac{\partial f}{\partial x^s}$ and N, we obtain

$$\frac{\partial^3 f}{\partial x^i \partial x^j \partial x^k} = \frac{\partial^3 f}{\partial x^j \partial x^k \partial x^i}.$$

$\square$

Problem 4.2.3 The following assertions are equivalent

(i) $\dfrac{\partial^2 N}{\partial x^i \partial x^j} = \dfrac{\partial^2 N}{\partial x^j \partial x^i}$

(ii) $\dfrac{\partial h_{jk}}{\partial x^i} + \Gamma^s_{jk} \cdot h_{si} = \dfrac{\partial h_{ki}}{\partial x^j} + \Gamma^s_{ki} \cdot h_{sj}$

Solution.

$h^s_i = g^{sk} \cdot h_{ki}$ and Weingarten's formulas

$$-\frac{\partial N}{\partial x^i} = h^s_i \cdot \frac{\partial f}{\partial x^s},$$

imply

$$-\frac{\partial N}{\partial x^i} = h_{ik} \cdot g^{ks} \cdot \frac{\partial f}{\partial x^s}.$$

Deriving with respect to x^j, it results

$$-\frac{\partial^2 N}{\partial x^i \partial x^j} = \left(\frac{\partial h_{ik}}{\partial x^j} \cdot g^{ks} + h_{ik} \cdot \frac{\partial g^{ks}}{\partial x^j}\right) \cdot \frac{\partial f}{\partial x^s} + h_{ik} \cdot g^{ks} \cdot \frac{\partial^2 f}{\partial x^s \partial x^j} =$$

$$= \left(\frac{\partial h_{ik}}{\partial x^j} \cdot g^{ks} + h_{ik} \cdot \frac{\partial g^{ks}}{\partial x^j}\right) \cdot \frac{\partial f}{\partial x^s} + h_{ik} \cdot g^{kr} \cdot \left(\Gamma^s_{rj} \frac{\partial f}{\partial x^s} + h_{rj} \cdot N\right) =$$

$$= \left(\frac{\partial h_{ik}}{\partial x^j} \cdot g^{ks} + h_{ik} \cdot \frac{\partial g^{ks}}{\partial x^j} + h_{ik} \cdot g^{kr} \cdot \Gamma^s_{rj}\right) \cdot \frac{\partial f}{\partial x^s} + h_{ik} \cdot g^{kr} \cdot h_{rj} \cdot N.$$

From the equality

$$\frac{\partial^2 N}{\partial x^i \partial x^j} = \frac{\partial^2 N}{\partial x^j \partial x^i}$$

using

$$h_{ik} \cdot g^{kr} \cdot h_{rj} = h_{jk} \cdot g^{kr} \cdot h_{ri},$$

we obtain the following equality

$$\frac{\partial h_{ik}}{\partial x^j} \cdot g^{ks} + h_{ik} \cdot \frac{\partial g^{ks}}{\partial x^j} + h_{ik} \cdot g^{kr} \cdot \Gamma^s_{rj} = \frac{\partial h_{jk}}{\partial x^i} \cdot g^{ks} + h_{jk} \cdot \frac{\partial g^{ks}}{\partial x^i} + h_{jk} \cdot g^{kr} \cdot \Gamma^s_{ri}.$$

Summing after multiplying by g_{sp} implies

$$\frac{\partial h_{ip}}{\partial x^j} + h_{ik} \cdot g_{sp} \cdot \frac{\partial g^{ks}}{\partial x^j} + \frac{1}{2} h_{ik} \cdot g^{kr} \cdot \left(\frac{\partial g_{rp}}{\partial x^j} + \frac{\partial g_{ip}}{\partial x^r} - \frac{\partial g_{jr}}{\partial x^p}\right) =$$

$$= \frac{\partial h_{jp}}{\partial x^i} + h_{jk} \cdot g_{sp} \cdot \frac{\partial g^{ks}}{\partial x^i} + \frac{1}{2} h_{jk} \cdot g^{kr} \cdot \left(\frac{\partial g_{rp}}{\partial x^i} + \frac{\partial g_{ip}}{\partial x^r} - \frac{\partial g_{ir}}{\partial x^p}\right).$$

Then

$$\frac{\partial h_{ip}}{\partial x^j} + h_{ik} \cdot \left(g_{sp} \cdot \frac{\partial g^{ks}}{\partial x^j} + g^{ks} \cdot \frac{\partial g_{sp}}{\partial x^j}\right) - \frac{1}{2} h_{ik} \cdot g^{kr} \cdot \left(\frac{\partial g_{jr}}{\partial x^p} + \frac{\partial g_{rp}}{\partial x^j} - \frac{\partial g_{jp}}{\partial x^r}\right) =$$

$$= \frac{\partial h_{jp}}{\partial x^i} + h_{jk} \cdot \left(g_{sp} \cdot \frac{\partial g^{ks}}{\partial x^i} + g^{ks} \cdot \frac{\partial g_{ps}}{\partial x^j}\right) - \frac{1}{2} h_{jk} \cdot g^{kr} \cdot \left(\frac{\partial g_{ir}}{\partial x^p} + \frac{\partial g_{rp}}{\partial x^i} - \frac{\partial g_{jp}}{\partial x^r}\right).$$

Using the derivative of the relation $g_{sp} \cdot g^{sk} = \delta_{pk}$, that is $g_{sp} \cdot \dfrac{\partial g^{sk}}{\partial x^i} + \dfrac{\partial g_{sp}}{\partial x^i} \cdot g^{sk} = 0$, it results

$$\frac{\partial h_{ip}}{\partial x^j} - h_{ik} \cdot g^{kr} \cdot \Gamma_{jp,r} = \frac{\partial h_{jp}}{\partial x^i} - h_{jk} \cdot g^{kr} \cdot \Gamma_{ip,r}.$$

Therefore the Codazzi–Mainardi equations

$$\frac{\partial h_{ij}}{\partial x^k} + \Gamma_{ij}^s \cdot h_{sk} = \frac{\partial h_{ik}}{\partial x^j} + \Gamma_{ik}^s \cdot h_{sj}$$

are obtained.

$(ii) \Rightarrow (i)$

Almost obvious using the way back in the previous computations. $\square$

Theorem 4.2.4 *Riemann symbols R_{ijkl} have the properties*

$$R_{ijkl} = -R_{ijlk};$$
$$R_{ijkl} = -R_{jikl};$$
$$R_{ijkl} = R_{jilk};$$
$$R_{ijkl} = R_{klij};$$
$$R_{ijkl} + R_{iklj} + R_{iljk} = 0 \ (Bianchi's \ first \ identity).$$

Proof Using previous Gauss' equations

$$R_{ijkl} = h_{ik} \cdot h_{jl} - h_{il} \cdot h_{jk}$$

and some replacements of indexes. $\square$

A consequence of the first relation is $R_{2111} = -R_{2111}$, that is $R_{2111} = 0$. Same way $R_{1222} = 0$, or generally, if three indexes coincide then $R_{jiii} = 0$. We may also observe the relations

$$R_{1212} = -R_{2112} = -R_{1221} = R_{2121}.$$

In the same way $R_{iikl} = -R_{iikl}$, i.e. $R_{iikl} = 0$.

Theorem 4.2.5 (Theorema Egregium) *The Gaussian curvature of a surface depends on the coefficients of the metric only.*

Proof

$$K(x) = \frac{\det\left(h_{ij}(x)\right)}{\det\left(g_{ij}(x)\right)} = \frac{h_{11}(x) \cdot h_{22}(x) - h_{12}^2(x)}{\det\left(g_{ij}(x)\right)} = \frac{R_{1212}(x)}{\det\left(g_{ij}(x)\right)}.$$

$\square$

The previous theorem shows that the Gauss curvature belongs to the intrinsic Geometry of the surface.

As we discussed earlier, this particular result allows us to think about Differential Geometry in a more general frame, for example, considering sets which are not necessary embedded in a space with an extra-dimension. The Differential Geometry of such a set will be described only by a "metric tensor", i.e. a matrix g_{ij}, which plays the role of the "first fundamental form" we used in the case of surfaces. Therefore, the quadratic form we defined earlier,

$$ds^2 = g_{ij}(x)dx^i dx^j,$$

is the only thing we need to develop Differential Geometry on sets without extra dimensions.

Example 4.2.6 THE PSEUDOSPHERE

In some examples we noticed the existence of surfaces with null Gaussian curvature as the plane, the circular cylinder, the circular cone. We also highlighted spheres as surfaces of constant positive Gaussian curvature.

Now, we intend to describe a surface with constant negative curvature.

This surface is called a *pseudosphere* and was described by Eugenio Beltrami in his 1868 paper on models of hyperbolic geometries [32].

In order to discuss a pseudosphere, let us suppose firstly to know the equation of a curve called *tractrix*. The tractrix is imagined as "a curve whose tangent are all equal length"; let us explain this definition. At a given point A of the tractrix we consider the tangent. The tangent intersects the tractrix asymptote at a second point, B. AB is the segment of constant length. If the initial point is $(1, 0)$ and the asymptote is the $y-$axis, the length becomes 1.

Identifying the tractrix equation $y = y(x)$ means to select the point where the tangent line

$$Y - y(x) = y'(x) \cdot (X - x)$$

intersects the line $X = 0$. It results $Y = y(x) - x \cdot y'(x)$. The constant length from the definition $(Y - y(x))^2 + x^2 = 1$, $x \in (0, 1)$ leads to the tractrix equation

$$x^2 \cdot (y'(x))^2 + x^2 = 1, \ x \in (0, 1), \ y'(x) < 0.$$

From the tractrix equation $-y'(x) = \dfrac{\sqrt{1 - x^2}}{x}$, it results $-\dfrac{dy}{dx} = \dfrac{\sqrt{1 - x^2}}{x}$, which is equivalent to

$$\int dy = -\int \frac{\sqrt{1 - x^2}}{x} dx.$$

For $t := \sqrt{1 - x^2}$ we have

$$\int \frac{t^2}{1 - t^2} dt = -t - \frac{1}{2} \ln \frac{1 - t}{1 + t},$$

therefore

$$y(x) = -\int \frac{\sqrt{1 - x^2}}{x} dx = -\sqrt{1 - x^2} - \ln \frac{1 - \sqrt{1 - x^2}}{x} + C$$

with C determined by the condition $y(1) = 0$, that is $C = 0$.

Finally we can consider the equation of the symmetric tractrix with respect to x−axis

$$y(x) = \sqrt{1 - x^2} + \ln \frac{1 - \sqrt{1 - x^2}}{x}.$$

The pseudosphere is obtained when the tractrix is rotated around y−axis and its equation is

$$f(x, y) = \left(x, y, \sqrt{1 - (x^2 + y^2)} + \ln \frac{1 - \sqrt{1 - (x^2 + y^2)}}{\sqrt{x^2 + y^2}} \right) \quad x, y \in (-1, 1).$$

We prefer the parameterization: if $x^1 \in (0, 2\pi)$, $x^2 \in \left(0, \frac{\pi}{2} \right)$,

$$f(x^1, x^2) = \left(\cos x^1 \cdot \sin x^2, \sin x^1 \cdot \sin x^2, \cos x^2 + \ln \left(\tan \left(\frac{x^2}{2} \right) \right) \right),$$

which produces the metric

$$ds^2 = \left(\sin x^2 \right)^2 \left(dx^1 \right)^2 + \left(\cot x^2 \right)^2 \left(dx^2 \right)^2.$$

We compute the curvature using the Theorema Egregium:

$$\Gamma_{11,1} = \Gamma_{12,2} = \Gamma_{21,2} = \Gamma_{22,1} = 0$$

$$\Gamma_{12,1} = \Gamma_{21,1} = \Gamma_{11,2} = \sin x^2 \cdot \cos x^2; \quad \Gamma_{22,2} = -\cot x^2 \cdot \frac{1}{(\sin x^2)^2}.$$

Then

$$\Gamma_{11}^1 = \Gamma_{12}^2 = \Gamma_{21}^2 = \Gamma_{22}^1 = 0$$

$$\Gamma_{12}^1 = \Gamma_{21}^1 = \cot x^2; \quad \Gamma_{11}^2 = \frac{(\sin x^2)^3}{\cos x^2}; \quad \Gamma_{22}^2 = -\frac{1}{\sin x^2 \cdot \cos x^2}.$$

It results

$$R^1_{212} = \frac{\partial \Gamma^1_{22}}{\partial x^1} - \frac{\partial \Gamma^1_{21}}{\partial x^2} + \Gamma^1_{22}\Gamma^1_{11} + \Gamma^2_{22}\Gamma^1_{21} - \Gamma^1_{21}\Gamma^1_{12} - \Gamma^2_{21}\Gamma^1_{22} = -(\cot x^2)^2$$

and

$$R_{1212} = g_{1s}R^s_{212} = g_{11}R^1_{212} = -(\cos x^2)^2; \ det\,(g_{ij}) = (\cos x^2)^2,$$

that is $K = -1$.   □

Consider the Ricci symbols $R_{ij} : U \longrightarrow \mathbb{R}$ defined by $R_{ij} := R^s_{isj} = R^1_{i1j} + R^2_{i2j}$.

Theorem 4.2.7 (Einstein) *For every surface $f : U \longrightarrow \mathbb{R}^3$, the Ricci tensor is proportional to the metric tensor via the Gauss curvature, i.e.*

$$R_{ij}(x) = K(x) \cdot g_{ij}(x).$$

Proof As before, we cancel x, therefore we start to compute the Ricci symbol R_{11}.

$$R_{11} = R^s_{1s1} = R^1_{111} + R^2_{121} = 0 + R^2_{121} = R^2_{121}.$$

But

$$R^2_{121} = g^{2s} \cdot R_{s121},$$

that is

$$R_{11} = g^{21} \cdot R_{1121} + g^{22} \cdot R_{2121} = 0 + g^{22} \cdot R_{2121} =$$

$$= g^{22} \cdot R_{2121} = \frac{g_{11}}{det\,(g_{ij})} \cdot R_{1212} = \frac{R_{1212}}{det\,(g_{ij})} \cdot g_{11} = K \cdot g_{11}.$$

In a similar way, starting from the Ricci symbol R_{22}, we have

$$R_{22} = R^s_{2s2} = R^1_{212} + R^2_{222} = R^1_{212} + 0 = R^1_{212}.$$

Since

$$R^1_{212} = g^{1s} \cdot R_{s212},$$

it results

$$R_{22} = g^{11} \cdot R_{1212} + g^{12} \cdot R_{2212} = g^{11} \cdot R_{1212} + 0 = \frac{g_{22}}{det\,(g_{ij})} \cdot R_{1212} = K \cdot g_{22}.$$

For R_{12}, we have:

$$R_{12} = R^s_{1s2} = R^1_{112} + R^2_{122} = g^{1s} \cdot R_{s112} + g^{2s} \cdot R_{s122} =$$

$$= g^{11} \cdot R_{1112} + g^{12} \cdot R_{2112} + g^{21} \cdot R_{1122} + g^{22} \cdot R_{2122} =$$

$$= 0 + g^{12} \cdot R_{2112} + 0 + 0 = -g^{12} R_{1212} = -\frac{-g_{21}}{\det(g_{ij})} \cdot R_{1212} = \frac{R_{1212}}{\det(g_{ij})} \cdot g_{21} = K \cdot g_{12}.$$

In the same way, we can prove that $R_{21} = K \cdot g_{21}$.   □

A first consequence of the Einstein theorem is related to the symmetry of Ricci's symbol for surfaces. For a given surface, it is $R_{ij} = R_{ji}$. A general result about the symmetry of the Ricci symbols and their geometric nature is presented in the next chapter.

Let $c = f \circ x : I \longrightarrow \mathbb{R}^3$ be a curve on the surface f, $f : U \longrightarrow \mathbb{R}^3$ and let $X : I \longrightarrow \mathbb{R}^3$ be a differentiable map such that $X(t) \in T_{c(t)}f$, i.e.

$$X(t) = X^k(t) \cdot \frac{\partial f}{\partial x^k}(x(t)) \in T_{(f \circ x)(x)}f.$$

$\frac{dX}{dt}(t)$ is a vector field along $c(t) = (f \circ x)(t)$, which, in general, does not belong to $T_{c(t)}$. We consider the normal projection $pr_t : T_{c(t)}\mathbb{R}^3 \longrightarrow T_{c(t)}f$ and we denote by

$$\frac{\nabla X(t)}{dt} := pr_t \frac{dX}{dt}(t)$$

the *covariant derivative of the field* X.

Theorem 4.2.8 *The covariant derivative of the vector field* X *is*

$$\frac{\nabla X(t)}{dt} := \left[\dot{X}^k(t) + \Gamma^k_{ij}(x(t)) \cdot X^i(t) \cdot \dot{x}^j(t) \right] \frac{\partial f}{\partial x^k}(x(t)).$$

Proof From $X(t) = X^k(t) \cdot \frac{\partial f}{\partial x^k}(x(t))$, we have

$$\frac{dX(t)}{dt} = \dot{X}^k(t) \cdot \frac{\partial f}{\partial x^k}(x(t)) + X^k(t) \cdot \frac{\partial^2 f}{\partial x^j \partial x^k}(x(t)) \cdot \dot{x}^j(t).$$

Using Gauss' formulas

$$\frac{\partial^2 f}{\partial x^j \partial x^k} = \Gamma^i_{jk} \cdot \frac{\partial f}{\partial x^i} + h_{jk} \cdot N$$

after arranging the dummy indexes, we obtain

$$\frac{dX(t)}{dt} = \left[\dot{X}^k(t) + \Gamma^k_{ij}(x(t)) \cdot X^i(t) \cdot \dot{x}^j(t)\right] \frac{\partial f}{\partial x^k}(x(t)) +$$

$$+ X^k(t) \cdot x^j(t) \cdot h_{kj}(x(t)) \cdot N(x(t)).$$

The projection onto the tangent plane makes the normal component to vanish, therefore

$$\frac{\nabla X(t)}{dt} = pr_t \frac{dX(t)}{dt} = \left[\dot{X}^k(t) + \Gamma^k_{ij}(x(t)) \cdot X^i(t) \cdot \dot{x}^j(t)\right] \frac{\partial f}{\partial x^k}(x(t)).$$

$\square$

Definition 4.2.9 The parallel transport along a curve $c = f \circ x : I \longrightarrow \mathbb{R}^3$ of the vector field $X : I \longrightarrow \mathbb{R}^3$ is described by the condition $\dfrac{\nabla X(t)}{dt} = \vec{0}$.

Therefore, the equations of the parallel transport are

$$\dot{X}^k(t) + \Gamma^k_{ij}(x(t)) \cdot X^i(t) \cdot \dot{x}^j(t) = 0, \ k \in \{1, 2\}.$$

The parallel transport equations can be completely determined if we consider an initial condition. It is enough to have a point p of the curve and the initial vector V_p at p. Then, the system of equations has, as unique solution, the vector field X such that, at $p = c(t_0)$, it is $X(t_0) = V_p$.

Let us underline the following point. The system of differential equations

$$\dot{X}^k(t) + \Gamma^k_{ij}(x(t)) \cdot X^i(t) \cdot \dot{x}^j(t) = 0, \quad k \in \{1, 2\}$$

which describes the parallel transport, shows that the vector field

$$X(t) = X^i(t) \cdot \frac{\partial f}{\partial x^i}(x(t))$$

is completely determined if we know X at a given point of the curve $c(t) = (f \circ x)(t)$.

Example 4.2.10 The case of parallel transport along any curve of the plane.

We may consider, without loose of generality, that the algebraic equation of the plane is $z = 0$. Then the surface f is $f(x^1, x^2) = (x^1, x^2, 0)$, the metric is $ds^2 = (dx^1)^2 + (dx^2)^2$, all $\Gamma_{ij,k} = 0$, all $\Gamma^k_{ij} = 0$, and the equations of the parallel transport are $\dot{X}^k(t) = 0, k \in \{1, 2\}$. It results, by integration, that the vector field X is a constant one, i.e. $X(t) = (a, b)$.

If we are looking at the support lines of this vector field along a given line, we see Euclidean parallel lines, therefore we understand the meaning of the parallel transport above.   □

In the particular case when the tangent field to the curve c, $\dot{c}\,(t)$, is parallel transported along the curve, then the curve c, by definition, is called *a geodesic of the surface f*.

The equations of a geodesic are $\dfrac{\nabla \dot{c}\,(t)}{dt} = \vec{0}$. In fact, as above, there are two equations. Since $\dot{c}\,(t) = \widehat{(f \circ x)}\,(t) = \dot{x}^i\,(t)\,\dfrac{\partial f}{\partial x^i}\,(x\,(t))$, the equations are

$$\ddot{x}^k\,(t) + \Gamma_{ij}^k\,(x\,(t)) \cdot \dot{x}^i\,(t) \cdot \dot{x}^j\,(t) = 0, \quad k \in \{1, 2\}.$$

The system of equations for geodesics is completely determined if we consider an initial condition. It is enough to have both:

- a point p of the geodesic;
- the initial vector $v_p = (\dot{x}^1(t_0), \dot{x}^2(t_0))$ at $p = c(t_0)$.

Example 4.2.11 Let us show that the geodesics of the plane $z = x^3 = 0$ are lines.

Since $f(x^1, x^2) = (x^1, x^2, 0)$, all $\Gamma_{ij,k} = 0$, all $\Gamma_{ij}^k = 0$, and the equations of the geodesics are $\ddot{x}^k\,(t) = 0, k \in \{1, 2\}$.
It results, by integration, the curve $c(t) = (v_1 t + x_0^1, v_2 t + x_0^2, 0)$. The curve has constant speed $\sqrt{(v_1)^2 + (v_2)^2}$ (Fig. 4.2).

Example 4.2.12 Find the geodesics of the cylinder $f(x^1, x^2) = (R \cos x^1, R \sin x^1, x^2)$.

Hint. The geodesics are helices of the cylinder

$$c(t) = (R \cos(v_1 t + \alpha_0), R \sin(v_1 t + \alpha_0), v_2 t + b),$$

and the speed along geodesic is constant, $||\dot{c}(t)|| = R$ (Fig. 4.3).

Fig. 4.2 Line geodesic

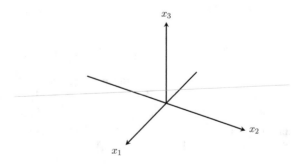

Fig. 4.3 Geodesic on a cylinder

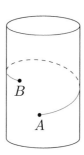

Example 4.2.13 Study the geodesics of the metric

$$ds^2 = R^2 dx^2 + R^2 \sin^2 x\, dy^2$$

of a sphere.

Hint. We may think at a sphere centered in O, with the length of the radius, R. We have $g_{11} = R^2$, $g_{22} = R^2 \sin^2 x_1$, $g_{12} = g_{21} = 0$, $g^{11} = \dfrac{1}{R^2}$, $g^{22} = \dfrac{1}{R^2 \sin^2 x_1}$, $g^{12} = g^{21} = 0$. Then, considering $x = x^1$, $y = x^2$, the Christoffel symbols are

$$\Gamma_{11,1} = \Gamma_{11,2} = \Gamma_{12,1} = \Gamma_{21,1} = \Gamma_{22,2} = 0, \ \Gamma_{12,2} = \Gamma_{21,2} = R^2 \sin x \cos x = -\Gamma_{22,1}$$

$$\Gamma^1_{11} = \Gamma^2_{11} = \Gamma^1_{12} = \Gamma^1_{21} = \Gamma^2_{22} = 0, \ \Gamma^2_{12} = \Gamma^2_{21} = \cot x, \ \Gamma^1_{22} = \sin x \cos x.$$

The geodesic equation, written for the first variable, is

$$\ddot{x} + \sin x \cos x \cdot \dot{y} = 0.$$

For the second variable y, the geodesic equation is

$$\ddot{y} + 2 \cot x \cdot \dot{x}\dot{y} = 0.$$

We may observe that $x = \dfrac{\pi}{2}$, $y = s$ is a solution, therefore $c(s) = (R \cos s, R \sin s, 0)$ is a geodesic of the sphere, in fact, it is the great circle obtained by the intersection of the plane $z = 0$ with the sphere. If we rotate the sphere around its center, another great circle becomes a geodesic. So the geodesics are the great circles of the sphere.

A comment is important at this point: under a change of coordinates, a rotation of the sphere around the origin means that "a geodesic is mapped into a geodesic". This result is proved in next chapter but we can use it now to better understand what is happened before. It results that all great circles are geodesics of the sphere. Now, since a geodesic is determined by a point and by a direction, it results that the geodesics of the sphere are only the great circles. The terminology is related to the fact that among all circles obtained from the intersection of a sphere with planes, the maximum radius is for the circles obtained at the intersection with a plane through the center of the sphere.   □

In all examples, the tangent vector to geodesic has constant length. This is a general property for geodesics and the proof of this fact is also provided in the next chapter.□

The intrinsic Geometry of the surface depends on the coefficients of the first fundamental form. We need to know how Cristoffel, Riemann, Ricci symbols and geodesics are transformed under a change of coordinates. In the next chapter, we will discuss these important topics.

4.3   Geometry of Surfaces in a 3D-Minkowski Space

The Minkowski three dimensional space, denoted by M^3, can be thought as the vector space $\mathbb{R}^3$ over the field $\mathbb{R}$ endowed with the *Minkowski product*

$$\langle a, b \rangle_M := a_0 b_0 - a_1 b_1 - a_2 b_2,$$

where $a = (a_0, a_1, a_2)$, $b = (b_0, b_1, b_2)$.

The theory below, (see [30]), can also be developed for the Minkowski product

$$\langle a, b \rangle_M := a_0 b_0 + a_1 b_1 - a_2 b_2.$$

It is a good exercise for the reader to follow the below steps regarding the first Minkowski product for developing the theory corresponding to this second possible Minkowski product.

The components of a vector $a = (a_0, a_1, a_2)$ appear because of the vectorial structure. With respect to the basis $\vec{i} = (1, 0, 0)$, $\vec{j} = (0, 1, 0)$, $\vec{k} = (0, 0, 1)$ we have the coordinates, so that we can assign them to a point A. We can write $A(a_0, a_1, a_2)$ and this point can be seen as the endpoint of the vector a whose origin is at the point $(0, 0, 0)$. We observe that, in a Minkowski space M^3, there are vectors like $u = (x, x, 0)$ such that

$$\langle u, u \rangle_M = 0.$$

All vectors a having this property,

$$\langle a, a \rangle_M = 0,$$

are called *lightlike vectors* or *null vectors*.

Also, it exists vectors as $v = (x, x, 1)$ such that

$$\langle u, u \rangle_M < 0;$$

All these vectors are called *space-like vectors*.

And there are vectors as $w = (x + 1, \sqrt{x}, \sqrt{x})$ such that

$$\langle u, u \rangle_M > 0$$

which are called *time-like vectors*. The null vectors determine a cone if we are looking at the equation $t^2 - x^2 - y^2 = 0$, (only from the Euclidean point of view. Why?). We may refer to this equation as the equation of null vectors in a Minkowski 3-space. The origin $(0, 0, 0)$ and the points from the exterior of this cone determine space-like vectors.

The origin and the points from the interior of the cone determine time-like vectors.

According to the situations seen before, the length of the vector a is, by definition, its norm, that is

$$||a||_M := \sqrt{|\langle a, a \rangle_M|} = \sqrt{|a_0^2 - a_1^2 - a_2^2|}.$$

Minkowski perpendicular vectors correspond to null Minkowski product, i.e. a and b are *Minkowski perpendicular (or Minkowski orthogonal)* if $\langle a, b \rangle_M = 0$.

As in the Euclidean plane, we observe that a frame generated by the vectors $(1, 0, 0)$, $(0, 1, 0)$ and $(0, 0, 1)$ is a Minkowski orthogonal frame. The length of a vector a becomes the "Minkowski distance" between the origin $O(0, 0, 0)$ and the point $A(a_0, a_1, a_2)$. In fact this is not a distance in the mathematical sense. The triangle rule is not working in general, as we saw when we discussed about pure time-like triangles in Minkowski two dimensional spaces. We continue to use the term "Minkowski distance" keeping in mind our remark before. The *Minkowski distance* between two points $A(a_0, a_1, a_2)$, $B(b_0, b_1, b_2)$ is given by the formula

$$d_M(A, B) := ||a - b||_M = \sqrt{|\langle a - b, a - b \rangle_M|},$$

therefore

$$d_M(A, B) = \sqrt{|(a_0 - b_0)^2 - (a_1 - b_1)^2 - (a_2 - b_2)^2|}.$$

The *Minkowski crossproduct* of two vectors is given by the formula

$$a \times_M b := (a_1 b_2 - a_2 b_1, a_0 b_2 - a_2 b_0, a_1 b_0 - a_0 b_1).$$

Exactly as in the case of the cross product in an Euclidean space, it is easier to remember it from the formal developing of the determinant

$$\begin{vmatrix} \vec{i} & -\vec{j} & -\vec{k} \\ a_0 & a_1 & a_2 \\ b_0 & b_1 & b_2 \end{vmatrix}.$$

Since $\langle a \times_M b, a \rangle_M = 0$ and $\langle a \times_M b, b \rangle_M = 0$, the vector $a \times_M b$ is Minkowski orthogonal to the plane determined by the vectors a and b.

These assertions are obvious to prove, since they can be reduced by algebraic computations. The first orthogonality means

$$(a_1b_2 - a_2b_1)a_0 - (a_0b_2 - a_2b_0)a_1 - (a_1b_0 - a_0b_1)a_2 = 0,$$

etc.

We can observe that the vectors $(1, 0, 0)$ and $(0, 1, 0)$ lead to the orthogonal vector $(0, 0, -1)$, that is the frame determined by the three vectors is negative oriented.

In general,

$$\begin{vmatrix} a_0 & a_1 & a_2 \\ b_0 & b_1 & b_2 \\ a_1b_2 - a_2b_1 & a_0b_2 - a_2b_0 & a_1b_0 - a_0b_1 \end{vmatrix} =$$

$$= (a_1b_2 - a_2b_1)^2 - (a_0b_2 - a_2b_0)^2 - (a_1b_0 - a_0b_1)^2,$$

that is we obtain $\langle a \times_M b, a \times_M b \rangle_M$. The frame orientation depends on the nature of the vector $a \times_M b$.

Surfaces in the Minkowski three dimensional space M^3 are defined exactly as the surfaces in the Euclidean space E^3: they are smooth mappings of an open set $U \subset \mathbb{R}^2$ into $\mathbb{R}^3$ with an extra property: at each point $f(x)$ of the surface, there exists a tangent plane.

Of course, the tangent plane is generated by the vectors $\left\{ \dfrac{\partial f}{\partial x^1}(x), \dfrac{\partial f}{\partial x^2}(x) \right\}$.

The only difference with respect to the Euclidean surfaces is the equation of the tangent plane generated by $\dfrac{\partial f}{\partial x^1}(x) \times_M \dfrac{\partial f}{\partial x^2}(x)$, i.e. it is

$$\begin{vmatrix} X - f^1(x) & -Y + f^2(x) & -Z + f^3(x) \\ \dfrac{\partial f^1}{\partial x^1}(x) & \dfrac{\partial f^2}{\partial x^1}(x) & \dfrac{\partial f^3}{\partial x^1}(x) \\ \dfrac{\partial f^1}{\partial x^2}(x) & \dfrac{\partial f^2}{\partial x^2}(x) & \dfrac{\partial f^3}{\partial x^2}(x) \end{vmatrix} = 0.$$

The tangent plane is denoted by $T_{f(x)}f$ and any vector $X(x)$ which belongs to $T_{f(x)}f$ can be written in the form

$$X(x) = X^1(x) \frac{\partial f}{\partial x^1}(x) + X^2(x) \frac{\partial f}{\partial x^2}(x)$$

where the coefficient $X^1, X^2 : U \longrightarrow \mathbb{R}$ are smooth maps. The line which is Minkowski perpendicular to the tangent plane at the point $f(x)$ is called a Minkowski normal line to the surface and has the equation

$$\frac{X - f^1(x)}{\begin{vmatrix} \dfrac{\partial f^2}{\partial x^1}(x) & \dfrac{\partial f^3}{\partial x^1}(x) \\ \dfrac{\partial f^2}{\partial x^2}(x) & \dfrac{\partial f^3}{\partial x^2}(x) \end{vmatrix}} = \frac{-Y + f^2(x)}{\begin{vmatrix} \dfrac{\partial f^1}{\partial x^1}(x) & \dfrac{\partial f^3}{\partial x^1}(x) \\ \dfrac{\partial f^1}{\partial x^2}(x) & \dfrac{\partial f^3}{\partial x^2}(x) \end{vmatrix}} = \frac{-Z + f^3(x)}{\begin{vmatrix} \dfrac{\partial f^2}{\partial x^1}(x) & \dfrac{\partial f^1}{\partial x^1}(x) \\ \dfrac{\partial f^2}{\partial x^2}(x) & \dfrac{\partial f^1}{\partial x^2}(x) \end{vmatrix}}.$$

The vector $\dfrac{\partial f}{\partial x^1}(x) \times_M \dfrac{\partial f}{\partial x^2}(x)$ generates the Minkowski–Gauss normal unitary vector

$$n(x) := \frac{\dfrac{\partial f}{\partial x^1}(x) \times_M \dfrac{\partial f}{\partial x^2}(x)}{\left\| \dfrac{\partial f}{\partial x^1}(x) \times_M \dfrac{\partial f}{\partial x^2}(x) \right\|_M}.$$

The nature of the normal vector rises supplementary problems when we are considering Minkowski surfaces.

We choose $N = \varepsilon n,\ \varepsilon \in \{-1, 1\}$ such that he frame $\{\dfrac{\partial f}{\partial x^1}(x), \dfrac{\partial f}{\partial x^2}(x), N(x)\}$ is positive oriented. This one is called a *Minkowski–Gauss frame*, or simply, a *Minkowski frame* attached to the surface f in $f(x)$. At each point of a surface, this frame is a vector basis in $T_x U \times R$.

The Differential Geometry of such a surface is described by the properties of the coefficients of equations determined by the partial derivatives of the vectors of this frame, exactly as it happens in Euclidean spaces. All the other properties are like in the Euclidean frame.

Definition 4.3.1 The map $I_x^M(\cdot, \cdot) : T_x \mathbb{R}^2 \times T_x \mathbb{R}^2 \longrightarrow \mathbb{R}$, defined by

$$I_x^M(X, Y) := \langle df_x X, df_x Y \rangle_M$$

is called the Minkowski first fundamental form of the surface f at the point $f(x)$ in the Minkowski frame.

The matrix of this bilinear map with respect to the basis $\{e_1, e_2\}$ in $T_x U$ is described by the coefficients $(g_{ij}(x))_{i,j=1,2}$,

$$g_{11}^M(x) = I_x^M(e_1, e_1) = \langle df_x e_1, df_x e_1 \rangle_M = \left\langle \frac{\partial f}{\partial x^1}(x), \frac{\partial f}{\partial x^1}(x) \right\rangle_M$$

$$g_{12}^M(x) = I_x^M(e_1, e_2) = g_{21}^M(x) = \langle df_x e_1, df_x e_2 \rangle_M = \left\langle \frac{\partial f}{\partial x^1}(x), \frac{\partial f}{\partial x^2}(x) \right\rangle_M$$

$$g_{22}^M(x) = I_x^M(e_2, e_2) = \langle df_x e_2, df_x e_2 \rangle_M = \left\langle \frac{\partial f}{\partial x^2}(x), \frac{\partial f}{\partial x^2}(x) \right\rangle_M.$$

The way these coefficients are described by the Minkowski product of M^3 points out the way in which the ambient Minkowski 3-space endows with its Geometry each tangent plane to the surface.

Therefore, if $X(x) = X^1(x)e_1 + X^2(x)e_2,\ Y(x) = Y^1(x)e_1 + Y^2(x)e_2 \in T_x U$, then

$$I_x^M (X, Y) := \langle df_x X, df_x Y \rangle_M$$

and $I_x^M (X, Y)$ is

$$g_{11}^M(x)X^1(x)Y^1(x) + g_{12}^M(x)X^1(x)Y^2(x) + g_{21}^M(x)X^2(x)Y^1(x) + g_{22}^M(x)X^2(x)Y^2(x).$$

The Einstein summation convention leads to a simplified formula

$$I_x^M (X, Y) := \langle df_x X, df_x Y \rangle_M = g_{ij}^M(x)X^i(x)Y^j(x).$$

The changes of coordinates and the isometries of a Minkowski 3-space preserves the Minkowski first fundamental form.

Then, the curves on a surface in a Minkowski 3-space has the same two important properties: the tangent vector $\dot{c}(t)$ belongs to the tangent plane to the surface, $T_{c(t)}f$, and, the Minkowski length of this tangent vector, depends on the coefficients g_{ij}^M of the Minkowski first fundamental form but also on the type of the vector, i.e. we can prove

Theorem 4.3.2 *If c is a curve on the surface $f : U \longrightarrow \mathbb{R}^3$ from the Minkowski 3-space M^3, then*

(i) $\dot{c}(t) = \dot{x}^1(t) \dfrac{\partial f}{\partial x^1}(x(t)) + \dot{x}^2(t) \dfrac{\partial f}{\partial x^2}(x(t)) \in T_{f(x(t))}f, \quad \forall t \in I.$

(ii) $\|\dot{c}(t)\|_M^2 = g_{11}^M \cdot \left(\dot{x}^1(t)\right)^2 + 2g_{12}^M \cdot \dot{x}^1(t) \cdot \dot{x}^2(t) + g_{22}^M \cdot \left(\dot{x}^2(t)\right)^2$

if the right member is positive; otherwise we have to consider a $-$ sign in front of the entire right member.

Formula (ii) highlights a quadratic form denoted by ds^2 which acts on a vector $v = (v^1, v^2)$ after the rule

$$ds^2(v, v) = g_{11}^M (x(t)) \cdot \left(v^1(t)\right)^2 + 2g_{12}^M (x(t)) \cdot v^1(t) \cdot v^2(t) + g_{22}^M (x(t)) \cdot \left(v^2(t)\right)^2.$$

Taking into account that $dx^i(v) = v^i$, the previous formula of the quadratic form can be written as

$$ds^2 = g_{11}^M (x(t)) \cdot \left(dx^1\right)^2 + 2g_{12}^M (x(t)) \cdot dx^1 \cdot dx^2 + g_{22}^M (x(t)) \cdot \left(dx^2\right)^2$$

or, using Einstein notation,

$$ds^2 = g_{ij}^M(x)dx^i dx^j.$$

This quadratic form induced by the first fundamental form is called *a metric for the surface f in the Minkowski 3-space M^3*.

The only difference with respect to the Euclidean case is the "Minkowski" way to determine the coefficients g_{ij} and g_{ij}^M. This can be understood through two examples.

Example 4.3.3 Consider the plane $z = 0$ described by $f : U = \overset{\circ}{U} \subset \mathbb{R}^2 \longrightarrow \mathbb{R}^3$, having the form

$$f(x, y) = (x, y, 0).$$

We determine the metric in the case the plane "f is a surface in the Euclidean 3-space E^3".

In both cases the generators of the tangent plane are:

$$\frac{\partial f}{\partial x} = (1, 0, 0) ; \quad \frac{\partial f}{\partial y} = (0, 1, 0) ;$$

In the first case, the coefficients will be determined using the Euclidean inner product

$$\langle a, b \rangle = a_0 b_0 + a_1 b_1 + a_2 b_2,$$

therefore

$$g_{11} = 1; \ g_{22} = 1; \ g_{12} = g_{21} = 0.$$

The metric of the plane f, seen as a surface in the Euclidean 3-space, is

$$ds^2 = (dx)^2 + (dy)^2.$$

Let us now determine the metric of the plane in the case "it is a surface in the Minkowski 3-space M^3". In the Minkowski 3-space, the coefficients of the metric are determined using the Minkowski product

$$\langle a, b \rangle_M = a_0 b_0 - a_1 b_1 - a_2 b_2.$$

It results

$$g_{11}^M = 1; \ g_{22}^M = -1; \ g_{12}^M = g_{21}^M = 0;$$

and the metric of the same plane f, but now seen as a surface in the Minkowski 3-space, is

$$ds^2 = (dx)^2 - (dy)^2.\ \square$$

Example 4.3.4 Consider a surface $f : U = \overset{\circ}{U} \subset \mathbb{R}^2 \longrightarrow \mathbb{R}^3$, having the form

$$f(x, y) = (x, y, u(x, y)).$$

Let us determine the metric in the case "f is a surface in the Euclidean 3-space E^3", and in the case "f is a surface in the Minkowski 3-space M^3". In both cases, the generators of the tangent plane are:

$$\frac{\partial f}{\partial x} = \left(1, 0, \frac{\partial u}{\partial x}\right); \quad \frac{\partial f}{\partial y} = \left(0, 1, \frac{\partial u}{\partial y}\right).$$

In the first case, the coefficients will be determined using the Euclidean inner product

$$\langle a, b \rangle = a_0 b_0 + a_1 b_1 + a_2 b_2,$$

therefore

$$g_{11} = 1 + \left(\frac{\partial u}{\partial x}\right)^2; \quad g_{22} = 1 + \left(\frac{\partial u}{\partial y}\right)^2; \quad g_{12} = g_{21} = \left(\frac{\partial u}{\partial x}\right)\left(\frac{\partial u}{\partial y}\right).$$

The metric of the surface f, seen as a surface in the Euclidean 3-space, is

$$ds^2 = \left(1 + \left(\frac{\partial u}{\partial x}\right)^2\right)(dx)^2 + 2\left(\frac{\partial u}{\partial x}\right)\left(\frac{\partial u}{\partial y}\right)dxdy + \left(1 + \left(\frac{\partial u}{\partial y}\right)^2\right)(dy)^2.$$

In the Minkowski 3-space, the coefficients of the metric are determined using the Minkowski product

$$\langle a, b \rangle_M = a_0 b_0 - a_1 b_1 - a_2 b_2.$$

Therefore

$$g_{11}^M = 1 - \left(\frac{\partial u}{\partial x}\right)^2; \quad g_{22}^M = -\left(1 + \left(\frac{\partial u}{\partial y}\right)^2\right); \quad g_{12}^M = g_{21}^M = -\left(\frac{\partial u}{\partial x}\right)\left(\frac{\partial u}{\partial y}\right);$$

and the metric of the same surface f, but now seen as a surface in the Minkowski 3-space, is

$$ds^2 = \left(1 - \left(\frac{\partial u}{\partial x}\right)^2\right)(dx)^2 - 2\left(\frac{\partial u}{\partial x}\right)\left(\frac{\partial u}{\partial y}\right)dxdy - \left(1 + \left(\frac{\partial u}{\partial y}\right)^2\right)(dy)^2.$$

$\square$

Let us consider the Gauss frame $\left\{\dfrac{\partial f}{\partial x^1}(x), \dfrac{\partial f}{\partial x^2}(x), N(x)\right\}$ at each point $f(x)$ on the surface $f : U \subset \mathbb{R}^2 \longrightarrow \mathbb{R}^3$, seen as a surface in the Minkowski 3-space.

Since the length of the Minkowski–Gauss normal vector can be -1 or 1, it results both $\left\langle \dfrac{\partial N}{\partial x^1}(x), N(x)\right\rangle_M = 0$ and $\left\langle \dfrac{\partial N}{\partial x^2}(x), N(x)\right\rangle_M = 0$, i.e. the vectors $\dfrac{\partial N}{\partial x^1}(x)$

and $\dfrac{\partial N}{\partial x^2}(x)$ are Minkowski orthogonal to the Minkowski–Gauss vector $N(x)$ at each point of the surface. Therefore $\left\{\dfrac{\partial N}{\partial x^1}(x), \dfrac{\partial N}{\partial x^2}(x)\right\} \subset T_{f(x)}f$.

Here it appears a main difference with respect to the Euclidean case.

We consider only the case when $\langle N, N \rangle_M = -1$ and we leave to the reader the other case.

So, we analyze the case when the normal to the surface is a space-like vector. Let us explain how we have to think in this case. We intend to have the same Gauss formulas,

$$\frac{\partial^2 f}{\partial x^i \partial x^k}(x) = \Gamma^s_{ik}(x) \cdot \frac{\partial f}{\partial x^s}(x) + N(x) \cdot h_{ik}(x).$$

But now $\langle N, N \rangle_M = -1$, therefore the formula for the coefficients of the second fundamental form has to change. It becomes

$$h^M_{ij}(x) = -\left\langle N(x), \frac{\partial^2 f}{\partial x^i \partial x^j}(x)\right\rangle_M = \left\langle \frac{\partial N}{\partial x^i}(x), \frac{\partial f}{\partial x^j}(x)\right\rangle_M.$$

This means that, in order to preserve the formula which connects the second fundamental form coefficients by the first fundamental form coefficients via the *Minkowski-Weingarten matrix* of coefficients, $h^M_{ij} = h^s_i\, g^M_{sj}$, we need to consider a modified formula for Minkowski-Weingarten coefficients, that is

$$\frac{\partial N}{\partial x^i} = h^s_i \frac{\partial f}{\partial x^s}.$$

Using Gauss' formulas and Weingarten's formulas, we obtain *Minkowski–Gauss' equations* in the modified form

$$R_{ijkl} = -\left(h^M_{ik} h^M_{jl} - h^M_{il} h^M_{jk}\right).$$

And finally, considering

$$K_M = \frac{R_{1212}}{\det g^M_{ij}},$$

we can define the *Minkowski–Gauss curvature* by the formula

$$K_M := -\frac{\det h^M_{ij}}{\det g^M_{ij}} = -\det h^j_i.$$

The rest is the same. To see an example where this theory works, let us take into account the computations of Minkowski–Gauss curvature of the affine sphere $x^2 - y^2 - z^2 = -a^2$ in the last chapter of the book. Without considering the formulas with the change imposed by the space-like nature of the Minkowski–Gauss normal vectors, the Minkowski–Gauss curvature would be obtained with two different signs, instead to obtain only the value $-\dfrac{1}{a^2}$.

Therefore, for a given Minkowski metric

$$ds^2 = g_{ij}^M(x)dx^i dx^j$$

we can construct:

- the Christoffel symbols of first kind

$$\Gamma_{ij,k} = \frac{1}{2}\left(\frac{\partial g_{ik}^M}{\partial x^j} + \frac{\partial g_{jk}^M}{\partial x^i} - \frac{\partial g_{ij}^M}{\partial x^k}\right);$$

- the Christoffel symbols of second kind

$$\Gamma_{jk}^i = {}^M g^{is}\Gamma_{jk,s} = \frac{1}{2}\,{}^M g^{is}\left(\frac{\partial g_{js}^M}{\partial x^k} + \frac{\partial g_{ks}^M}{\partial x^j} - \frac{\partial g_{jk}^M}{\partial x^s}\right)$$

where ${}^M g^{ij}$ are the components of the inverse matrix of the metric, i.e.

$$ {}^M g^{is} g_{sj}^M = \delta_j^i $$

as in the Euclidean case.
- The Riemann symbols of second kind are

$$R_{ijk}^h = \frac{\partial \Gamma_{ik}^h}{\partial x^j} - \frac{\partial \Gamma_{ij}^h}{\partial x^k} + \Gamma_{mj}^h \Gamma_{ik}^m - \Gamma_{mk}^h \Gamma_{ij}^m.$$

- The Riemann symbols, first kind $R_{ijkl} = g_{is}^M R_{jkl}^s$.
- The Ricci symbols $R_{ij} = R_{isj}^s$ are obtained from the Riemann symbols of second kind R_{imj}^s by contracting the indexes $s = m$.
- The Minkowski parallel transport equations are induced in the same way by the covariant derivative and they are

$$\dot{X}^k(t) + \Gamma_{ij}^k(x(t)) \cdot X^i(t) \cdot \dot{x}^j(t) = 0, \ k \in \{1, 2\}.$$

- Geodesics, i.e. curves $f(x(t))$ such that $x(t) = (x^1(t), x^2(t))$, satisfies the equations

$$\frac{d^2 x^r}{dt^2} + \Gamma_{pq}^r \frac{dx^p}{dt} \frac{dx^q}{dt} = 0.$$

All theorems proved in the Euclidean space are available in the Minkowski case. They have the same statement and the same proof.

- Gauss' formulas are :

$$\frac{\partial^2 f}{\partial x^j \partial x^k} = \Gamma^s_{jk} \cdot \frac{\partial f}{\partial x^s} + N \cdot h_{jk}$$

- Gauss' equations are (in the case when the normal is a space-like vector):

$$R_{ijkl} = -(h^M_{ik} h^M_{jl} - h^M_{il} h^M_{jk})$$

- The properties of the Riemann symbols of first kind are

$$R_{ijkl} = -R_{ijlk};$$
$$R_{ijkl} = -R_{jikl};$$
$$R_{ijkl} = R_{jilk};$$
$$R_{ijkl} = R_{klij};$$
$$R_{ijkl} + R_{iklj} + R_{iljk} = 0 \ (Bianchi's \ first \ identity).$$

- Codazzi–Mainardi's equations are:

$$\frac{\partial h_{ij}}{\partial x^k} + \Gamma^s_{ij} \cdot h_{sk} = \frac{\partial h^M_{ik}}{\partial x^j} + \Gamma^s_{ik} \cdot h^M_{sj}.$$

- The Theorema Egregium is:

$$K(x) = \frac{R_{1212}(x))}{det(g^M_{ij}(x))}.$$

- Einstein's theorem is: For a Minkowski surface, $R_{ij} = K \cdot g_{ij}$.

We may conclude:
For a surface in a Minkowski 3D-dimensional space, the Minkowski product induces its Minkowski first fundamental form with coefficients (g^M_{ij}). Therefore it induces its metric

$$ds^2 = g^M_{ij}(x) dx^i dx^j.$$

We can measure lengths and angles for vectors belonging to tangent planes to the surface, length of curves who belong to surfaces and areas of regions included in surfaces exactly as we have done in the 3D-Euclidean space.

Now, if the normal N is a space-like vector, the partial derivatives of N from the Minkowski–Gauss frame allow us to discuss about the *Minkowski–Gaussian curvature of a surface at a point,*

$$K_M(x) = -det(h^i_j(x)) = -\frac{det(h^M_{ij}(x))}{det(g^M_{ij}(x))}.$$

This "new" Gaussian curvature seems to be dependent on the embedding in the ambient 3-Minkowski space, but after we prove Minkowski–Gauss' formulas and Minkowski-Ricci's equations, we step into the intrinsic theory of Minkowski surfaces where Theorema Egregium

$$K(x) = \frac{R_{1212}(x))}{det(g_{ij}^M(x))}$$

offers another perspective: the Minkowski surface can be seen as a piece of a plane (x^1, x^2) endowed with a metric, and this metric only determines the Minkowski–Gauss curvature. The lines of this Geometry are the geodesic which satisfy the equations

$$\frac{d^2 x^r}{dt^2} + \Gamma_{pq}^r \frac{dx^p}{dt} \frac{dx^q}{dt} = 0, \ r \in \{1, 2\}.$$

The surface is no longer needed. The Geometry becomes the Geometry of the metric.

4.4 A Short Story of a Person Embedded in a Surface

For a person embedded in a surface, the surface is her/his Universe. The person cannot see the surface in which is embedded as the image of a function $f : U \to \mathbb{R}^3$, in the same way we cannot see from "outside" the four dimensional Universe in which we live.

Suppose the person is interested in developing a theory to see how much one can understand about the Universe she/he lives.

The way has to be as it is in this book.

I. At the beginning the person will try to develop geometric concepts, because the person would like to create images which correspond to the surrounding environment.

To do this, the person has to think about geometries in an axiomatic way, providing abstract definitions for points, lines, plane and even the space, the relations between them being established through axioms. We have to accept that the intuition of the person is an Euclidean one, exactly as our intuition is. Therefore the axiomatic system seems to be as Hilbert's one. Once this context is established, the person can start to prove theorems, to introduce new concepts until the basic part of the Absolute Geometry we have presented before can be highlighted.

At a moment, the person will succeed to prove Legendre's Theorem related to the sum of angles of a triangle, "$\angle A + \angle B + \angle C \leq 2R$, where R is the value of a right angle."

Using the defect of a triangle, a concept introduced as a result of Legendre's Theorem, an important consequence appears: "if there exists one triangle with the sum of angles $2R$, all the triangles have the same property."

And more, "if there exists one triangle with the sum of angles strictly less than $2R$, all the triangles have the same property."

You can understand the person surprise when statements as the previous ones can be proved. It means that two geometries exist, even if, at the beginning, the person thought in the existence of one only! In one of these two geometries, for all triangles, we have

$$\angle A + \angle B + \angle C = 2R$$

in the other,

$$\angle A + \angle B + \angle C < 2R.$$

An axiom will make the first situation possible, the denial of it (seen as an axiom) will provide the second situation. All the geometric results obtained using the initial axiomatic system and the first axiom are similar to our Euclidean Geometry, while the Geometry provided by the initial axiomatic system together the second axiom is similar to our Non-Euclidean Geometry.

Now, the person knows two theoretical geometries developed in an axiomatic frame. The person will be interested in seeing models for each Geometry.

Algebra will be used to construct the Euclidean plane and the Euclidean space. Even spaces with more than three dimensions can be constructed in this way. In fact, vector spaces, trigonometric functions, inner products and groups which preserve the inner product, norms and the Euclidean distance d_E are the ingredients for the Euclidean models in a plane or in a space. An algebraic language is created and used to explain the Euclidean geometric results. The person will understand later that this is the frame for the basic Physics of its space. The lines of the Euclidean Geometry are provided by sets of points of the plane, or of the space, denoted by l, such that for every three points of l in the order A, B, C the equality

$$d_E(A, B) + d_E(B, C) = d_E(A, C)$$

happens.

A mimetic algebraic construction, in which the role of inner product is taken by the Minkowski product, provides the Minkowski Geometry. Three types of vectors appear and the hyperbolic trigonometric functions sinh and cosh replace the Euclidean trigonometric functions in the "rotation" matrix of this Geometry. Minkowski-Pythagoras' Theorem has different forms related to the hypotenuse vector type. There are triangles where we cannot discus about the sum of angles. Since here it does not exist a distance in the mathematical sense of this word, the Minkowski Geometry is different from the Euclidean and the Non-Euclidean Geometries studied. This special Geometry is the frame for Special Relativity, our person will understand this later.

Next, developing both the geometric inversion and the projective transformations, the person will construct the Poincaré Disk as a model for the Non-Euclidean Geometry.

A special distance d is highlighted and the disk appears unbounded with respect to this distance. The lines of this Geometry are orthogonal arcs to the circle which bound the disk. For three points on a line in this Geometry in the order A, B, C we have $d(A, B) + d(B, C) = d(A, C)$ i.e. the orthogonal arc is a geodesic with respect to the distance d. We can see that through a given point from the interior of the disk which does not belong to a given orthogonal arc a, at least two non-intersecting orthogonal arcs a_1 and a_2 can be constructed such that $a \cap a_1 = \emptyset$, $a \cap a_2 = \emptyset$. The sum of angles of a triangle in this Poincaré disk model is strictly less than two right angles. But all these things are theoretical, how to acts in its reality?

The person can think at another way to study the geometric objects. May be replacing algebra with calculus can help. And more, the person could think at the convenient Geometry of its Universe.

II. This is the moment when the person defines its environment as a surface and all the Geometry is created by calculus. The person needs three dimensions in which it can be supposed the existence of the image of the surface. The three dimensional space leave its Geometry in the tangent planes at each point of the surface f.

The metric coefficients appear from the product between the vectors $\dfrac{\partial f}{\partial x^i}$ and $\dfrac{\partial f}{\partial x^j}$, therefore the metric $ds^2 = g_{ij}(x)dx^i dx^j$ appears. The coefficients are produced by the type of space in which exists the image of the surface.

The product can be the inner product if the three dimensional space is an Euclidean one;

Or, the product can be the Minkowski product if the three dimensional space is a Minkowski one.

How the person can choose? If some physical facts can impose, say, a maximum speed for all objects are moving, the person will choose the Minkowski type product as we will see later in the text. If not, the person will think at the Euclidean product. Once the product is established, the theory is known: the metric is established, length of curves, the angle between curves, the area of a domain included in the surface can be computed only with respect the coefficients of the metric. Formulas as Gauss' ones

$$\frac{\partial^2 f}{\partial x^j \partial x^k} = \Gamma^s_{jk} \cdot \frac{\partial f}{\partial x^s} + N \cdot h_{jk}$$

involve Christoffel symbols, first kind

$$\Gamma_{ij,k} = \frac{1}{2}\left(\frac{\partial g_{ik}}{\partial x^j} + \frac{\partial g_{jk}}{\partial x^i} - \frac{\partial g_{ij}}{\partial x^k}\right),$$

Christoffel symbols, second kind

$$\Gamma^i_{jk} = g^{is}\Gamma_{jk,s} = \frac{1}{2}g^{is}\left(\frac{\partial g_{js}}{\partial x^k} + \frac{\partial g_{ks}}{\partial x^j} - \frac{\partial g_{jk}}{\partial x^s}\right)$$

where g^{ij} are the components of the inverse matrix of the metric, i.e.

$$g^{is}g_{sj} = \delta^i_j.$$

The Geometry depends on the Riemannian symbols second kind

$$R^h_{ijk} = \frac{\partial \Gamma^h_{ik}}{\partial x^j} - \frac{\partial \Gamma^h_{ij}}{\partial x^k} + \Gamma^h_{mj}\Gamma^m_{ik} - \Gamma^h_{mk}\Gamma^m_{ij},$$

by the Riemannian symbols first kind $R_{ijkl} = g_{is}R^s_{jkl}$,
and by the Ricci symbols $R_{ij} = R^s_{isj}$, which are obtained from the Riemannian symbols second kind R^s_{imj} by contracting the indexes $s = m$.

The lines of the Universe of the person are described by the equations

$$\frac{d^2x^i}{dt^2} + \Gamma^i_{jk}\frac{dx^j}{dt}\frac{dx^k}{dt} = 0, \; i \in \{1, 2\}.$$

The Gaussian curvature can be written in the form

$$K(x) = \frac{R_{1212}(x))}{det(g_{ij}(x))},$$

showing its intrinsic geometric nature. The extra dimension necessary to create and to understand this Geometry can be forgotten.

But still a problem remains: who is f, that is how to determine precisely the correct coefficients? This is not mathematics again, it depends by the "physical reality" of its space. Suppose that somehow the person realizes that the curvature can help him to understand more about the Geometry of its Universe. The complete revelation for our person is the moment when he understands that Einstein's theorem offers the equations of its Universe. Let us write them again,

$$R_{ij} = K \cdot g_{ij}.$$

It is the geometric way to go further.

In this two dimensional case, the Universe is shaped by its Gaussian curvature. But in fact, you will see that is not about the shape, its about the metric.

The next example clarifies this aspect.

A very normal question for our person is to try to find something about its Universe if the Universe equations are

$$R_{ij} = 0.$$

We analyze a simple case in the Euclidean space. The person chooses the metric as

$$ds^2 = a(x^1)(dx^1)^2 + (dx^2)^2.$$

From Einstein equations it results that $K(x) = 0$. If we compute the Christoffel symbols we have:

$$\Gamma_{11,1} = \frac{1}{2}\dot{a}(x^1); \ \Gamma_{ij,k} = 0, \ i, j, k \neq 1,$$

$$\Gamma_{11}^1 = \frac{1}{2}\frac{\dot{a}(x^1)}{a(x^1)}; \ \Gamma_{jk}^i = 0, \ i, j, k \neq 1,$$

$$R_{1212} = g_{1s}R_{212}^s = g_{11}R_{212}^1 = 0,$$

because $R_{212}^1 = 0$. Therefore, for any function $a(x^1)$ the chosen metric fulfills the property $R_{ij} = 0$. If $a(x^1) = b$, where b is a constant, the person can not decide if the Universe is a plane or a cylinder or some other surface having the metric above. The shape of the Universe can not be determined. Even the geodesic equations are the same, because $\Gamma_{jk}^i = 0$. But, if the person can somehow prove that there are closed geodesic, then a cylinder-Universe is a good possibility and the plane Universe is out of the possibilities.

It is a progress, the Geometry of a person embedded is more consistent now, but the person has to work to cancel the supplementary dimension. Think at the fact that the properties of first kind Riemann symbols

$$R_{ijkl} = -R_{ijlk};$$
$$R_{ijkl} = -R_{jikl};$$
$$R_{ijkl} = R_{jilk};$$
$$R_{ijkl} = R_{klij};$$
$$R_{ijkl} + R_{iklj} + R_{iljk} = 0 \ (the \ first \ Bianchi's \ identity).$$

were deduced using Gauss' equations.

Therefore the person has to find some other proofs for the above formulas, proofs in which the "extra dimension" is not involved. The same for the geodesics which were defined taking into account the projection along the normal to the surface. These things will be seen in the next chapter.

Chapter 5
Basic Differential Geometry

Geometria substantia rerum.

Abstract As we saw in the previous chapter, multi-index quantities exist in Differential Geometry. We highlighted that we did not see yet the mathematical nature of the first fundamental form; or the nature of Riemann symbols, Ricci symbols, or some important properties for geodesics. In this chapter, we take into account this nature and how general are the concepts introduced when we studied surfaces and curves on surfaces. We are interested in proving that the coefficients of the metric, the Riemann symbols, the Ricci symbols or the geodesics remain invariant when we deal with a change of coordinates. The substance of the General Relativity is related to the invariance under changes of coordinates and to the tensor structure of objects that have to present the same form in any reference frame. These two main properties can be related to the deep meaning of General Relativity which has the Equivalence Principle as the physical starting point.

5.1 Covariant and Contravariant Vectors and Tensors. The Christoffel Symbols

Differential Geometry deals with a set M endowed with a coordinate system $(x^0, x^1, \ldots, x^n)$, $x^i \in \mathbb{R}$. In our notation, the system of coordinates starts from x^0 instead x^1 to make a distinction between the first coordinate which, in Physics, represents time and the other three coordinates represent the spatial coordinates, often denoted by Greek letters, α, β and γ.[1] We may keep this notation even if the number of coordinates is greater than 4.

The dimension of our set M is $n + 1$ without insisting on the structure of M. The reader has to imagine M as an open set of $\mathbb{R}^{n+1}$ whose Geometry depends on a metric defined on it. Let us observe that there is no extra dimensions to study M. In geometric examples below, we choose to use the coordinate system starting with x^1, or we use directly the letters x, y, etc. because the physical meaning will be discussed in the later chapters.

A fundamental object in Differential Geometry is the tensor.

In simple words, a tensor is a multi-index quantity which, under a change of coordinates, is transforming linearly with respect to the indexes. In Differential Geometry (and Physics), the components of a tensor depend smoothly on the points of the space, in our case M. So, tensors are functions having derivatives of all order everywhere in M. Let us suppose we have a quantity $T_{l_1 l_2 \ldots l_p}^{i_1 i_2 \ldots i_k}(x)$ in a given system of coordinates $(x^0, x^1, \ldots, x^n)$ and let us consider a change of coordinates

$$x^i = x^i(\bar{x}) = x^i(\bar{x}^0, \bar{x}^1, \ldots, \bar{x}^n), \ i \in \{0, 1, \ldots, n\}.$$

In a simpler form, it is

$$x^i = x^i(\bar{x}^j), \ i, j \in \{0, 1, \ldots, n\}.$$

The inverse transformation of coordinates is $\bar{x}^i = \bar{x}^i(x^j)$, $i, j \in \{0, 1, \ldots, n\}$. It is worth noticing that the Einstein notation $a_i b^i = \sum_{i=1}^{n} a_i b^i$ can be used in this form or in its multi-index form. With these positions in mind, we can denote by $\bar{T}_{q_1 q_2 \ldots q_p}^{j_1 j_2 \ldots j_k}(\bar{x})$, the quantity $T_{l_1 l_2 \ldots l_p}^{i_1 i_2 \ldots i_k}(x)$ written with respect the new coordinates.

Definition 5.1.1 $T_{l_1 l_2 \ldots l_p}^{i_1 i_2 \ldots i_k}$ is a tensor contravariant of rank k and covariant of rank p, or simply a (k, p) tensor, if, under the previous change of coordinates $x^i = x^i(\bar{x}^j)$, the formula of $\bar{T}_{q_1 q_2 \ldots q_p}^{j_1 j_2 \ldots j_k}$ is

$$\bar{T}_{q_1 q_2 \ldots q_p}^{j_1 j_2 \ldots j_k}(\bar{x}) = T_{l_1 l_2 \ldots l_p}^{i_1 i_2 \ldots i_k}(x) \frac{\partial x^{l_1}}{\partial \bar{x}^{q_1}} \frac{\partial x^{l_2}}{\partial \bar{x}^{q_2}} \ldots \frac{\partial x^{l_p}}{\partial \bar{x}^{q_p}} \frac{\partial \bar{x}^{j_1}}{\partial x^{i_1}} \frac{\partial \bar{x}^{j_2}}{\partial x^{i_2}} \ldots \frac{\partial \bar{x}^{j_k}}{\partial \bar{x}^{i_k}}.$$

Example 5.1.2 $a_k(x)$ is a covariant tensor of rank 1, or a covariant vector, if, under a change of coordinates $x^i = x^i(\bar{x}^j)$, $\bar{a}_i(\bar{x})$ is defined as

$$\bar{a}_i(\bar{x}) = a_k(x) \frac{\partial x^k}{\partial \bar{x}^i};$$

---

[1]It is worth noticing that spacetime coordinates can be indicated with Latin indexes $i, j, \ldots = 0, 1, 2, 3, \ldots$ while space coordinates can be indicated with Greek indexes $\alpha, \beta, \ldots = 1, 2, 3 \ldots$. However, in literature, there is also the opposite choice, that is $\alpha, \beta, \ldots = 0, 1, 2, 3 \ldots$ for spacetime coordinates and $i, j, \ldots = 1, 2, 3, \ldots$ for purely spatial coordinates. Here, we will adopt the first notation.

Example 5.1.3 $a_{kl}(x)$ is a covariant tensor of rank 2, if after a change of coordinates $x^i = x^i(\bar{x}^j)$, $\bar{a}_{ij}(\bar{x})$ is defined as

$$\bar{a}_{ij}(\bar{x}) = a_{kl}(x)\frac{\partial x^k}{\partial \bar{x}^i}\frac{\partial x^l}{\partial \bar{x}^j};$$

Example 5.1.4 $a_{pqr}(x)$ is a covariant tensor of rank 3, if after a change of coordinates $x^i = x^i(\bar{x}^j)$, $\bar{a}_{ijk}(\bar{x})$ is defined as

$$\bar{a}_{ijk}(\bar{x}) = a_{pqr}(x)\frac{\partial x^p}{\partial \bar{x}^i}\frac{\partial x^q}{\partial \bar{x}^j}\frac{\partial x^r}{\partial \bar{x}^k};$$

Example 5.1.5 $T_{i_1 i_2 \dots i_p}(x)$ is a covariant tensor of rank p, if after a change of coordinates $x^i = x^i(\bar{x}^j)$, $\bar{T}_{j_1 j_2 \dots j_p}(\bar{x})$ is defined as

$$\bar{T}_{j_1 j_2 \dots j_p}(\bar{x}) = T_{i_1 i_2 \dots i_p}(x)\frac{\partial x^{i_1}}{\partial \bar{x}^{j_1}}\frac{\partial x^{i_2}}{\partial \bar{x}^{j_2}}\dots\frac{\partial x^{i_p}}{\partial \bar{x}^{j_p}}.$$

Let us observe that in the definition of (k, p) tensor above the contravariant indexes change using the inverse transformation of coordinates. The matrix of change of coordinates is called Jacobian matrix. The transformation is regular if the determinant of Jacobian matrix is always finite and different from zero. Otherwise, the transformation is singular. In general, if an object transforms under any change of coordinates with a non-singular Jacobian determinant, the object is a tensor. In this case the transformation is called linear. The rank of the tensor determines the number of Jacobian matrixes concurring into the transformation. For example, a rank 2 tensor transforms, under a coordinate changes, by the multiplication of two Jacobian matrices, one for each index.[2]

Example 5.1.6 $a^k(x)$ is a contravariant tensor of rank 1, or simply a contravariant vector, if at a change of coordinates $x^i = x^i(\bar{x}^j)$, $\bar{a}^i(\bar{x})$ is defined as

$$\bar{a}^i(\bar{x}) = a^k(x)\frac{\partial \bar{x}^i}{\partial x^k}.$$

Observe that we have used the inverse change of coordinates formula.

Example 5.1.7 $T^{i_1 i_2 \dots i_k}(x)$ is a contravariant tensor of rank k, if under a change of coordinates $x^i = x^i(\bar{x}^j)$, $\bar{T}^{j_1 j_2 \dots j_k}(\bar{x})$ is defined as

$$\bar{T}^{j_1 j_2 \dots j_k}(\bar{x}) = T^{i_1 i_2 \dots i_k}(x)\frac{\partial \bar{x}^{j_1}}{\partial x^{i_1}}\frac{\partial \bar{x}^{j_2}}{\partial x^{i_2}}\dots\frac{\partial \bar{x}^{j_k}}{\partial x^{i_k}}.$$

---

[2]It is worth noticing that General Relativity, from a mathematical point of view, is the physical theory whose objects are invariant under the group of linear transformations in four dimension, i.e. $GL(4)$.

The Geometry on M is assigned by the following objects:

$\cdot\ ds^2 = g_{ij}(x^0, x^1, \ldots, x^n) dx^i dx^j$ is a *metric* determined by the matrix

$$\mathbf{G} = (g_{ij}), \quad g_{ij} = g_{ij}(x^0, x^1, \ldots, x^n), \quad i, \ j \in \{0, 1, \ldots, n\},$$

which is a rank-2 tensor which can be recovered by the extension of a *metric* tensor. It has the following properties:

1. $g_{ij}(x)$ is a smooth function of x on M.
2. at each point x, the metric is symmetric, i.e. $g_{ij}(x) = g_{ji}(x)$
3. at each point x, it exists the inverse $\mathbf{G}^{-1} = (g^{ij})$; using the Einstein notation, the inverse is described by the relations: $g^{ij} g_{jk} = \delta_k^i$, $g_{js} g^{sk} = \delta_j^k$.

Imagine the attached bilinear form with coefficients g_{ij}, denoted $S(u, v) = g_{ij} u^i v^j$.

This one can have the property $S(u, u) > 0$, $\forall u$, $u = (u^0, \ldots, u^n)$. In this case the metric is called a *Riemannian metric*.

Otherwise is called a *non-Riemannian metric*; some textbooks use the equivalent terminology: *semi-Riemannian* or *pseudo-Riemannian*.

The *signature of a metric* is defined as the signature of the corresponding quadratic form. For Minkowski metric, the signature we use is $(+--)$.[3]

As an example, the Euclidean metric $ds^2 = (dx^0) + (dx^1)^2$ is a Riemannian metric, its signature is $(++)$ while the Minkowski metric $ds^2 = (dx^0) - (dx^1)^2$ is a non-Riemannian metric with $(+-)$ signature.

- At each point x, it exists a *tangent space* of M, denoted by $T_x M$, whose coordinates are $(\dot{x}^0, \dot{x}^1, \ldots, \dot{x}^n)$. Consider a curve $x(t)$ in M, and the vector

$$\dot{x}(t) = \frac{dx}{dt} = (\dot{x}^0(t), \dot{x}^1(t), \ldots, \dot{x}^n(t)).$$

This vector belongs to the tangent space and, under a change of coordinates $x^i = x^i(\bar{x}^j)$, we have

$$\dot{x}^i(t) = \frac{dx^i}{dt} = \frac{dx^i}{d\bar{x}^j} \frac{d\bar{x}^j}{dt} = \dot{\bar{x}}^j(t) \frac{dx^i}{d\bar{x}^j},$$

that is

$$\dot{\bar{x}}^j(t) = \dot{x}^i(t) \frac{d\bar{x}^j}{dx^i}.$$

Therefore, a tangent vector to a curve is a contravariant vector. Considering vectors, we prefer to write $V = (V^0, V^1, \ldots, V^n)$ in the simpler form $V = V^k$, or only V^k as we did it before. That is, a vector can be seen through its components.

---

[3]In several textbook, the signature $(-+++)$ is adopted. In this case, time-like and space-like vectors have opposite sign.

- Christoffel symbols of first kind:

$$\Gamma_{ij,k} := \frac{1}{2} \left(\frac{\partial g_{ik}}{\partial x^j} + \frac{\partial g_{jk}}{\partial x^i} - \frac{\partial g_{ij}}{\partial x^k} \right)$$

- Christoffel symbols of second kind:

$$\Gamma^i_{jk} := g^{is} \Gamma_{jk,s} = \frac{1}{2} g^{is} \left(\frac{\partial g_{js}}{\partial x^k} + \frac{\partial g_{ks}}{\partial x^j} - \frac{\partial g_{jk}}{\partial x^s} \right)$$

- Riemannian curvature tensor, mixed:

$$R^h_{ijk} := \frac{\partial \Gamma^h_{ik}}{\partial x^j} - \frac{\partial \Gamma^h_{ij}}{\partial x^k} + \Gamma^h_{mj} \Gamma^m_{ik} - \Gamma^h_{mk} \Gamma^m_{ij}$$

The fact that the previous multi-index quantity is a tensor is proved later in this chapter. The same for all lower multi-index quantities.
- Riemannian curvature tensor, covariant: $R_{ijkl} := g_{is} R^s_{jkl}$
- Ricci tensor: $R_{ij} = R^s_{isj}$ which is obtained from the curvature tensor R^s_{imj} by contracting the indexes $s = m$.
- Geodesics, i.e. curves $c(t) = (x^1(t), x^2(t), \ldots, x^n(t))$ which satisfy the equations

$$\frac{d^2 x^r}{dt^2} + \Gamma^r_{pq} \frac{dx^p}{dt} \frac{dx^q}{dt} = 0.$$

We will see that the Christoffel symbols are not tensor, while g_{ij}, R^i_{jkl}, R_{ijkl}, R_{ij} and geodesics are tensors and their form is preserved under any change of coordinates.

Proposition 5.1.8 *A change of coordinates* $x^r = x^r(\overline{x}^h)$, $r, h \in \{0, 1, \ldots, n\}$ *transforms the metric under the rule* $\overline{G}_{\overline{x}} = (dM_{\overline{x}})^t \cdot G_x \cdot (dM_{\overline{x}})$.

Proof Suppose that $M : \overline{U} \to U$ is the previous change of coordinates which transforms $(\overline{x}^0, \ldots, \overline{x}^n)$ into the coordinates $(x^0, \ldots, x^n)$. The first metric is described by the matrix $\overline{G} = (\overline{g}_{ij}(\overline{x})$ and the second metric is described by the matrix $G = (g_{rs}(x))$.

We first suggest why the formula should be as it is in the statement before. Consider a quadratic form $\sum_{i,j=1}^n \alpha_{ij} y^i y^j$ written in its matrix form

$$\mathbf{y}^t \cdot \boldsymbol{\alpha} \cdot \mathbf{y},$$

where $\mathbf{y}$ is a column vector such that its transposed is $\mathbf{y}^t = (y^1, \ldots, y^n)$ and the matrix α is $\boldsymbol{\alpha} = (\alpha_{ij})$, $i \in \{1, \ldots, n\}$, $j \in \{1, \ldots, n\}$.
The change of coordinates $\mathbf{y} = \mathbf{A} \cdot \mathbf{x}$ leads to

$$\mathbf{y}^t \cdot \boldsymbol{\alpha} \cdot \mathbf{y} = (\mathbf{A} \cdot \mathbf{x})^t \cdot \boldsymbol{\alpha} \cdot (\mathbf{A} \cdot \mathbf{x}),$$

that is $(\mathbf{x}^t \cdot \mathbf{A}^t) \cdot \boldsymbol{\alpha} \cdot (\mathbf{A} \cdot \mathbf{x})$, i.e.

$$\mathbf{x}^t \cdot (\mathbf{A}^t \cdot \boldsymbol{\alpha} \cdot \mathbf{A}) \cdot \mathbf{x}.$$

So, the transformed quadratic form has its matrix $\mathbf{B}$ given by the formula $\mathbf{B} = \mathbf{A}^t \cdot \boldsymbol{\alpha} \cdot \mathbf{A}$.

In our case, using Einstein's rule, the second metric is

$$g_{rs}(\mathbf{x})dx^r dx^s = \sum_{r,s=0}^{n} g_{rs}(\mathbf{x})dx^r dx^s.$$

Since the change of coordinates in terms of differentials is $\mathbf{dx} = \left(\dfrac{\partial x^i}{\partial \bar{x}^j}\right)\mathbf{d\bar{x}}$, i.e.
$dM_{\bar{\mathbf{x}}} = \left(\dfrac{\partial x^i}{\partial \bar{x}^j}\right)$, the previous formula $\mathbf{B} = \mathbf{A}^t \cdot \boldsymbol{\alpha} \cdot \mathbf{A}$ for $\mathbf{B} = \bar{G}_{\bar{x}}, \alpha = G_{\mathbf{x}}, \mathbf{A} = dM_{\bar{\mathbf{x}}}$ leads to

$$\bar{G}_{\bar{x}} = (dM_{\bar{\mathbf{x}}})^t \cdot G_{\mathbf{x}} \cdot (dM_{\bar{\mathbf{x}}}).$$

□

Of course, we have a similar formula for the inverse of our initial coordinates transform.

Corollary 5.1.9 *A change of coordinates* $\bar{x}^r = \bar{x}^r(x^h), \quad r \in \{0, 1, \ldots, n\}$, $h \in \{0, 1, \ldots, n\}$ *transforms the metric according to the rule* $G_x = (dM_{\bar{x}(x)})^t \cdot \bar{G}_{\bar{x}(x)} \cdot (dM_{\bar{x}(x)})$.

We will see this formula acting later to transform the Poincaré metric of the disk into the Poincaré metric of the half-plane.

Corollary 5.1.10 *The formula* $\bar{G}_{\bar{x}} = (dM_{\bar{\mathbf{x}}})^t \cdot G_x \cdot (dM_{\bar{\mathbf{x}}})$ *is described in coordinates by* $\bar{g}_{ij} = g_{kl}\dfrac{\partial x^k}{\partial \bar{x}^i}\dfrac{\partial x^l}{\partial \bar{x}^j}$, *that is the metric tensor* g_{ij} *is a covariant tensor of rank* 2.

Proof To prove the result, we are looking at the geometric meaning of the previous obtained formula. Between two vectors $\bar{u}, \bar{v}$ of the tangent space at $\bar{x}$ and their images u, v in the tangent space at x, we have the connection $u = dM_{\bar{x}}\bar{u}, \; v = dM_{\bar{x}}\bar{v}$ and

$$\bar{u}^t \cdot \bar{G} \cdot \bar{v} = u^t \cdot G \cdot v = (dM_{\bar{x}}\bar{u})^t \cdot G \cdot dM_{\bar{x}}\bar{v} = \bar{u}^t \cdot dM_{\bar{x}}^t \cdot G \cdot dM_{\bar{x}} \cdot \bar{v},$$

i.e.

$$\bar{u}^t \cdot [\bar{G} - dM_{\bar{x}}^t \cdot G \cdot dM_{\bar{x}}] \cdot \bar{v} = 0.$$

Let us now write this in coordinates:

$$\bar{u} = \frac{d\bar{x}^i}{dt}, \quad \bar{v} = \frac{d\bar{x}^j}{dt}, \quad u = \frac{dx^k}{dt} = \frac{\partial x^k}{\partial \bar{x}^i}\frac{d\bar{x}^i}{dt}, \quad v = \frac{dx^l}{dt} = \frac{\partial x^l}{\partial \bar{x}^j}\frac{d\bar{x}^j}{dt}.$$

Taking into account the equality of metrics before and after transformation, we have

$$\bar{g}_{ij}\bar{u}\bar{v} = \bar{g}_{ij}\frac{d\bar{x}^i}{dt}\frac{d\bar{x}^j}{dt} = g_{kl}\frac{\partial x^k}{\partial \bar{x}^i}\frac{d\bar{x}^i}{dt}\frac{\partial x^l}{\partial \bar{x}^j}\frac{d\bar{x}^j}{dt} = g_{kl}uv.$$

Therefore

$$\left[\bar{g}_{ij} - g_{kl}\frac{\partial x^k}{\partial \bar{x}^i}\frac{\partial x^l}{\partial \bar{x}^j}\right]\frac{d\bar{x}^i}{dt}\frac{d\bar{x}^j}{dt} = 0$$

and since $\dfrac{d\bar{x}^i}{dt}, \dfrac{d\bar{x}^j}{dt}$ are arbitrary, we obtain

$$\bar{g}_{ij} = g_{kl}\frac{\partial x^k}{\partial \bar{x}^i}\frac{\partial x^l}{\partial \bar{x}^j}$$

$\square$

It is obvious that the inverse transformation leads to

Corollary 5.1.11 *The formula* $G_x = (dM_{\bar{x}(x)})^t \cdot \bar{G}_{\bar{x}(x)} \cdot (dM_{\bar{x}(x)})$ *is described in coordinates by* $g_{ij} = \bar{g}_{kl}\dfrac{\partial \bar{x}^k}{\partial x^i}\dfrac{\partial \bar{x}^l}{\partial x^j}.$

Corollary 5.1.12 *The inverse matrix of the metric tensor* G^{-1} *is also a contravariant tensor of rank 2, i.e.* $\bar{g}^{ij} = g^{kl}\dfrac{\partial \bar{x}^i}{\partial x^k}\dfrac{\partial \bar{x}^j}{\partial x^l}.$

In the above examples, we used both this kind of change of coordinates and a direct way suggested from calculus.

Theorem 5.1.13 *A change of coordinates* $x^r = x^r(\bar{x}^h), r \in \{0, 1, \ldots, n\}$, $h \in \{0, 1, \ldots, n\}$ *transforms the Christoffel symbols of first kind under the rule*

$$\bar{\Gamma}_{ij,k} = \Gamma_{rs,p}\frac{\partial x^r}{\partial \bar{x}^i}\frac{\partial x^s}{\partial \bar{x}^j}\frac{\partial x^p}{\partial \bar{x}^k} + g_{rs}\frac{\partial^2 x^r}{\partial \bar{x}^i \partial \bar{x}^j}\frac{\partial x^s}{\partial \bar{x}^k}.$$

Proof Let start from

$$\Gamma_{ij,k} = \frac{1}{2}\left(\frac{\partial g_{ik}}{\partial x^j} + \frac{\partial g_{jk}}{\partial x^i} - \frac{\partial g_{ij}}{\partial x^k}\right)$$

and

$$\overline{g}_{jk} = g_{rs} \frac{\partial x^r}{\partial \overline{x}^j} \frac{\partial x^s}{\partial \overline{x}^k}.$$

We have

$$\frac{\overline{g}_{jk}}{\partial \overline{x}^i} = \frac{g_{rs}}{\partial \overline{x}^p} \frac{\partial x^p}{\partial \overline{x}^i} \frac{\partial x^r}{\partial \overline{x}^j} \frac{\partial x^s}{\partial \overline{x}^k} + g_{rs} \frac{\partial^2 x^r}{\partial \overline{x}^i \partial \overline{x}^j} \frac{\partial x^s}{\partial \overline{x}^k} + g_{rs} \frac{\partial x^r}{\partial \overline{x}^j} \frac{\partial^2 x^s}{\partial \overline{x}^i \partial \overline{x}^k},$$

$$\frac{\overline{g}_{ki}}{\partial \overline{x}^j} = \frac{g_{rs}}{\partial \overline{x}^p} \frac{\partial x^p}{\partial \overline{x}^j} \frac{\partial x^r}{\partial \overline{x}^k} \frac{\partial x^s}{\partial \overline{x}^i} + g_{rs} \frac{\partial^2 x^r}{\partial \overline{x}^j \partial \overline{x}^k} \frac{\partial x^s}{\partial \overline{x}^i} + g_{rs} \frac{\partial x^r}{\partial \overline{x}^k} \frac{\partial^2 x^s}{\partial \overline{x}^j \partial \overline{x}^i},$$

$$\frac{\overline{g}_{ij}}{\partial \overline{x}^k} = \frac{g_{rs}}{\partial \overline{x}^p} \frac{\partial x^p}{\partial \overline{x}^k} \frac{\partial x^r}{\partial \overline{x}^i} \frac{\partial x^s}{\partial \overline{x}^j} + g_{rs} \frac{\partial^2 x^r}{\partial \overline{x}^k \partial \overline{x}^i} \frac{\partial x^s}{\partial \overline{x}^j} + g_{rs} \frac{\partial x^r}{\partial \overline{x}^i} \frac{\partial^2 x^s}{\partial \overline{x}^k \partial \overline{x}^j}.$$

Since

$$r \to s \to p \to r \implies \frac{g_{rs}}{\partial \overline{x}^p} \frac{\partial x^p}{\partial \overline{x}^i} \frac{\partial x^r}{\partial \overline{x}^j} \frac{\partial x^s}{\partial \overline{x}^k} = \frac{g_{sp}}{\partial \overline{x}^r} \frac{\partial x^r}{\partial \overline{x}^i} \frac{\partial x^p}{\partial \overline{x}^k} \frac{\partial x^s}{\partial \overline{x}^j},$$

$$r \to p \to s \to r \implies \frac{g_{rs}}{\partial \overline{x}^p} \frac{\partial x^p}{\partial \overline{x}^j} \frac{\partial x^r}{\partial \overline{x}^k} \frac{\partial x^s}{\partial \overline{x}^i} = \frac{g_{pr}}{\partial \overline{x}^s} \frac{\partial x^r}{\partial \overline{x}^i} \frac{\partial x^p}{\partial \overline{x}^k} \frac{\partial x^s}{\partial \overline{x}^j},$$

after we add, considering the first two equalities with plus and the last one with minus, we obtain

$$\overline{\Gamma}_{ij,k} = \Gamma_{rs,p} \frac{\partial x^r}{\partial \overline{x}^i} \frac{\partial x^s}{\partial \overline{x}^j} \frac{\partial x^p}{\partial \overline{x}^k} + g_{rs} \frac{\partial^2 x^r}{\partial \overline{x}^i \partial \overline{x}^j} \frac{\partial x^s}{\partial \overline{x}^k}$$

$$\square$$

Theorem 5.1.14 *A change of coordinates* $x^r = x^r(\overline{x}^h), r \in \{0, 1, , \ldots, n\}$, $h \in \{0, 1, \ldots, n\}$ *transforms the Christoffel symbols of second kind under the rule*

$$\frac{\partial^2 x^k}{\partial \overline{x}^i \partial \overline{x}^j} = -\Gamma^k_{rs} \frac{\partial x^r}{\partial \overline{x}^i} \frac{\partial x^s}{\partial \overline{x}^j} + \overline{\Gamma}^r_{ij} \frac{\partial x^k}{\partial \overline{x}^r}$$

Proof We start from the equalities

$$\begin{cases} \overline{g}^{ij} \overline{g}_{jk} = \delta^i_k \\ \overline{g}_{jk} = g_{rs} \dfrac{\partial x^r}{\partial \overline{x}^j} \dfrac{\partial x^s}{\partial \overline{x}^k} \end{cases} \implies \overline{g}^{ij} g_{rs} \frac{\partial x^r}{\partial \overline{x}^j} \frac{\partial x^s}{\partial \overline{x}^k} = \delta^i_k$$

which are multiplied by $g^{pq} \dfrac{\partial \overline{x}^k}{\partial x^p}$. It results

$$\overline{g}^{ij} g_{rs} \frac{\partial x^r}{\partial \overline{x}^j} \frac{\partial x^s}{\partial \overline{x}^k} g^{pq} \frac{\partial \overline{x}^k}{\partial x^p} = \delta^i_k g^{pq} \frac{\partial \overline{x}^k}{\partial x^p}$$

i.e.

$$\overline{g}^{ij} g_{rs} \frac{\partial x^r}{\partial \overline{x}^j} g^{pq} \frac{\partial x^s}{\partial x^p} = g^{pq} \frac{\partial \overline{x}^i}{\partial x^p}.$$

The left side makes sense only if $s = p$, therefore we have

$$\overline{g}^{ij} g^{pq} g_{pr} \frac{\partial x^r}{\partial \overline{x}^j} = g^{pq} \frac{\partial \overline{x}^i}{\partial x^p}$$

that is

$$\overline{g}^{ij} \frac{\partial x^q}{\partial \overline{x}^j} = g^{pq} \frac{\partial \overline{x}^i}{\partial x^p}.$$

Multiplying

$$\overline{\Gamma}_{ij,k} = \Gamma_{rs,p} \frac{\partial x^r}{\partial \overline{x}^i} \frac{\partial x^s}{\partial \overline{x}^j} \frac{\partial x^p}{\partial \overline{x}^k} + g_{rs} \frac{\partial^2 x^r}{\partial \overline{x}^i \partial \overline{x}^j} \frac{\partial x^s}{\partial \overline{x}^k}$$

by $\overline{g}^{mk}$, we obtain

$$\overline{\Gamma}^m_{ij} = \overline{g}^{mk} \overline{\Gamma}_{ij,k} = \Gamma_{rs,p} \frac{\partial x^r}{\partial \overline{x}^i} \frac{\partial x^s}{\partial \overline{x}^j} \cdot \overline{g}^{mk} \frac{\partial x^p}{\partial \overline{x}^k} + g_{rs} \frac{\partial^2 x^r}{\partial \overline{x}^i \partial \overline{x}^j} \cdot \overline{g}^{mk} \frac{\partial x^s}{\partial \overline{x}^k},$$

that is

$$\overline{\Gamma}^m_{ij} = \Gamma_{rs,p} g^{pq} \frac{\partial \overline{x}^m}{\partial x^q} \cdot \frac{\partial x^r}{\partial \overline{x}^i} \frac{\partial x^s}{\partial \overline{x}^j} + g_{rs} \cdot g^{sq} \frac{\partial \overline{x}^m}{\partial x^q} \cdot \frac{\partial^2 x^r}{\partial \overline{x}^i \partial \overline{x}^j},$$

and finally

$$\overline{\Gamma}^m_{ij} = \Gamma^q_{rs} \frac{\partial \overline{x}^m}{\partial x^q} \cdot \frac{\partial x^r}{\partial \overline{x}^i} \frac{\partial x^s}{\partial \overline{x}^j} + \frac{\partial \overline{x}^m}{\partial x^r} \cdot \frac{\partial^2 x^r}{\partial \overline{x}^i \partial \overline{x}^j}.$$

After multiplying by $\dfrac{\partial x^k}{\partial \overline{x}^m}$, it results

$$\overline{\Gamma}^m_{ij} \frac{\partial x^k}{\partial \overline{x}^m} = \Gamma^q_{rs} \frac{\partial x^k}{\partial \overline{x}^m} \frac{\partial \overline{x}^m}{\partial x^q} \cdot \frac{\partial x^r}{\partial \overline{x}^i} \frac{\partial x^s}{\partial \overline{x}^j} + \frac{\partial \overline{x}^m}{\partial x^r} \frac{\partial x^k}{\partial \overline{x}^m} \cdot \frac{\partial^2 x^r}{\partial \overline{x}^i \partial \overline{x}^j},$$

which can be arranged in the form

$$\frac{\partial^2 x^k}{\partial \overline{x}^i \partial \overline{x}^j} = -\Gamma^k_{rs} \frac{\partial x^r}{\partial \overline{x}^i} \frac{\partial x^s}{\partial \overline{x}^j} + \overline{\Gamma}^r_{ij} \frac{\partial x^k}{\partial \overline{x}^r}.$$

□

5.2 Covariant Derivative for Vectors and Tensors. Geodesics

Theorem 5.2.1 *Consider a curve $x(t)$ and a contravariant vector $V^k(t)$ along the curve. Then, the vector with the components*

$$\frac{dV^k}{dt} + \Gamma^k_{ij} V^j \frac{dx^i}{dt}$$

is a contravariant vector attached to this curve.

Proof Consider the contravariant vector V having the components $V^r(x)$. Being a contravariant vector, under a change of coordinates $\bar{x}^i = \bar{x}^i(x^j)$, $i, j \in \{0, 1, \ldots, n\}$, its components become $\overline{V}^r(\bar{x})$,

$$\overline{V}^r(\bar{x}) = V^j(x)\frac{\partial \bar{x}^r}{\partial x^j}.$$

If we consider the partial derivative with respect to x^i, we obtain

$$\frac{\partial \overline{V}^r}{\partial \bar{x}^p}\frac{\partial \bar{x}^p}{\partial x^i} = \frac{\partial V^j}{\partial x^i}\frac{\partial \bar{x}^r}{\partial x^j} + V^j\frac{\partial^2 \bar{x}^r}{\partial x^i \partial x^j}.$$

We multiply the last relation by $\dfrac{\partial x^k}{\partial \bar{x}^r}$ and it results

$$\frac{\partial \overline{V}^r}{\partial \bar{x}^p}\frac{\partial \bar{x}^p}{\partial x^i}\frac{\partial x^k}{\partial \bar{x}^r} = \frac{\partial V^j}{\partial x^i}\frac{\partial \bar{x}^r}{\partial x^j}\frac{\partial x^k}{\partial \bar{x}^r} + V^j\frac{\partial^2 \bar{x}^r}{\partial x^i \partial x^j}\frac{\partial x^k}{\partial \bar{x}^r} = \frac{\partial V^k}{\partial x^i} + V^j\frac{\partial^2 \bar{x}^r}{\partial x^i \partial x^j}\frac{\partial x^k}{\partial \bar{x}^r},$$

which can be written in the form

$$\frac{\partial V^k}{\partial x^i} = \frac{\partial \overline{V}^r}{\partial \bar{x}^p}\frac{\partial \bar{x}^p}{\partial x^i}\frac{\partial x^k}{\partial \bar{x}^r} - V^j\frac{\partial^2 \bar{x}^r}{\partial x^i \partial x^j}\frac{\partial x^k}{\partial \bar{x}^r}.$$

From Christoffel symbols of second kind

$$\Gamma^k_{ij} = \overline{\Gamma}^r_{pq}\frac{\partial \bar{x}^p}{\partial x^i}\frac{\partial \bar{x}^q}{\partial x^j}\frac{\partial x^k}{\partial \bar{x}^r} + \frac{\partial^2 \bar{x}^p}{\partial x^i \partial x^j}\frac{\partial x^k}{\partial \bar{x}^p},$$

we deduce

$$V^j\Gamma^k_{ij} = V^j\overline{\Gamma}^r_{pq}\frac{\partial \bar{x}^p}{\partial x^i}\frac{\partial \bar{x}^q}{\partial x^j}\frac{\partial x^k}{\partial \bar{x}^r} + V^j\frac{\partial^2 \bar{x}^p}{\partial x^i \partial x^j}\frac{\partial x^k}{\partial \bar{x}^p} = \overline{V}^q\overline{\Gamma}^r_{pq}\frac{\partial \bar{x}^p}{\partial x^i}\frac{\partial x^k}{\partial \bar{x}^r} + V^j\frac{\partial^2 \bar{x}^p}{\partial x^i \partial x^j}\frac{\partial x^k}{\partial \bar{x}^p}.$$

If we add the two relations obtained above we have

$$\frac{\partial V^k}{\partial x^i} + V^j \Gamma^k_{ij} = \left(\frac{\partial \overline{V}^r}{\partial \overline{x}^p} + \overline{V}^q \overline{\Gamma}^r_{pq} \right) \frac{\partial \overline{x}^p}{\partial x^i} \frac{\partial x^k}{\partial \overline{x}^r}.$$

Considering the coordinates x and $\bar{x}$ as functions of t we can write the last equality in the form

$$\frac{\partial V^k}{\partial x^i} \frac{dx^i}{dt} + V^j \Gamma^k_{ij} \frac{dx^i}{dt} = \left(\frac{\partial \overline{V}^r}{\partial \overline{x}^p} + \overline{V}^q \overline{\Gamma}^r_{pq} \right) \frac{\partial \overline{x}^p}{\partial x^i} \frac{\partial x^k}{\partial \overline{x}^r} \frac{dx^i}{dt},$$

or equivalently

$$\frac{dV^k}{dt} + \Gamma^k_{ij} V^j \frac{dx^i}{dt} = \left(\frac{d\overline{V}^r}{dt} + \overline{\Gamma}^r_{pq} \overline{V}^q \frac{d\overline{x}^p}{dt} \right) \frac{\partial x^k}{\partial \overline{x}^r},$$

which ends the proof. ☐

The above formula

$$\frac{\partial V^k}{\partial x^i} + V^j \Gamma^k_{ij} = \left(\frac{\partial \overline{V}^r}{\partial \overline{x}^p} + \overline{V}^q \overline{\Gamma}^r_{pq} \right) \frac{\partial \overline{x}^p}{\partial x^i} \frac{\partial x^k}{\partial \overline{x}^r}$$

shows that the expression $\dfrac{\partial V^k}{\partial x^i} + V^j \Gamma^k_{ij}$ remains invariant at a change of coordinates, therefore we have

Definition 5.2.2 For the contravariant vector $V^k(x)$, the covariant derivative is the $(1, 1)$ tensor with the components

$$\frac{\partial V^k}{\partial x^i} + V^j \Gamma^k_{ij}.$$

We denote the covariant derivative of the contravariant vector $V(x) = V^k(x)$ by

$$V^k_{;i} := \frac{\partial V^k}{\partial x^i} + V^j \Gamma^k_{ij}$$

Other possible notations for $V^k_{;i}$ are $\dfrac{\nabla V^k}{\partial x^i}$ or $\dfrac{\nabla V}{\partial x^i}$.

Definition 5.2.3 For the contravariant vector $V^k(t)$ the covariant derivative is the contravariant vector

$$\frac{dV^k}{dt} + \Gamma^k_{ij} V^j \frac{dx^i}{dt}.$$

We denote this in the form

$$V^k_{;} := \frac{dV^k}{dt} + \Gamma^k_{ij} V^j \frac{dx^i}{dt}$$

Other possible notations for $V^k_{;}$ are $\dfrac{\nabla V^k}{dt}$ or $\dfrac{\nabla V}{dt}$.

Definition 5.2.4 The contravariant vector $V^k(t)$ is parallel transported along the curve $x(t)$ if

$$\frac{dV^k}{dt} + \Gamma^k_{ij} V^j \frac{dx^i}{dt} = 0, \ k \in \{0, 1, 2, \ldots, n\}.$$

Definition 5.2.5 The curve $x = x(t) = (x^0(t), x^1(t), \ldots, x^n(t))$ is a geodesic if its contravariant tangent vector $\dot{x}(t)$ is parallel transported along the curve.

Proposition 5.2.6 *A curve* $x(t) = (x^0(t), x^1(t), \ldots, x^n(t))$ *whose components satisfy the equations*

$$\frac{d^2 x^k}{dt^2} + \Gamma^k_{ij} \frac{dx^i}{dt} \frac{dx^j}{dt} = 0, \ k \in \{0, 1, \ldots, n\}$$

is a geodesic of M.

Proof We replace V^k by $\dfrac{dx^k}{dt}$ in the formula

$$\frac{dV^k}{dt} + \Gamma^k_{ij} V^j \frac{dx^i}{dt} = 0,$$

and we obtain

$$\frac{d^2 x^k}{dt^2} + \Gamma^k_{ij} \frac{dx^i}{dt} \frac{dx^j}{dt} = 0.$$

$\square$

In the following statement we prove that a change of coordinates transforms a geodesic into a geodesic.

Theorem 5.2.7 *The change of coordinates* $x^r = x^r(\overline{x}^h)$, $r \in \{0, 1, \ldots, n\}, h \in \{0, 1, \ldots, n\}$ *transforms the equations*

$$\frac{d^2 \overline{x}^h}{dt^2} + \overline{\Gamma}^h_{ij} \frac{d\overline{x}^i}{dt} \frac{d\overline{x}^j}{dt} = 0$$

for the curve $\overline{c}(t) = (\overline{x}^0(t), \overline{x}^1(t), \ldots, \overline{x}^n(t))$ *into the equations*

$$\frac{d^2 x^r}{dt^2} + \Gamma^r_{pq} \frac{dx^p}{dt} \frac{dx^q}{dt} = 0$$

for the curve $c(t) = (x^0(t), x^1(t), \ldots, x^n(t))$.

Proof From

$$\frac{d^2 \overline{x}^h}{dt^2} + \overline{\Gamma}^h_{ij} \frac{d\overline{x}^i}{dt} \frac{d\overline{x}^j}{dt} = \left(\frac{d^2 x^r}{dt^2} + \Gamma^r_{pq} \frac{dx^p}{dt} \frac{dx^q}{dt} \right) \frac{\partial \overline{x}^h}{\partial x^r}$$

if

$$\frac{d^2 \overline{x}^h}{dt^2} + \overline{\Gamma}^h_{ij} \frac{d\overline{x}^i}{dt} \frac{d\overline{x}^j}{dt} = 0$$

then

$$\frac{d^2 x^r}{dt^2} + \Gamma^r_{pq} \frac{dx^p}{dt} \frac{dx^q}{dt} = 0.$$

$\square$

Problem 5.2.8 Prove directly, without considering the theory above, that, for a curve $x(t) = x^r(t)$, the vector

$$\frac{d^2 x^h}{dt^2} + \Gamma^h_{ij} \frac{dx^i}{dt} \frac{dx^j}{dt}$$

is a contravariant vector attached to this curve.

Solution. We wish to prove that a change of coordinates

$$x^r = x^r(\overline{x}^h), r \in \{0, 1, \ldots, n\}, h \in \{0, 1, \ldots, n\},$$

transforms the previous vector under the rule

$$\frac{d^2 \overline{x}^h}{dt^2} + \overline{\Gamma}^h_{ij} \frac{d\overline{x}^i}{dt} \frac{d\overline{x}^j}{dt} = \left(\frac{d^2 x^r}{dt^2} + \Gamma^r_{pq} \frac{dx^p}{dt} \frac{dx^q}{dt} \right) \frac{\partial \overline{x}^h}{\partial x^r}.$$

Let us start from $\dfrac{dx^r}{dt} = \dfrac{\partial x^r}{\partial \overline{x}^h} \dfrac{d\overline{x}^h}{dt}$. Then

$$\frac{d^2 x^r}{dt^2} = \frac{\partial^2 x^r}{\partial \overline{x}^h \partial \overline{x}^j} \frac{d\overline{x}^h}{dt} \frac{d\overline{x}^j}{dt} + \frac{\partial x^r}{\partial \overline{x}^h} \frac{d^2 \overline{x}^h}{dt^2}.$$

We arrange in the form

$$\frac{\partial x^r}{\partial \overline{x}^h} \frac{d^2 \overline{x}^h}{dt^2} = \frac{d^2 x^r}{dt^2} - \frac{\partial^2 x^r}{\partial \overline{x}^h \partial \overline{x}^j} \frac{d\overline{x}^h}{dt} \frac{d\overline{x}^j}{dt}$$

and we multiply by $\dfrac{\partial \overline{x}^h}{\partial x^r}$. It results

$$\frac{d^2 \overline{x}^h}{dt^2} = \frac{d^2 x^r}{dt^2}\frac{\partial \overline{x}^h}{\partial x^r} - \frac{\partial^2 x^r}{\partial \overline{x}^h \partial \overline{x}^j}\frac{d\overline{x}^h}{dt}\frac{d\overline{x}^j}{dt}\frac{\partial \overline{x}^h}{\partial x^r}$$

and we add $\overline{\Gamma}^h_{ij}\dfrac{d\overline{x}^i}{dt}\dfrac{d\overline{x}^j}{dt}$ in both members.

$$\frac{d^2 \overline{x}^h}{dt^2} + \overline{\Gamma}^h_{ij}\frac{d\overline{x}^i}{dt}\frac{d\overline{x}^j}{dt} = \frac{d^2 x^r}{dt^2}\frac{\partial \overline{x}^h}{\partial x^r} - \frac{\partial^2 x^r}{\partial \overline{x}^h \partial \overline{x}^j}\frac{d\overline{x}^h}{dt}\frac{d\overline{x}^j}{dt}\frac{\partial \overline{x}^h}{\partial x^r} + \overline{\Gamma}^h_{ij}\frac{d\overline{x}^i}{dt}\frac{d\overline{x}^j}{dt}$$

Let us use the formula which transforms the Christoffel second type symbols written in the form

$$\frac{\partial^2 x^p}{\partial \overline{x}^i \partial \overline{x}^j}\frac{\partial \overline{x}^h}{\partial x^p} + \Gamma^r_{pq}\frac{\partial x^p}{\partial \overline{x}^i}\frac{\partial x^q}{\partial \overline{x}^j}\frac{\partial \overline{x}^h}{\partial x^r} = \overline{\Gamma}^h_{ij}$$

in the right member of our last equality.

$$\frac{d^2 \overline{x}^h}{dt^2} + \overline{\Gamma}^h_{ij}\frac{d\overline{x}^i}{dt}\frac{d\overline{x}^j}{dt} =$$

$$= \frac{d^2 x^r}{dt^2}\frac{\partial \overline{x}^h}{\partial x^r} - \frac{\partial^2 x^r}{\partial \overline{x}^h \partial \overline{x}^j}\frac{d\overline{x}^h}{dt}\frac{d\overline{x}^j}{dt}\frac{\partial \overline{x}^h}{\partial x^r} + \left(\frac{\partial^2 x^p}{\partial \overline{x}^i \partial \overline{x}^j}\frac{\partial \overline{x}^h}{\partial x^p} + \Gamma^r_{pq}\frac{\partial x^p}{\partial \overline{x}^i}\frac{\partial x^q}{\partial \overline{x}^j}\frac{\partial \overline{x}^h}{\partial x^r}\right)\frac{d\overline{x}^i}{dt}\frac{d\overline{x}^j}{dt}.$$

Finally we obtain

$$\frac{d^2 \overline{x}^h}{dt^2} + \overline{\Gamma}^h_{ij}\frac{d\overline{x}^i}{dt}\frac{d\overline{x}^j}{dt} = \left(\frac{d^2 x^r}{dt^2} + \Gamma^r_{pq}\frac{dx^p}{dt}\frac{dx^q}{dt}\right)\frac{\partial \overline{x}^h}{\partial x^r}.$$

$\square$

Another important fact about geodesics is

Theorem 5.2.9 *If $c(t)$ is a geodesic, then $||\dot{c}(t)||$ is a constant.*

Proof Recall first that the length of a vector in a given metric $ds^2 = gijdx^i dx^j$ obviously depends on the type of the vector. Therefore, in our case, the formula is

$$||\dot{c}(t)||^2 := \overset{-}{+}\, g_{ij}(x^0(t), x^1(t), \ldots, x^n(t))\frac{dx^i(t)}{dt}\frac{dx^j(t)}{dt}.$$

We continue using the $+$ sign, in the other case the computations are the same. Having in mind that

$$\frac{d^2 x^i}{dt^2} = -\Gamma^i_{lm}\frac{dx^l}{dt}\frac{dx^m}{dt}$$

we start to compute $\dfrac{d}{dt}\left(||\dot{c}(t)||^2\right)$. We obtain

$$\frac{d}{dt}\left(||\dot{c}(t)||^2\right) = \frac{\partial g_{ij}}{\partial x^k}\frac{dx^k}{dt}\frac{dx^i}{dt}\frac{dx^j}{dt} + g_{ij}\frac{d^2x^i}{dt^2}\frac{dx^j}{dt} + g_{ij}\frac{dx^i}{dt}\frac{d^2x^j}{dt^2} =$$

$$= \frac{\partial g_{ij}}{\partial x^k}\frac{dx^k}{dt}\frac{dx^i}{dt}\frac{dx^j}{dt} + 2g_{ij}\frac{d^2x^i}{dt^2}\frac{dx^j}{dt} =$$

$$= \frac{\partial g_{ij}}{\partial x^k}\frac{dx^k}{dt}\frac{dx^i}{dt}\frac{dx^j}{dt} - 2g_{ij}\Gamma^i_{lm}\frac{dx^l}{dt}\frac{dx^m}{dt}\frac{dx^j}{dt} = \frac{\partial g_{ij}}{\partial x^k}\frac{dx^k}{dt}\frac{dx^i}{dt}\frac{dx^j}{dt} - 2\Gamma_{lm,j}\frac{dx^l}{dt}\frac{dx^m}{dt}\frac{dx^j}{dt} =$$

$$= \frac{\partial g_{ij}}{\partial x^k}\frac{dx^k}{dt}\frac{dx^i}{dt}\frac{dx^j}{dt} - \left(\frac{\partial g_{lj}}{\partial x^m} + \frac{\partial g_{mj}}{\partial x^l} - \frac{\partial g_{lm}}{\partial x^l}\right)\frac{dx^l}{dt}\frac{dx^m}{dt}\frac{dx^j}{dt} = 0.$$

This happens because after we relabel the summation indexes in the last three terms, in the expression, we have in fact the term $\dfrac{\partial g_{ij}}{\partial x^k}\dfrac{dx^k}{dt}\dfrac{dx^i}{dt}\dfrac{dx^j}{dt}$ written four times, two with the sign plus and two with minus. It results $\dfrac{d}{dt}\left(||\dot{c}(t)||^2\right) = 0$, i.e. $||\dot{c}(t)||^2 = b$, where b is a constant. □

The concept of covariant derivative allows us to obtain the same result later. The fact that, along a geodesic, the length of the tangent vector at geodesic is a constant one, allows us to replace the parameter t with $s = \dfrac{t}{\sqrt{|b|}}$. The geodesic $g = g(s)$ has the property $||\dot{c}(s)|| = 1$.

Definition 5.2.10 A curve which fulfills such a property, i.e. $||\dot{c}(s)|| = 1$, is called a canonically parameterized curve.

These last two theorems are used later to understand how geodesics of the disk are transformed by inversion in geodesics of the Poincaré half-plane. And more, how geodesics of the disk are transformed by inversion in geodesics outside the disk. All these facts will allow us to better understand the connections among Non-Euclidean Geometry basic models.

5.3 Riemann Mixed, Riemann Curvature Covariant, Ricci and Einstein Tensors

The Riemann symbols in the case of surfaces were obtained by considering the partial derivative of Gauss formulas. The way we introduced the parallel transport of contravariant vectors, without using an extra-dimension, allows us to think at a way to introduce the Riemann mixed curvature tensor without an extra-dimension.
The key is the parallel transport of contravariant vectors along infinitesimal vectors.
Suppose the infinitesimal vector is $A = (\delta x^0, \delta x^1, \ldots, \delta x^n)$ and let V be the vector we parallel transport along the infinitesimal vector A. If we act in an Euclidean

space where $\Gamma_{ij}^k = 0$, at the end we have only the vector $A + V$ of components $A^k + V^k$. But, in general, $\Gamma_{ij}^k \neq 0$ and the parallel transport highlights the vector of components $A^k + V^k + \delta V^k$ where

$$\delta V^k = -\Gamma_{ij}^k V^j \delta x^i.$$

If the components of V are dx^k, i.e. $V = (dx^0, dx^1, .., dx^n)$, the previous formula becomes

$$\delta(dx^k) = -\Gamma_{ij}^k dx^j \delta x^i$$

and each component of the parallel transported vector V along A becomes $\delta x^k + dx^k + \delta(dx^k)$, that is

$$\delta x^k + dx^k - \Gamma_{ij}^k dx^j \delta x^i.$$

If we consider the parallel transport of the infinitesimal vector A along the infinitesimal vector V, at the end we have the components $V^k + A^k + dA^k$, where

$$dA^k = -\Gamma_{ij}^k A^j dx^i.$$

Therefore, at the end of the parallel transport of the vector A along the vector V we obtain

$$dx^k + \delta x^k - \Gamma_{ij}^k \delta x^j dx^i.$$

In the Euclidean space, the condition $A^k + V^k = V^k + A^k$, $k \in \{0, 1, \ldots, n\}$ describes a parallelogram. Here, the parallelogram is described by the condition

$$A^k + (V^k + \delta V^k) = V^k + (A^k + dA^k), \ k \in \{0, 1, \ldots, n\},$$

that is

$$\cancel{\delta x^k} + \cancel{dx^k} - \Gamma_{ij}^k dx^j \delta x^i = \cancel{dx^k} + \cancel{\delta x^k} - \Gamma_{ij}^k \delta x^j dx^i.$$

This condition may be written in the form

$$\Gamma_{ij}^k dx^j \delta x^i = \Gamma_{ji}^k \delta x^i dx^j.$$

The last equality is true because the Christoffel symbols are symmetric, i.e. $\Gamma_{ij}^k = \Gamma_{ji}^k$.

Let us take into account the parallelogram considered above and a vector $W = (W^0, W^1, \ldots, W^n)$. We consider the parallel transport of W along the first two sides. Along A we first obtain the vector X of coordinates $X^k := A^k + (W^k + \delta W^k)$. Then, this vector is parallel transported along V. We obtain the vector of coordinates $V^k + (X^k + dX^k)$. Therefore the parallel transport of W along the first two sides leads to the vector of coordinates

$$T_1^k := V^k + [A^k + (W^k + \delta W^k) + d(A^k + (W^k + \delta W^k))] \quad (1)$$

The parallel transport of W along V leads to the vector Y of coordinates

$$Y^k := V^k + (W^k + dW^k)$$

and the parallel transport of Y along A leads to the vector of coordinates

$$A^k + (Y^k + \delta Y^k).$$

Therefore the parallel transport of W along the other two sides leads to the vector of coordinates

$$T_2^k := A^k + [V^k + (W^k + dW^k) + \delta(V^k + (W^k + dW^k))] \quad (2)$$

We continue to work in coordinates. The relation which allows us to consider the initial parallelogram is $dA^k = \delta V^k$. If we denote by

$$R := T_2^k - T_1^k,$$

it results

$$R = \delta(dW^k) - d(\delta W^k).$$

If we compute

$$-\delta(dW^k) = \delta(\Gamma_{ij}^k W^i dx^j) = \delta\Gamma_{ij}^k W^i dx^j + \Gamma_{ij}^k(\delta W^i)dx^j + \Gamma_{ij}^k W^i \delta(dx^j),$$

we obtain

$$-\delta(dW^k) = \frac{\partial \Gamma_{ij}^k}{\partial x^l}\delta x^l W^i dx^j - \Gamma_{ij}^k \Gamma_{ab}^i W^a \delta x^b dx^j - \Gamma_{ij}^k \Gamma_{ab}^j W^i dx^a \delta x^b.$$

Arranging the indexes

$$\delta(dW^k) = \left(-\frac{\partial \Gamma_{ij}^k}{\partial x^l} + \Gamma_{sj}^k \Gamma_{il}^s + \Gamma_{si}^k \Gamma_{jl}^s\right) W^i dx^j \delta x^l,$$

and in the same way

$$-d(\delta W^k) = \left(\frac{\partial \Gamma_{il}^k}{\partial x^j} - \Gamma_{sl}^k \Gamma_{ij}^s - \Gamma_{si}^k \Gamma_{lj}^s\right) W^i dx^j \delta x^l.$$

Therefore, after canceling $\Gamma_{si}^k \Gamma_{lj}^s$, one obtains

$$\delta(dW^k) - d(\delta W^k) = \left(\frac{\partial \Gamma_{il}^k}{\partial x^j} - \frac{\partial \Gamma_{ij}^k}{\partial x^l} + \Gamma_{sj}^k \Gamma_{il}^s - \Gamma_{sl}^k \Gamma_{ij}^s\right) W^i dx^j \delta x^l = R_{ijl}^k W^i dx^j \delta x^l.$$

The second type Riemann symbol is highlighted. If the vector $R = \delta(dW^k) - d(\delta W^k)$ is 0, the two vectors are coincident, it happens as in the Euclidean Geometry. If not, a curvature of M appears.

It remains to prove the tensorial nature of R^i_{jkl}.

Theorem 5.3.1 *A change of coordinates transforms the Riemann mixed curvature tensor and the Riemann curvature covariant tensor after the formulas*

$$(1)\ \bar{R}^a_{bgd}\frac{\partial \overline{x}^i}{\partial x^a} = R^i_{jkl}\frac{\partial x^j}{\partial \overline{x}^b}\frac{\partial x^k}{\partial \overline{x}^g}\frac{\partial x^l}{\partial \overline{x}^d},$$

$$(2)\ \bar{R}_{ebgd} = R_{rjkl}\frac{\partial x^r}{\partial \overline{x}^e}\frac{\partial x^j}{\partial \overline{x}^b}\frac{\partial x^k}{\partial \overline{x}^g}\frac{\partial x^l}{\partial \overline{x}^d}.$$

For the Ricci tensor, we have

$$(3)\ \bar{R}_{bg} = R_{jl}\frac{\partial x^j}{\partial \overline{x}^b}\frac{\partial x^l}{\partial \overline{x}^g}.$$

Proof For (1), we consider the partial derivative $\dfrac{\partial}{\partial \overline{x}^l}$ of the expression

$$\frac{\partial^2 x^k}{\partial \overline{x}^i \partial \overline{x}^j} = -\Gamma^k_{rs}\frac{\partial x^r}{\partial \overline{x}^i}\frac{\partial x^s}{\partial \overline{x}^j} + \overline{\Gamma}^r_{ij}\frac{\partial x^k}{\partial \overline{x}^r}.$$

It results

$$\frac{\partial^3 x^k}{\partial \overline{x}^l \partial \overline{x}^i \partial \overline{x}^j} =$$

$$= -\frac{\partial \Gamma^k_{rs}}{\partial x^p}\frac{\partial x^p}{\partial \overline{x}^l}\frac{\partial x^r}{\partial \overline{x}^i}\frac{\partial x^s}{\partial \overline{x}^j} - \Gamma^k_{rs}\left(\frac{\partial^2 x^r}{\partial \overline{x}^l \partial \overline{x}^i}\frac{\partial x^s}{\partial \overline{x}^j} + \frac{\partial x^r}{\partial \overline{x}^i}\frac{\partial^2 x^s}{\partial \overline{x}^l \partial \overline{x}^j}\right) + \frac{\partial \overline{\Gamma}^r_{ij}}{\partial \overline{x}^l}\frac{\partial x^k}{\partial \overline{x}^r} + \overline{\Gamma}^r_{ij}\frac{\partial^2 x^k}{\partial \overline{x}^l \partial \overline{x}^r}.$$

We switch l and j indexes in the previous formula

$$\frac{\partial^3 x^k}{\partial \overline{x}^j \partial \overline{x}^i \partial \overline{x}^l} =$$

$$= -\frac{\partial \Gamma^k_{rs}}{\partial x^p}\frac{\partial x^p}{\partial \overline{x}^j}\frac{\partial x^r}{\partial \overline{x}^i}\frac{\partial x^s}{\partial \overline{x}^l} - \Gamma^k_{rs}\left(\frac{\partial^2 x^r}{\partial \overline{x}^j \partial \overline{x}^i}\frac{\partial x^s}{\partial \overline{x}^l} + \frac{\partial x^r}{\partial \overline{x}^i}\frac{\partial^2 x^s}{\partial \overline{x}^j \partial \overline{x}^l}\right) + \frac{\partial \overline{\Gamma}^r_{il}}{\partial \overline{x}^j}\frac{\partial x^k}{\partial \overline{x}^r} + \overline{\Gamma}^r_{il}\frac{\partial^2 x^k}{\partial \overline{x}^j \partial \overline{x}^r}.$$

Since

$$\frac{\partial^3 x^k}{\partial \overline{x}^l \partial \overline{x}^i \partial \overline{x}^j} = \frac{\partial^3 x^k}{\partial \overline{x}^j \partial \overline{x}^i \partial \overline{x}^l}$$

we put the two equalities together and we separate the quantities having bar on second kind Christoffel symbols by the ones without bar. We also cancel the equal quantities and the left member, here denoted as LM, becomes

$$LM = \frac{\partial \Gamma_{rs}^k}{\partial x^p} \frac{\partial x^p}{\partial \overline{x}^j} \frac{\partial x^r}{\partial \overline{x}^i} \frac{\partial x^s}{\partial \overline{x}^l} - \frac{\partial \Gamma_{rs}^k}{\partial x^p} \frac{\partial x^p}{\partial \overline{x}^l} \frac{\partial x^r}{\partial \overline{x}^i} \frac{\partial x^s}{\partial \overline{x}^j} + \Gamma_{rs}^k \frac{\partial^2 x^r}{\partial \overline{x}^j \partial \overline{x}^i} \frac{\partial x^s}{\partial \overline{x}^l} - \Gamma_{rs}^k \frac{\partial^2 x^r}{\partial \overline{x}^l \partial \overline{x}^i} \frac{\partial x^s}{\partial \overline{x}^j}.$$

We divide the left member LM in two parts. In the first part, we interchange p and s. Then

$$\frac{\partial \Gamma_{rs}^k}{\partial x^p} \frac{\partial x^p}{\partial \overline{x}^j} \frac{\partial x^r}{\partial \overline{x}^i} \frac{\partial x^s}{\partial \overline{x}^l} - \frac{\partial \Gamma_{rs}^k}{\partial x^p} \frac{\partial x^p}{\partial \overline{x}^l} \frac{\partial x^r}{\partial \overline{x}^i} \frac{\partial x^s}{\partial \overline{x}^j} = \frac{\partial \Gamma_{rp}^k}{\partial x^s} \frac{\partial x^s}{\partial \overline{x}^j} \frac{\partial x^r}{\partial \overline{x}^i} \frac{\partial x^p}{\partial \overline{x}^l} - \frac{\partial \Gamma_{rs}^k}{\partial x^p} \frac{\partial x^p}{\partial \overline{x}^l} \frac{\partial x^r}{\partial \overline{x}^i} \frac{\partial x^s}{\partial \overline{x}^j} =$$

$$= \left(\frac{\partial \Gamma_{rp}^k}{\partial x^s} - \frac{\partial \Gamma_{rs}^k}{\partial x^p} \right) \frac{\partial x^s}{\partial \overline{x}^j} \frac{\partial x^r}{\partial \overline{x}^i} \frac{\partial x^p}{\partial \overline{x}^l}$$

In the second part of the left member, we use the formulas which explain how the second type Christoffel symbols are transformed under a change of coordinates:

$$\Gamma_{rs}^k \frac{\partial^2 x^r}{\partial \overline{x}^j \partial \overline{x}^i} \frac{\partial x^s}{\partial \overline{x}^l} - \Gamma_{rs}^k \frac{\partial^2 x^r}{\partial \overline{x}^l \partial \overline{x}^i} \frac{\partial x^s}{\partial \overline{x}^j} =$$

$$= \Gamma_{rs}^k \left(-\Gamma_{ab}^r \frac{\partial x^a}{\partial \overline{x}^j} \frac{\partial x^b}{\partial \overline{x}^i} + \overline{\Gamma}_{ji}^a \frac{\partial x^r}{\partial \overline{x}^a} \right) \frac{\partial x^s}{\partial \overline{x}^l} - \Gamma_{rs}^k \left(-\Gamma_{ab}^r \frac{\partial x^a}{\partial \overline{x}^l} \frac{\partial x^b}{\partial \overline{x}^i} + \overline{\Gamma}_{li}^a \frac{\partial x^r}{\partial \overline{x}^a} \right) \frac{\partial x^s}{\partial \overline{x}^j} =$$

$$= -\Gamma_{rs}^k \Gamma_{ab}^r \frac{\partial x^a}{\partial \overline{x}^j} \frac{\partial x^b}{\partial \overline{x}^i} \frac{\partial x^s}{\partial \overline{x}^l} + \Gamma_{rs}^k \Gamma_{ab}^r \frac{\partial x^a}{\partial \overline{x}^l} \frac{\partial x^b}{\partial \overline{x}^i} \frac{\partial x^s}{\partial \overline{x}^j} + \Gamma_{rs}^k \overline{\Gamma}_{ji}^a \frac{\partial x^r}{\partial \overline{x}^a} \frac{\partial x^s}{\partial \overline{x}^l} - \Gamma_{rs}^k \overline{\Gamma}_{li}^a \frac{\partial x^r}{\partial \overline{x}^a} \frac{\partial x^s}{\partial \overline{x}^j}$$

We rearrange the dummy indexes such that the product of the three ratios becomes

$$\frac{\partial x^s}{\partial \overline{x}^j} \frac{\partial x^r}{\partial \overline{x}^i} \frac{\partial x^p}{\partial \overline{x}^l}.$$

The left member LM becomes:

$$\left(\frac{\partial \Gamma_{rp}^k}{\partial x^s} - \frac{\partial \Gamma_{rs}^k}{\partial x^p} + \Gamma_{as}^k \Gamma_{pr}^a - \Gamma_{ap}^k \Gamma_{rs}^a \right) \frac{\partial x^s}{\partial \overline{x}^j} \frac{\partial x^r}{\partial \overline{x}^i} \frac{\partial x^p}{\partial \overline{x}^l} + \Gamma_{rs}^k \overline{\Gamma}_{ji}^a \frac{\partial x^r}{\partial \overline{x}^a} \frac{\partial x^s}{\partial \overline{x}^l} - \Gamma_{rs}^k \overline{\Gamma}_{li}^a \frac{\partial x^r}{\partial \overline{x}^a} \frac{\partial x^s}{\partial \overline{x}^j}.$$

The final form of the left member is

$$LM = R_{rsp}^k \frac{\partial x^s}{\partial \overline{x}^j} \frac{\partial x^r}{\partial \overline{x}^i} \frac{\partial x^p}{\partial \overline{x}^l} + \Gamma_{rs}^k \overline{\Gamma}_{ji}^a \frac{\partial x^r}{\partial \overline{x}^a} \frac{\partial x^s}{\partial \overline{x}^l} - \Gamma_{rs}^k \overline{\Gamma}_{li}^a \frac{\partial x^r}{\partial \overline{x}^a} \frac{\partial x^s}{\partial \overline{x}^j}.$$

In same way, the right member becomes

$$RM = \frac{\partial \bar{\Gamma}^r_{il}}{\partial \bar{x}^j} \frac{\partial x^k}{\partial \bar{x}^r} - \frac{\partial \bar{\Gamma}^r_{ij}}{\partial \bar{x}^l} \frac{\partial x^k}{\partial \bar{x}^r} + \bar{\Gamma}^r_{il} \frac{\partial^2 x^k}{\partial \bar{x}^j \partial \bar{x}^r} - \bar{\Gamma}^r_{ij} \frac{\partial^2 x^k}{\partial \bar{x}^l \partial \bar{x}^r} =$$

$$\left(\frac{\partial \bar{\Gamma}^r_{il}}{\partial \bar{x}^j} - \frac{\partial \bar{\Gamma}^r_{ij}}{\partial \bar{x}^l} \right) \frac{\partial x^k}{\partial \bar{x}^r} + \bar{\Gamma}^r_{il} \left(-\Gamma^k_{ab} \frac{\partial x^a}{\partial \bar{x}^j} \frac{\partial x^b}{\partial \bar{x}^r} + \bar{\Gamma}^s_{jr} \frac{\partial x^k}{\partial \bar{x}^s} \right) + \bar{\Gamma}^r_{ij} \left(-\Gamma^k_{ab} \frac{\partial x^a}{\partial \bar{x}^l} \frac{\partial x^b}{\partial \bar{x}^r} + \bar{\Gamma}^s_{lr} \frac{\partial x^k}{\partial \bar{x}^s} \right)$$

$$= \left(\frac{\partial \bar{\Gamma}^s_{il}}{\partial \bar{x}^j} - \frac{\partial \bar{\Gamma}^s_{ij}}{\partial \bar{x}^l} + \bar{\Gamma}^r_{il} \bar{\Gamma}^s_{jr} - \bar{\Gamma}^r_{ij} \bar{\Gamma}^s_{lr} \right) \frac{\partial x^k}{\partial \bar{x}^s} - \bar{\Gamma}^r_{il} \Gamma^k_{ab} \frac{\partial x^a}{\partial \bar{x}^j} \frac{\partial x^b}{\partial \bar{x}^r} + \bar{\Gamma}^r_{ij} \Gamma^k_{ab} \frac{\partial x^a}{\partial \bar{x}^l} \frac{\partial x^b}{\partial \bar{x}^r} =$$

$$= \bar{R}^s_{ijl} \frac{\partial x^k}{\partial \bar{x}^s} - \bar{\Gamma}^r_{il} \Gamma^k_{ab} \frac{\partial x^a}{\partial \bar{x}^j} \frac{\partial x^b}{\partial \bar{x}^r} + \bar{\Gamma}^r_{ij} \Gamma^k_{ab} \frac{\partial x^a}{\partial \bar{x}^l} \frac{\partial x^b}{\partial \bar{x}^r}.$$

Comparing the final forms of the left and right members after reducing the equal terms, we obtain the formula

$$R^k_{rsp} \frac{\partial x^r}{\partial \bar{x}^i} \frac{\partial x^s}{\partial \bar{x}^j} \frac{\partial x^p}{\partial \bar{x}^l} = \bar{R}^m_{ijl} \frac{\partial x^k}{\partial \bar{x}^m}.$$

(2) In (1), we multiply the right member by $g_{ri} \dfrac{\partial x^r}{\partial \bar{x}^e}$ and the left member by the same quantity written in the form $\bar{g}_{se} \dfrac{\partial \bar{x}^s}{\partial x^i}$.

This is possible because the formula $g_{ri} \dfrac{\partial x^r}{\partial \bar{x}^e} = \bar{g}_{se} \dfrac{\partial \bar{x}^s}{\partial x^i}$ comes from

$$\bar{g}_{es} = \bar{g}_{se} = g_{ri} \frac{\partial x^r}{\partial \bar{x}^e} \frac{\partial x^i}{\partial \bar{x}^s},$$

formula which was proved before. Then

$$\bar{g}_{se} \bar{R}^a_{bgd} \frac{\partial x^i}{\partial x^a} \frac{\partial \bar{x}^s}{\partial x^i} = g_{ri} R^i_{jkl} \frac{\partial x^r}{\partial \bar{x}^e} \frac{\partial x^j}{\partial \bar{x}^b} \frac{\partial x^k}{\partial \bar{x}^g} \frac{\partial x^l}{\partial \bar{x}^d}$$

In the left member, $s := a$ leads to

$$\bar{R}_{ebgd} = R_{rjkl} \frac{\partial x^r}{\partial \bar{x}^e} \frac{\partial x^j}{\partial \bar{x}^b} \frac{\partial x^k}{\partial \bar{x}^g} \frac{\partial x^l}{\partial \bar{x}^d}.$$

(3) We start from $\bar{R}_{bg} = \bar{R}^a_{bad}$. Then $\bar{R}_{bg} \dfrac{\partial x^i}{\partial \bar{x}^a} = \bar{R}^a_{bad} \dfrac{\partial x^i}{\partial \bar{x}^a} = R^i_{jil} \dfrac{\partial x^j}{\partial \bar{x}^b} \dfrac{\partial x^i}{\partial \bar{x}^a} \dfrac{\partial x^l}{\partial \bar{x}^g}$, therefore

$$\bar{R}_{bg} = R_{jl} \frac{\partial x^j}{\partial \bar{x}^b} \frac{\partial x^l}{\partial \bar{x}^g}.$$

We may say that Riemann mixed curvature tensor is a $(1, 3)$ tensor, the Riemann curvature covariant tensor is $(0, 4)$ tensor and the Ricci tensor is a covariant $(0, 2)$ tensor.   □

When we worked with surfaces embedded in 3-space (Euclidean or Minkowski) using the Gauss equations, we proved that the Riemann symbols R_{ijkl} have the properties

$$R_{ijkl} = -R_{ijlk};$$
$$R_{ijkl} = -R_{jikl};$$
$$R_{ijkl} = R_{klij};$$
$$R_{ijkl} + R_{iklj} + R_{iljk} = 0.$$

The keys of demonstration are the Gauss equations which depend on the second fundamental form. Can we prove such formulas in this abstract framework? The answer is yes, but we need to discuss first the covariant derivative. Let us remember the definition of the covariant derivative of a contravariant vector as the $(1, 1)$ tensor defined by $V^k_{;i} = \dfrac{\partial V^k}{\partial x^i} + V^j \Gamma^k_{ij}$.

Definition 5.3.2 The covariant derivative of a (k, p) tensor $T^{i_1 i_2 .. i_k}_{l_1 l_2 .. l_p}(x)$ is the $(k, p + 1)$ tensor defined by

$$T^{i_1 i_2 .. i_k}_{l_1 l_2 .. l_p; j}(x) =$$

$$= \frac{T^{i_1 i_2 .. i_k}_{l_1 l_2 .. l_p}}{\partial x^j} + T^{m i_2 .. i_k}_{l_1 l_2 .. l_p} \Gamma^{i_1}_{mj} + .. + T^{i_1 i_2 .. i_{k-1} m}_{l_1 l_2 ... l_p} \Gamma^{i_k}_{mj} - T^{i_1 i_2 ... i_k}_{m l_2 .. l_p} \Gamma^m_{l_1 j} - .. - T^{i_1 i_2 .. i_k}_{l_1 l_2 .. l_{p-1} m} \Gamma^m_{l_p j}.$$

Some particular cases: the covariant derivative of a covariant vector is the $(0, 2)$ covariant tensor

$$V_{i;j} = \frac{\partial V_i}{\partial x^j} - V_k \Gamma^k_{ij}.$$

The covariant derivative of a covariant $(0, 2)$ tensor is the $(0, 3)$ covariant tensor

$$a_{ij;k} = \frac{\partial a_{ij}}{\partial x^k} - a_{sj} \Gamma^s_{ik} - a_{si} \Gamma^s_{jk}.$$

The covariant derivative of a contravariant $(2, 0)$ tensor is the $(2, 1)$ tensor

$$b^{ij}_{;k} = \frac{\partial b^{ij}}{\partial x^k} + b^{lj} \Gamma^i_{lk} + b^{il} \Gamma^j_{kl}.$$

The covariant derivative of a $(1, 1)$ tensor is the $(1, 2)$ tensor $a^j_{i;k} = \dfrac{\partial a^j_i}{\partial x^k} + a^m_i \Gamma^j_{mk} - a^j_m \Gamma^m_{ik}$.

For two tensors T_L^I and S_P^J, where I, L, J, P are multi-index quantities, it makes sense the tensor $T_L^I S_P^J$ and the product rule of covariant derivative, that is

$$(T_L^I S_P^J)_{;m} = (T_L^I)_{;m} S_P^J + T_L^I (S_P^J)_{;m}.$$

For the metric tensor g_{ij} we can prove

Theorem 5.3.3 $g_{ij;k} = 0$

Proof We have

$$g_{ij;k} = \frac{\partial g_{ij}}{\partial x^k} - g_{sj} \Gamma_{ik}^s - g_{si} \Gamma_{jk}^s = \frac{\partial g_{ij}}{\partial x^k} - g_{sj} g^{sl} \Gamma_{ik,l} - g_{si} g^{sl} \Gamma_{jk,l} = \frac{\partial g_{ij}}{\partial x^k} - \Gamma_{ik,j} - \Gamma_{jk,i} = 0.$$

$\square$

A very important consequence appears.
Consider two contravariant vectors V^k and W^j and their "dot product" via the metric tensor, $\langle V^k, W^j \rangle = g_{kj} V^k W^j$.
Suppose that V^k and W^j are parallel transported along a curve $x(t)$, that is

$$\frac{dV^k}{dt} + \Gamma_{ij}^k V^j \frac{dx^i}{dt} = 0,$$

and

$$\frac{dW^j}{dt} + \Gamma_{il}^j W^l \frac{dx^i}{dt} = 0,$$

or simply, $V_{;k}^k(t) = 0$ and $W_{;j}^j(t) = 0$. The covariant derivative of the metric tensor g_{ij} vanishes,

$$g_{ij;k} = \frac{\partial g_{ij}}{\partial x^k} - g_{sj} \Gamma_{ik}^s - g_{si} \Gamma_{jk}^s = 0$$

and this can be written as

$$g_{ij;}(t) = \left(\frac{\partial g_{ij}}{\partial x^k} - g_{sj} \Gamma_{ik}^s - g_{si} \Gamma_{jk}^s \right) \frac{dx^k}{dt} = 0.$$

The derivative with respect to t of the "dot product" is, in fact, the covariant derivative of $g_{kj} V^k W^j$. Applying the product rule for the covariant derivative, we obtain

$$\left(g_{kj} V^k W^j \right)_{;i} = g_{kj;i} V^k W^j + g_{kj} V_{;i}^k W^j + g_{kj} V^k W_{;i}^j = 0.$$

It results:

Theorem 5.3.4 *(i) The length of a vector is conserved if the vector is parallel trans-ported along a curve. (ii) The length of the tangent vector to a geodesic is a constant.*

In each geometric space where the meaning of

$$\frac{\langle V^k, W^j \rangle}{|V^k| \, |W^j|} = \frac{g_{kj} V^k W^j}{\sqrt{|g_{kj} V^k V^j|} \, \sqrt{|g_{kj} W^k W^j|}}$$

is related to an angle via a trigonometric function $f(\alpha)$, as we saw on surfaces both in Euclidean spaces or in Minkowski spaces, the following theorem holds:

Theorem 5.3.5 *(i) The angle between two vectors parallel transported along a curve is conserved.*
(ii) The angle between a vector parallel transported along a geodesic and the tangent vector to a geodesic is the same at each point of the geodesic.

For the inverse g^{ij} of the metric tensor g_{ij} we have

Theorem 5.3.6 $g^{ij}_{;k} = 0.$

Proof Consider the covariant derivative of the expression $g_{is} g^{sj} = \delta^j_i$. It results

$$\delta^j_{i;k} = \frac{\partial \delta^j_i}{\partial x^k} + \delta^m_i \Gamma^j_{mk} - \delta^j_m \Gamma^m_{ik} = 0$$

and

$$0 = (g_{is} g^{sj})_{;k} = g_{is;k} g^{sj} + g_{is} g^{sj}_{;k} = g_{is} g^{sj}_{;k},$$

i.e.

$$g^{mi} g_{is} g^{sj}_{;k} = g^{mj}_{;k} = 0.$$

$\square$

Theorem 5.3.7

$$R_{ijkl} = \frac{1}{2} \left(\frac{\partial^2 g_{il}}{\partial x^j \partial x^k} + \frac{\partial^2 g_{jk}}{\partial x^i \partial x^l} - \frac{\partial^2 g_{jl}}{\partial x^i \partial x^k} - \frac{\partial^2 g_{ik}}{\partial x^j \partial x^l} \right) + g_{mp} (\Gamma^m_{il} \Gamma^p_{jk} - \Gamma^m_{ik} \Gamma^p_{jl}).$$

Proof

$$R_{ijkl} = g_{is} R^s_{jkl} = g_{is} \left(\frac{\partial \Gamma^s_{jl}}{\partial x^k} - \frac{\partial \Gamma^s_{jk}}{\partial x^l} + \Gamma^s_{mk} \Gamma^m_{jl} - \Gamma^s_{ml} \Gamma^m_{jk} \right) =$$

$$= g_{is} \left(\frac{\partial \Gamma^s_{jl}}{\partial x^k} - \frac{\partial \Gamma^s_{jk}}{\partial x^l} \right) + g_{is} \left(\Gamma^s_{mk} \Gamma^m_{jl} - \Gamma^s_{ml} \Gamma^m_{jk} \right) =$$

$$= g_{is}\left(\frac{\partial g^{sa}}{\partial x^k}\Gamma_{jl,a} - \frac{\partial g^{sa}}{\partial x^l}\Gamma_{jk,a}\right) + \left(\frac{\partial \Gamma_{jl,i}}{\partial x^k} - \frac{\partial \Gamma_{jk,i}}{\partial x^l}\right) + g_{is}(\Gamma^s_{mk}\Gamma^m_{jl} - \Gamma^s_{ml}\Gamma^m_{jk}) =$$

$$= g_{is}\left(\frac{\partial g^{sa}}{\partial x^k}\Gamma_{jl,a} - \frac{\partial g^{sa}}{\partial x^l}\Gamma_{jk,a}\right) + \frac{1}{2}\left(\frac{\partial^2 g_{li}}{\partial x^k \partial x^j} + \frac{\partial^2 g_{jk}}{\partial x^l \partial x^i} - \frac{\partial^2 g_{jl}}{\partial x^k \partial x^i} - \frac{\partial^2 g_{ki}}{\partial x^l \partial x^j}\right) +$$

$$+ g_{is}(\Gamma^s_{mk}\Gamma^m_{jl} - \Gamma^s_{ml}\Gamma^m_{jk})$$

In the first part, we can replace $\dfrac{\partial g^{sa}}{\partial x^k}$ by $-\Gamma^s_{kr}g^{ra} - \Gamma^a_{kr}g^{sr}$ and $\dfrac{\partial g^{sa}}{\partial x^l}$ by $-\Gamma^s_{lr}g^{ra} - \Gamma^a_{lr}g^{sr}$.
It results

$$g_{is}\left(\frac{\partial g^{sa}}{\partial x^k}\Gamma_{jl,a} - \frac{\partial g^{sa}}{\partial x^l}\Gamma_{jk,a}\right) = g_{is}[\Gamma_{jl,a}(-\Gamma^s_{kr}g^{ra} - \Gamma^a_{kr}g^{sr}) + \Gamma_{jk,a}(\Gamma^s_{lr}g^{ra} + \Gamma^a_{lr}g^{sr})].$$

Continuing,

$$g_{is}\Gamma_{jl,a}(-\Gamma^s_{kr}g^{ra} - \Gamma^a_{kr}g^{sr}) = -g_{is}\Gamma^r_{jl}\Gamma^s_{kr} - \Gamma^a_{ki}\Gamma_{jl,a} =$$

$$= -\Gamma_{kr,i}\Gamma^r_{jl} - \Gamma^a_{ki}\Gamma_{jl,a} = -\Gamma^m_{ki}\Gamma_{jl,m} - \Gamma^m_{jl}\Gamma_{km,i}.$$

Analogously

$$g_{is}\Gamma_{jk,a}(\Gamma^s_{lr}g^{ra} + \Gamma^a_{lr}g^{sr}) = \Gamma^m_{li}\Gamma_{jk,m} + \Gamma^m_{jk}\Gamma_{lm,i},$$

i.e.

$$g_{is}\left(\frac{\partial g^{sa}}{\partial x^k}\Gamma_{jl,a} - \frac{\partial g^{sa}}{\partial x^l}\Gamma_{jk,a}\right) = -\Gamma^m_{ki}\Gamma_{jl,m} - \Gamma^m_{jl}\Gamma_{km,i} + \Gamma^m_{li}\Gamma_{jk,m} + \Gamma^m_{jk}\Gamma_{lm,i}.$$

Next step is to compute

$$g_{is}(\Gamma^s_{mk}\Gamma^m_{jl} - \Gamma^s_{ml}\Gamma^m_{jk}) - \Gamma^m_{ki}\Gamma_{jl,m} - \Gamma^m_{jl}\Gamma_{km,i} + \Gamma^m_{li}\Gamma_{jk,m} + \Gamma^m_{jk}\Gamma_{lm,i}$$

which means

$$\cancel{g_{is}g^{sa}\Gamma_{mk,a}\Gamma^m_{jl}} - \cancel{g_{is}g^{sa}\Gamma_{ml,a}\Gamma^m_{jk}} - \Gamma^m_{ki}\Gamma_{jl,m} - \cancel{\Gamma^m_{jl}\Gamma_{km,i}} + \Gamma^m_{li}\Gamma_{jk,m} + \cancel{\Gamma^m_{jk}\Gamma_{lm,i}},$$

that is

$$g_{mp}(\Gamma^m_{li}\Gamma^p_{jk} - \Gamma^m_{ki}\Gamma^p_{jl}).$$

The formula we obtained, after arranging the indexes, is

$$R_{ijkl} = \frac{1}{2}\left(\frac{\partial^2 g_{il}}{\partial x^j \partial x^k} + \frac{\partial^2 g_{jk}}{\partial x^i \partial x^l} - \frac{\partial^2 g_{jl}}{\partial x^i \partial x^k} - \frac{\partial^2 g_{ik}}{\partial x^j \partial x^l}\right) + g_{mp}(\Gamma_{il}^m \Gamma_{jk}^p - \Gamma_{ik}^m \Gamma_{jl}^p)$$

and allows to prove quickly the following formulas

$$R_{ijkl} = -R_{ijlk};$$
$$R_{ijkl} = -R_{jikl};$$
$$R_{ijkl} = R_{klij};$$
$$R_{ijkl} + R_{iklj} + R_{iljk} = 0.$$

in their intrinsic form.

The last identity, $R_{ijkl} + R_{iklj} + R_{iljk} = 0$ is known as Bianchi's first formula. □

Theorem 5.3.8 *The Ricci tensor is symmetric, that is $R_{ij} = R_{ji}$.*

Proof After multiplying the Bianchi first identity $R_{ijkl} + R_{iklj} + R_{iljk} = 0$ by g^{jl}, using the previous formulas, we obtain

$$g^{lj} R_{jilk} + g^{jl} R_{iklj} - g^{lj} R_{jkli} = 0,$$

that is

$$R_{ilk}^l + g^{jl} R_{iklj} - R_{kli}^l = 0,$$

or simply

$$R_{ik} + g^{jl} R_{iklj} - R_{ki} = 0.$$

If we show that $g^{jl} R_{iklj} = 0$, we complete the proof. We have

$$g^{jl} R_{iklj} = g^{lj} R_{iklj} = -g^{lj} R_{ikjl}.$$

If we take into account that j and l are dummy indexes, we have

$$g^{jl} R_{iklj} = -g^{jl} R_{iklj},$$

therefore $g^{jl} R_{iklj} = 0$, i.e.

$$R_{ik} = R_{ki}.$$

  □

Theorem 5.3.9 (Bianchi's second formula):

$$R_{ijk;l}^s + R_{ikl;j}^s + R_{ilj;k}^s = 0.$$

Proof We start from the covariant derivative formula $R^s_{ijk;l}$, i.e.

$$R^s_{ijk;l} = \frac{\partial R^s_{ijk}}{\partial x^l} + R^a_{ijk}\Gamma^s_{al} - R^s_{mjk}\Gamma^m_{il} - R^s_{imk}\Gamma^m_{jl} - R^s_{ijm}\Gamma^m_{kl},$$

$$R^s_{ikl;j} = \frac{\partial R^s_{ikl}}{\partial x^j} + R^a_{ikl}\Gamma^s_{aj} - R^s_{mkl}\Gamma^m_{ij} - R^s_{iml}\Gamma^m_{jk} - R^s_{ikm}\Gamma^m_{lj}$$

and

$$R^s_{ilj;k} = \frac{\partial R^s_{ilj}}{\partial x^k} + R^a_{ilj}\Gamma^s_{ak} - R^s_{mlj}\Gamma^m_{ik} - R^s_{imj}\Gamma^m_{lk} - R^s_{ilm}\Gamma^m_{kj}.$$

We add the three equalities and we use the obvious equality $R^s_{ijk} = -R^s_{ikj}$. It remains to prove that

$$\frac{\partial R^s_{ijk}}{\partial x^l} + \frac{\partial R^s_{ikl}}{\partial x^j} + \frac{\partial R^s_{ilj}}{\partial x^k} + R^a_{ijk}\Gamma^s_{al} + R^a_{ikl}\Gamma^s_{aj} + R^a_{ilj}\Gamma^s_{ak} = R^s_{mjk}\Gamma^m_{il} + R^s_{mkl}\Gamma^m_{ij} + R^s_{mlj}\Gamma^m_{ik}.$$

Let us focus on $\dfrac{\partial R^s_{ijk}}{\partial x^l} + \dfrac{\partial R^s_{ikl}}{\partial x^j} + \dfrac{\partial R^s_{ilj}}{\partial x^k}$ which means

$$\frac{\partial}{\partial x^l}\left(\frac{\partial \Gamma^s_{ik}}{\partial x^j} - \frac{\partial \Gamma^s_{ij}}{\partial x^k} + \Gamma^s_{aj}\Gamma^a_{ik} - \Gamma^s_{ak}\Gamma^a_{ij}\right) + \frac{\partial}{\partial x^j}\left(\frac{\partial \Gamma^s_{il}}{\partial x^k} - \frac{\partial \Gamma^s_{ik}}{\partial x^l} + \Gamma^s_{ak}\Gamma^a_{il} - \Gamma^s_{al}\Gamma^a_{ik}\right) +$$

$$+ \frac{\partial}{\partial x^k}\left(\frac{\partial \Gamma^s_{ij}}{\partial x^l} - \frac{\partial \Gamma^s_{il}}{\partial x^j} + \Gamma^s_{al}\Gamma^a_{ij} - \Gamma^s_{aj}\Gamma^a_{il}\right),$$

that is

$$\frac{\partial}{\partial x^l}\left(\Gamma^s_{aj}\Gamma^a_{ik} - \Gamma^s_{ak}\Gamma^a_{ij}\right) + \frac{\partial}{\partial x^j}\left(\Gamma^s_{ak}\Gamma^a_{il} - \Gamma^s_{al}\Gamma^a_{ik}\right) + \frac{\partial}{\partial x^k}\left(\Gamma^s_{al}\Gamma^a_{ij} - \Gamma^s_{aj}\Gamma^a_{il}\right).$$

If we continue computing and if we add the missing part

$$R^a_{ijk}\Gamma^s_{al} + R^a_{ikl}\Gamma^s_{aj} + R^a_{ilj}\Gamma^s_{ak}$$

we obtain

$$\frac{\partial \Gamma^s_{aj}}{\partial x^l}\Gamma^a_{ik} + \frac{\partial \Gamma^a_{ik}}{\partial x^l}\Gamma^s_{aj} - \frac{\partial \Gamma^s_{ak}}{\partial x^l}\Gamma^a_{ij} - \frac{\partial \Gamma^a_{ij}}{\partial x^l}\Gamma^s_{ak} + \frac{\partial \Gamma^s_{ak}}{\partial x^j}\Gamma^a_{il} + \frac{\partial \Gamma^a_{il}}{\partial x^j}\Gamma^s_{ak} - \frac{\partial \Gamma^s_{al}}{\partial x^j}\Gamma^a_{ik} - \frac{\partial \Gamma^a_{ik}}{\partial x^j}\Gamma^s_{al} +$$

$$+ \frac{\partial \Gamma^s_{al}}{\partial x^k}\Gamma^a_{ij} + \frac{\partial \Gamma^a_{ij}}{\partial x^k}\Gamma^s_{al} - \frac{\partial \Gamma^s_{aj}}{\partial x^k}\Gamma^a_{il} - \frac{\partial \Gamma^a_{il}}{\partial x^k}\Gamma^s_{aj} + R^a_{ijk}\Gamma^s_{al} + R^a_{ikl}\Gamma^s_{aj} + R^a_{ilj}\Gamma^s_{ak}$$

which can be successively arranged as

$$\Gamma^s_{al}\left(R^a_{ijk} - \frac{\partial \Gamma^a_{ik}}{\partial x^j} + \frac{\partial \Gamma^a_{ij}}{\partial x^k}\right) + \Gamma^s_{aj}\left(R^a_{ikl} - \frac{\partial \Gamma^a_{il}}{\partial x^k} + \frac{\partial \Gamma^a_{ik}}{\partial x^l}\right) + \Gamma^s_{ak}\left(R^a_{ilj} - \frac{\partial \Gamma^a_{ij}}{\partial x^l} + \frac{\partial \Gamma^a_{il}}{\partial x^j}\right) +$$

$$+\Gamma^a_{ik}\left(\frac{\partial \Gamma^s_{aj}}{\partial x^l} - \frac{\partial \Gamma^s_{al}}{\partial x^j}\right) + \Gamma^a_{ij}\left(\frac{\partial \Gamma^s_{al}}{\partial x^k} - \frac{\partial \Gamma^s_{ak}}{\partial x^l}\right) + \Gamma^a_{il}\left(\frac{\partial \Gamma^s_{ak}}{\partial x^j} - \frac{\partial \Gamma^s_{aj}}{\partial x^k}\right) =$$

$$= \Gamma^s_{al}\left(\Gamma^a_{mj}\Gamma^m_{ik} - \Gamma^a_{mk}\Gamma^m_{ij}\right) + \Gamma^s_{aj}\left(\Gamma^a_{mk}\Gamma^m_{il} - \Gamma^a_{ml}\Gamma^m_{ik}\right) + \Gamma^s_{ak}\left(\Gamma^a_{ml}\Gamma^m_{ij} - \Gamma^a_{mj}\Gamma^m_{il}\right) +$$

$$+\Gamma^m_{ik}\left(\frac{\partial \Gamma^s_{mj}}{\partial x^l} - \frac{\partial \Gamma^s_{ml}}{\partial x^j}\right) + \Gamma^m_{ij}\left(\frac{\partial \Gamma^s_{ml}}{\partial x^k} - \frac{\partial \Gamma^s_{mk}}{\partial x^l}\right) + \Gamma^m_{il}\left(\frac{\partial \Gamma^s_{mk}}{\partial x^j} - \frac{\partial \Gamma^s_{mj}}{\partial x^k}\right) =$$

$$\Gamma^m_{ik}\left(\frac{\partial \Gamma^s_{mj}}{\partial x^l} - \frac{\partial \Gamma^s_{ml}}{\partial x^j} + \Gamma^s_{al}\Gamma^a_{mj} - \Gamma^s_{aj}\Gamma^a_{ml}\right) + \Gamma^m_{ij}\left(\frac{\partial \Gamma^s_{ml}}{\partial x^k} - \frac{\partial \Gamma^s_{mk}}{\partial x^l} + \Gamma^s_{ak}\Gamma^a_{ml} - \Gamma^s_{al}\Gamma^a_{mk}\right)$$

$$+\Gamma^m_{il}\left(\frac{\partial \Gamma^s_{mk}}{\partial x^j} - \frac{\partial \Gamma^s_{mj}}{\partial x^k} + \Gamma^s_{aj}\Gamma^a_{mk} - \Gamma^s_{ak}\Gamma^a_{mj}\right) = R^s_{mlj}\Gamma^m_{ik} + R^s_{mkl}\Gamma^m_{ij} + R^s_{mjk}\Gamma^m_{il}.$$

□

Theorem 5.3.10 (The covariant derivative of Einstein's tensor formula) *If g_{ij} is the (0, 2) metric tensor, g^{ij} its inverse contravariant (2, 0) tensor, R_{ij} is the Ricci tensor, $R := R^s_s$ is the curvature scalar derived from the (1, 1) mixed tensor $R^i_j = g^{is}R_{sj}$, it is possible to define the Einstein tensor*

$$E_{ij} := R_{ij} - \frac{1}{2}R \cdot g_{ij},$$

then, we have

$$E_{ij;a} = 0.$$

Proof First of all, let us observe that R can be written as $g^{ij}R_{ij}$. Indeed,

$$g^{ij}R_{ij} = g^{ij}R_{ji} = R^i_i = R.$$

We have to prove that the covariant derivative of the Einstein tensor is null, i.e.

$$\left(R_{ij} - \frac{1}{2}R \cdot g_{ij}\right)_{;a} = 0.$$

We start from Bianchi's formula

$$R^s_{ijk;l} + R^s_{ikl;j} + R^s_{ilj;k} = 0.$$

We contract the indexes $j = s$. It results

$$R^s_{isk;l} + R^s_{ikl;s} + R^s_{ils;k} = 0$$

and we use $R_{ik;l} = R^s_{isk;l}$, $R^s_{ils;k} = -R^s_{isl;k} = -R_{il;k}$. We obtain

$$R_{ik;l} + R^s_{ikl;s} - R_{il;k} = 0.$$

Using the fact that the covariant derivative of g^{ij} is null, we can write

$$(g^{ia} R_{ik})_{;l} + (g^{ia} R^s_{ikl})_{;s} - (g^{ia} R_{il})_{;k} = 0,$$

i.e.

$$R^a_{k;l} + (g^{ia} R^s_{ikl})_{;s} - R^a_{l;k} = 0.$$

We contract $a = l$ and we have

$$R^a_{k;a} + (g^{ia} R^s_{ika})_{;s} - R^a_{a;k} = 0.$$

Now, $g^{ia} R^s_{ika} = g^{ia} g^{sb} R_{bika} = g^{ia} g^{sb} R_{ibak} = g^{sb} g^{ia} R_{ibak} = g^{sb} R^a_{bak} = R^s_k$ and we replace in the previous equality.
It results $R^a_{k;a} + R^s_{k;s} - R_{;k} = 0$, that is $2R^a_{k;a} - R_{;k} = 0$, which can be written in the form

$$\left(R^a_k - \frac{1}{2} \delta^a_k R \right)_{;a} = 0.$$

We use that the covariant derivative of the metric tensor is null, then we have

$$\left(g_{ma} R^a_k - \frac{1}{2} \delta^a_k g_{ma} R \right)_{;a} = 0,$$

that is

$$\left(R_{mk} - \frac{1}{2} R g_{mk} \right)_{;a} = 0.$$

$\Box$

One comment. The Einstein tensor $E_{ij} := R_{ij} - \frac{1}{2} R \cdot g_{ij}$ has the property

$$E_{ij;a} = 0.$$

The Einstein field equations in General Relativity are

$$R_{ij} - \frac{1}{2} R \cdot g_{ij} = k T_{ij},$$

where $k = \dfrac{8\pi G}{c^4}$ is a constant and T_{ij} is the so called stress-energy tensor, a tensor satisfying the same null covariant derivative equality,

$$T_{ij;a} = 0.$$

If we are looking at the left member we see somehow the Geometry of the space expressed in terms of tensors; in the right member there is a tensor which depends on mass and energy. The equality shows that the mass and the energy are creating the geometric structure of the spacetime. All these things will be better understood later when we construct all the "ingredients" of the theory.

Chapter 6
Non-Euclidean Geometries and Their Physical Interpretation

Gutta cavat lapidem.

Ovidius

At the end of this chapter, the big picture towards Relativity will emerge. Before discussing all the details and the proofs, we intend to sketch it now. The most known models for Non-Euclidean Geometry are the Poincaré disk model and the Poincaré half-plane model. Another related model, the exterior disk model, can be figured and presented. Two other models will be highlighted: the hemisphere model and the hyperboloid model. The first three models are connected among them by inversion. Two models have distances which can be described by a general principle of metrization; the distance between two points is

$$d(A, B) = \ln \frac{\max_{P \in K} g_{AB}(P)}{\min_{P \in K} g_{AB}(P)} = \ln \frac{\max_{P \in K} \frac{|PA|}{|PB|}}{\min_{P \in K} \frac{|PA|}{|PB|}},$$

where the set K and the set J have to be specified. In the case of the exterior of the disk, it is a good exercise for the reader to check that a similar construction works. Therefore all three models are endowed with a distance constructed by this special procedure.

The Poincaré Disk Model

Let us consider the circle $C(O, 1)$ having $O(0, 0)$ as a center and $r = 1$ as a radius. The interior of $C(O, 1)$,

$$D_n = int\, C(O, 1) = \{(x, y) \in E^2 | x^2 + y^2 < 1\}$$

is the Poincaré disk and in the same time the "plane" of the Non-Euclidean Geometry.

© The Editor(s) (if applicable) and The Author(s), under exclusive license to Springer Nature Switzerland AG 2020
W. Boskoff and S. Capozziello, *A Mathematical Journey to Relativity*, UNITEXT for Physics, https://doi.org/10.1007/978-3-030-47894-0_6

The points of the disk are the points of the Non-Euclidean Geometry. The lines of the Non-Euclidean Geometry are the orthogonal arcs to $C(O, 1)$. Why? Because it is easy to check that, for three points A, B, C in this order on an orthogonal arc, the following equality

$$d(A, C) = d(A, B) + d(B, C),$$

holds. It means that the orthogonal arcs are geodesics with respect to the distance d. An important particular case is related to diameters which are also geodesics in this Non-Euclidean Geometry inside the disk. $C(O, 1) = \partial D_n$ is the infinity domain of this Non-Euclidean Geometry. Why? According to the theory we presented, in the case when $A = O(0, 0)$ and $B = B(x, 0)$, the distance before becomes

$$d(O, B) = \frac{1}{2} \cdot \ln \frac{1-x}{1-0} : \frac{-1-x}{-1-0},$$

and when x approaches 1, $d(O, B)$ approaches to ∞.

In this Non-Euclidean Geometry, we see that, from a "point E" which does not belong to a "line l", there are, at least, "two lines d_1, d_2" i.e. two orthogonal arcs to $C(O, 1)$, such that

$$d_i \cap l = \emptyset, \ i = 1, 2.$$

What can we say more? The previous distance d induces a metric, which, in the case of the interior of the disk, is

$$ds^2 = \frac{4}{[1 - (x^2 + y^2)]^2} (dx^2 + dy^2).$$

The geodesics with respect this metric are the same orthogonal arcs to $C(O, 1)$. Therefore the geodesics of the metric coincide with the geodesic of the Poincaré distance. The Gaussian curvature of this metric is -1.

At this point, it is possible to offer a simple explanation regarding to the fact that this Poincaré Non-Euclidean Geometry inside the disk is also called Ponicaré Hyperbolic Geometry. In fact, the Poincaré distance can be expressed by a hyperbolic function, and this will happen for all the other non-Euclidean models. Let us look at the above formula

$$d(O, B) = \frac{1}{2} \cdot \ln \frac{1-x}{1-0} : \frac{-1-x}{-1-0}.$$

It results $\frac{1+x}{1-x} = e^{2d(O,B)}$, that is

$$d(O, B) = \tanh^{-1} x.$$

The Poincaré Half-Plane Model

The Poincaré half-plane model can be seen as simple as the disk model. The "plane" of the half-plane model is the set $H^2 := \{(x, y) \in E^2 | y > 0\}$. The infinity domain, in this case, is the line $y = 0$. The distance has the same form as in the case of disk model,

$$d(A, B) = \ln \frac{\max_{P \in K} g_{AB}(P)}{\min_{P \in K} g_{AB}(P)} = \ln \frac{\max_{P \in K} \frac{|PA|}{|PB|}}{\min_{P \in K} \frac{|PA|}{|PB|}},$$

where $K := \{(x, y), \ y = 0\}$ and $A, B \in J := H^2$. The previous distance induces the metric of the superior half-plane H^2, i.e. the Poincaré metric

$$ds^2 = \frac{1}{y^2}(dx^2 + dy^2).$$

The "lines" of the model are the semicircles centered on the $y = 0$ axis or they are half-lines Euclidean perpendicular to $y = 0$. It is easy to check that, for three points A, B, C in this order on a semicircle (or on an orthogonal line to the infinity domain), we have

$$d(A, C) = d(A, B) + d(B, C).$$

It means that the semicircles and the orthogonal half-lines to $y = 0$ are geodesics with respect to the distance d. As we saw already, they come by inversion from lines of the disk model. The same objects are geodesics with respect to the Poincaré half-plane metric.

Even if the two metrics can be deduced from the corresponding Poincaré distances using formula

$$ds^2 = \frac{1}{4}\left(\frac{1}{R_1} + \frac{1}{r_1}\right)^2 (dx_1^2 + dx_2^2),$$

they can be found starting from one of them and applying the inversion coordinate change formula for the coefficients g_{ij}.

Two "lines" of this model are non-secant lines if there is no point of intersection between them. Exactly as in the disk model, from a "point E" which does not belong to a "line l", there are at least "two lines d_1, d_2", such that $d_i \cap l = \emptyset$, $i = 1, 2$. Since the inversion preserves the angles between curves and, using the property of the sum of angles in the disk-model, we deduce that the sum of angles of a triangle, in this model, is strictly less than π.

For two points A, B on a line, $d(A, B)$ approaches ∞ when B approaches the endpoint S of the "line" which belongs to $y = 0$.

There exists another model of Non-Euclidean Geometry developed outside the disk. All the results and considerations available for this two models are available in the model on the set $ext C(O, 1) = \{(x, y) \in E^2 | x^2 + y^2 > 1\}$.

A stereographic projection is necessary to migrate from the half-plane model to the hemisphere model, another stereographic projection is used to migrate from the hemisphere model to the hyperboloid model. Shortly, the hemisphere model will be described by the set

$$H_+ = \{(y_1, y_2, y_3) \in S^2 \mid y_1^2 + y_2^2 + y_3^2 = 1, \ y_3 > 0\}$$

endowed with the metric

$$ds_{H_+}^2 := \frac{1}{y_3^2} \left(dy_1^2 + dy_2^2 + dy_3^2 \right)$$

and the hyperboloid model will be described by the set

$$\mathbb{H} := \{(x_1, x_2, x_3) \in E^3 \mid x_1^2 + x_2^2 - x_3^2 = -1, \ x_3 > 0\}$$

endowed with the metric
$$ds_{\mathbb{H}}^2 = dx_1^2 + dx_2^2 - dx_3^2.$$

As you can see, and this is remarkable, this last model describes, through a Minkowski metric, the Non-Euclidean Geometry. Some results of this chapter come from [11, 12].

6.1 Poincaré Distance and Poincaré Metric of the Disk

We proceed to prove all the facts asserted in "the big picture" before. Next theorem allows us to provide a metric starting from the special distance naturally attached to the Poincaré disk model.

Theorem 6.1.1 (Barbilian's Theorem) *Let K and J be two subsets of the Euclidean plane $\mathbb{R}^2$, and $K = \partial J$. Consider the influence $f(M, A) = |MA|$, where, by $|MA|$, we denote the Euclidean distance. Consider*

$$g_{AB}(M) = \frac{f(M, A)}{f(M, B)} = \frac{|MA|}{|MB|},$$

and consider the semi-distance induced on J by the metrization procedure

$$d(A, B) = \ln \frac{\max_{P \in K} g_{AB}(P)}{\min_{P \in K} g_{AB}(P)}.$$

Suppose furthermore that, for $M \in K$, the extrema $\max g_{AB}(M)$ and $\min g_{AB}(M)$ for any A and B in J are reached each for a single point in K. Then:

(i) *For any $A \in J$ and any line d passing through A there exist exactly two circles tangent to K and also to d at A.*

(ii) *The metric induced by the previous distance has the form*

$$ds^2 = \frac{1}{4} \left(\frac{1}{R_1} + \frac{1}{r_1} \right)^2 (dx_1^2 + dx_2^2),$$

where R and r are the radii of the circles described in (i).

Proof Consider $A\,(x_1, x_2)$ and $B\,(y_1, y_2)$ in J and $M\,(x^1, x^2)$ in $J \cup K$.

The circle determined by the relation $\dfrac{|MA|}{|MB|} = \sqrt{\lambda}$ has the equation

$$\left(x^1 - x_1\right)^2 + \left(x^2 - x_2\right)^2 - \lambda(\left(x^1 - y_1\right)^2 + \left(x^2 - y_2\right)^2) = 0.$$

Its radius $\mathcal{R}$ is

$$\mathcal{R}^2 = \frac{\lambda |AB|^2}{(1 - \lambda)^2}.$$

The maximum M_1 and the minimum m_1 values for the expression $\dfrac{|MA|^2}{|MB|^2}$ lead to the equalities

$$R_1^2 = \frac{M_1}{(1 - M_1)^2} |AB|^2, \quad r_1^2 = \frac{m_1}{(1 - m_1)^2} |AB|^2.$$

The first equality becomes

$$\left(\frac{1 + M_1}{1 - M_1} \right)^2 = \frac{|AB|^2 + 4R_1^2}{|AB|^2},$$

and taking into account that $M_1 \geq 1$, it results

$$M_1 = 1 + \frac{2|AB|}{-|AB| + \sqrt{|AB|^2 + 4R_1^2}}.$$

In the same way, using $m_1 \leq 1$, we have

$$m_1 = 1 - \frac{2|AB|}{|AB| + \sqrt{|AB|^2 + 4r_1^2}}.$$

If A and B are close enough, i.e. $B = A + dA$, the Euclidean distance $|AB|^2$ becomes the arc element

$$d\sigma^2 = dx_1^2 + dx_2^2.$$

The distance between the points A and $A + dA$ leads to the new arc element $d(A, A + dA)$ denoted by ds.

So,

$$ds = d(A, A + dA) = \frac{1}{2} \frac{M_1 - m_1}{m_1}.$$

We have the approximations

$$\frac{2d\sigma}{-d\sigma + \sqrt{d\sigma^2 + 4R_1^2}} = \frac{d\sigma}{R_1},$$

and

$$\frac{2d\sigma}{d\sigma + \sqrt{d\sigma^2 + 4r_1^2}} = \frac{d\sigma}{r_1}.$$

Final computation leads to

$$ds = \frac{1}{2} \left(\frac{1}{R_1} + \frac{1}{r_1} \right) d\sigma,$$

i.e. the metric corresponding to the previous distance is

$$ds^2 = \frac{1}{4} \left(\frac{1}{R_1} + \frac{1}{r_1} \right)^2 (dx_1^2 + dx_2^2).$$

□

Theorem 6.1.2 *Consider the circle Γ centered at origin and of radius R. Consider in the interior of the circle the Poincaré distance. Then, the associated metric, given by Barbilian's Theorem*

$$ds^2 = \frac{1}{4} \left(\frac{1}{R_1} + \frac{1}{r_1} \right)^2 (dx_1^2 + dx_2^2)$$

has the form

$$ds^2 = \frac{4R^2}{[R^2 - (x^2 + y^2)]^2} \cdot (dx^2 + dy^2).$$

Furthermore, the metric obtained by this procedure has the Gaussian curvature -1.

Proof In the case when Γ is a circle and J is its interior, we deal with a distance,

$$d(A, B) = \ln \frac{\max_{P \in \Gamma} g_{AB}(P)}{\min_{P \in \Gamma} g_{AB}(P)} = \ln \frac{\max_{P \in \Gamma} \frac{|PA|}{|PB|}}{\min_{P \in \Gamma} \frac{|PA|}{|PB|}},$$

called Poincaré distance of the disk. We would like to compute the metric of the disk induced by this distance and the previous theorem. Let A of coordinates (x_0, y_0), in the interior of Γ. Denote by $O_1(x_1, y_1)$ and $O_2(x_2, y_2)$ the centers of the two circles, each one tangent to the circle Γ and also tangent between them at A. Denote by m the slope of the tangent line Δ at A to both previous circles. Line $O_1 O_2$ has the equation

$$y - y_0 = -\frac{1}{m}(x - x_0).$$

Therefore, the points O_1 and O_2 have the coordinates $\left(x_i, y_0 - \frac{1}{m}(x_i - x_0)\right)$, for $i = 1, 2$. Furthermore,

$$R_i^2 = |O_i A|^2 = \frac{m^2 + 1}{m^2}(x_i - x_0)^2, \quad i = 1, 2.$$

Without losing of generality, we assume that $x_1 - x_0 \leq 0$ and $x_2 - x_0 \geq 0$, with the equality case reached when $\Delta \| Ox$. it is worth remarking that $x_1 - x_0 < 0$, if $m > 0$. Thus

$$|O_1 A| = \frac{\sqrt{m^2 + 1}}{m}(x_0 - x_1),$$

and

$$|O_2 A| = \frac{\sqrt{m^2 + 1}}{m}(x_2 - x_0).$$

Therefore, the circles have the centers $(x_1, y_0 - \frac{1}{m}(x_1 - x_0))$ and $(x_2, y_0 - \frac{1}{m}(x_2 - x_0))$, and the radii $R_1 = \frac{\sqrt{m^2 + 1}}{m}(x_0 - x_1)$ and $R_2 = \frac{\sqrt{m^2 + 1}}{m}(x_2 - x_0)$.

To obtain the coordinates of the point T_1', we recall that it lies at the intersection between the circle $x^2 + y^2 = R^2$ and the line

$$y = \frac{1}{x_1}\left[y_0 - \frac{1}{m}(x_1 - x_0)\right]x,$$

which passes through the collinear points O, O_1 and T_1'. Solving the system, we get the coordinates of T_1' as follows

$$\left(\frac{Rx_1}{\sqrt{x_1^2 + \left(y_0 - \frac{1}{m}(x_1 - x_0)\right)^2}}, \frac{R\left(y_0 - \frac{1}{m}(x_1 - x_0)\right)}{\sqrt{x_1^2 + \left(y_0 - \frac{1}{m}(x_1 - x_0)\right)^2}}\right).$$

By direct computation, we get

$$|O_1 T_1'| = R - \sqrt{x_1^2 + \left(y_0 - \frac{1}{m}(x_1 - x_0)\right)^2}.$$

Since the segments $O_1 T_1'$ and $O_1 A$ are radii of the circle of center O_1 and radius R_1, we set up the equalities

$$\frac{\sqrt{m^2 + 1}}{m}(x_0 - x_1) = R_1 = R - \sqrt{x_1^2 + (y_0 - \frac{1}{m}(x_1 - x_0))^2}.$$

It follows that

$$x_0 - x_1 = \frac{R_1 m}{\sqrt{m^2 + 1}},$$

therefore

$$(R - R_1)^2 = x_1^2 + \left(y_0 + \frac{R_1}{\sqrt{m^2 + 1}}\right)^2.$$

Since

$$x_1 = x_0 - \frac{R_1 m}{\sqrt{m^2 + 1}},$$

we have

$$(m^2 + 1)(R - R_1)^2 - (y_0\sqrt{m^2 + 1} + R_1)^2 = (x_0\sqrt{m^2 + 1} - R_1 m)^2$$

i.e.

$$R_1 = \frac{\sqrt{m^2 + 1}}{2} \cdot \frac{R^2 - x_0^2 - y_0^2}{R\sqrt{m^2 + 1} - x_0 m + y_0}.$$

In a similar way we obtain

$$R_2 = \frac{\sqrt{m^2 + 1}}{2} \cdot \frac{R^2 - x_0^2 - y_0^2}{R\sqrt{m^2 + 1} + x_0 m - y_0}.$$

It results the relation

$$\frac{1}{4}\left(\frac{1}{R_1} + \frac{1}{R_2}\right)^2 = \frac{4R^2}{(R^2 - x_0^2 - y_0^2)^2},$$

i.e. the *Poincaré metric of the disk* is

$$ds^2 = \frac{4R^2}{[R^2 - (x^2 + y^2)]^2} \cdot (dx^2 + dy^2).$$

By straightforward computation, we can easily see that the Gaussian curvature of this metric is $K(x, y) = -1$.

Therefore this metric generates the hyperbolic Geometry on the disk $D(O, R) = int\,\Gamma$. □

6.2 Poincaré Distance and Poincaré Metric of the Half-Plane

Consider the case when $K = \{(x, y) \in \mathbb{R}^2; y = 0\}$ and $J = \{(x, y) \in \mathbb{R}^2; y > 0\}$. Let $M \in K$ and $A(x_0, y_0) \in J$ and $B(x_1, y_1) \in J$. Consider the associated ratio

$$g_{AB}(M) = \frac{|MA|}{|MB|}.$$

We describe geometrically the points where the maximum and the minimum are reached. Let B_1 be the foot of the perpendicular drawn from B to x-axis. Consider the direct inversion with pole B and power $|BB_1|^2$.

Then, by this inversion, we have $B_1 \to B_1$ and $K \to C(BB_1)$, where we denote by $C(BB_1)$ the circle of diameter BB_1; we also have $A \to A'$ such that $|BA| \cdot |BA'| = |BB_1|^2$, where A' is a fixed point lying on the line AB.

Since any point $M \in K$ is mapped into $M' \in C(BB_1)$, we have that

$$|A'M'| = |BB_1|^2 \cdot \frac{|AM|}{|BA| \cdot |BM|};$$

Denote the constant ratio $k = \dfrac{|BB_1|^2}{|BA|}$, therefore $|A'M'| = k \cdot \dfrac{|AM|}{|BM|}$.

Let us remark that the ratio $\dfrac{|AM|}{|BM|}$ reaches its maximum or its minimum whenever $|A'M'|$ is maximum or minimum, respectively. Therefore, the antipodal points S_0 and S_0', bounding the diameter through A', lie diametrically opposite on $C(AA_1)$. Their inverse images, the points $\{M_0'\} = AS_0' \cap (y = 0)$, and $\{M_0\} = AS_0 \cap (y = 0)$, are the points where the minimum and the maximum of the ratio $\dfrac{|AM|}{|BM|}$ are reached, respectively.

Since $S_0 S_0'$ is orthogonal to the circle $C(AA_1)$ and since $B' \in S_0 S_0'$, we get that M_0 and M_0' are the endpoints of the arc twice orthogonal onto K passing through A and B. Thus, it makes sense to consider the distance

$$d(A, B) = \ln \frac{\max_{P \in K} g_{AB}(P)}{\min_{P \in K} g_{AB}(P)} = \ln \frac{\max_{P \in K} \frac{|PA|}{|PB|}}{\min_{P \in K} \frac{|PA|}{|PB|}},$$

for the half-plane J, since the two extrema needed in the logarithmic oscillation formula exist. So, in the half plane, it exists a Poincaré distance obtained exactly as in the case of the disk, that is

$$d(A, B) = \ln \frac{\max_{P \in K} \frac{|PA|}{|PB|}}{\min_{P \in K} \frac{|PA|}{|PB|}}.$$

Theorem 6.2.1 *Consider the previous Poincaré distance of the half plane. Then, the associated metric given by Barbilian's theorem*

$$ds^2 = \frac{1}{4} \left(\frac{1}{R_1} + \frac{1}{r_1} \right)^2 (dx_1^2 + dx_2^2)$$

has the form

$$ds^2 = \frac{1}{y^2} (dx^2 + dy^2).$$

(this important Riemannian metric is called *Poincaré metric of the half-plane.*)

Proof Let $A(x_0, y_0)$ and the arbitrary line through A given by $y - y_0 = m(x - x_0)$. Consider the circles of centers $O_1(x_1, y_1)$ and $O_2(x_2, y_2)$ tangents at A to the line and also tangents to K. Then we have:

$$y_1 - y_0 = -\frac{1}{m}(x_1 - x_0),$$

$$y_2 - y_0 = -\frac{1}{m}(x_2 - x_0),$$

$$y_1^2 = (x_1 - x_0)^2 + (y_1 - y_0)^2,$$

$$y_2^2 = (x_2 - x_0)^2 + (y_2 - y_0)^2, \quad y_1 < y_0, \ y_2 > y_0.$$

From the previous equations, we have

$$r_1 = y_1 = \frac{y_0 \sqrt{m^2 + 1}}{1 + \sqrt{m^2 + 1}}.$$

Similarly we obtain

$$r_2 = y_2 = \frac{y_0 \sqrt{m^2 + 1}}{-1 + \sqrt{m^2 + 1}},$$

therefore it results

$$ds^2 = \frac{1}{y^2}(dx^2 + dy^2).$$

☐

The Poincaré metric of the half-plane is a Riemann metric with constant Gaussian curvature -1. We left this as an exercise to the reader.

6.3 Connections Between the Models of Non-Euclidean Geometries

Suppose we are in the two dimensional Euclidean plane E^2 and let us consider the circle $C(T, 1)$ where the center T has the coordinates $(0, -1)$ and the length of the radius is 1. The point $A(0, -2)$ belongs to the circle, $B(x, y)$ belongs to the superior half-plane H^2, i.e. $y > 0$, and its inverse $B^*(x^*, y^*)$, with respect to the inversion I having A as pole and $\mu = 4$ as power, has the coordinates

$$x^*(x, y) = \frac{4x}{x^2 + (y + 2)^2},$$

$$y^*(x, y) = \frac{-2(x^2 + y^2 + 2y)}{x^2 + (y + 2)^2}.$$

We obtain this result using the fact that the algebraic equation of the line AB is

$$Y + 2 = \frac{y + 2}{x}X$$

and the collinear points A, B, B^* fulfill the condition $AB \cdot AB^* = 4$.
   Another easy computation shows that

$$(x^*)^2 + (y^* + 1)^2 < 1,$$

that is B^* belongs to the interior of the circle $C(T, 1)$. The mapping $M : H^2 \rightarrow int\,C(T, 1)$, which describes the change of coordinates $(x, y) \rightarrow (x^*, y^*)$, has the differential dM given by the formula

$$dM = \begin{pmatrix} \dfrac{\partial x^*}{\partial x} & \dfrac{\partial x^*}{\partial y} \\ \dfrac{\partial y^*}{\partial x} & \dfrac{\partial y^*}{\partial y} \end{pmatrix}$$

where

$$\frac{\partial x^*}{\partial x} = \frac{4[-x^2 + (y+2)^2]}{(x^2 + (y+2)^2)^2}; \quad \frac{\partial x^*}{\partial y} = \frac{-8x(y+2)}{[x^2 + (y+2)^2]^2};$$

$$\frac{\partial y^*}{\partial x} = \frac{-8x(y+2)}{[x^2 + (y+2)^2]^2}; \quad \frac{\partial y^*}{\partial y} = \frac{4[x^2 - (y+2)^2]}{[x^2 + (y+2)^2]^2}.$$

We observe that $dM = dM^t$.

The metric of the unit disk in coordinates (x^*, y^*) is

$$ds^2 = \frac{4}{[1 - (x^*)^2 - (y^*)^2]^2} \cdot ((dx^*)^2) + (dy^*)^2).$$

In our case, the disk is translated such that $O(0, 0) \rightarrow T(0, -1)$, i.e. $x^* \rightarrow x^*$, $y^* \rightarrow y^* + 1$. The metric becomes

$$ds^2 = \frac{4}{[1 - (x^*)^2 - (y^* + 1)^2]^2} \cdot ((dx^*)^2 + (dy^*)^2)$$

We transform the matrix

$$\bar{G}(x^*, y^*) = \begin{pmatrix} \dfrac{4}{[1 - (x^*)^2 - (y^* + 1)^2]^2} & 0 \\ 0 & \dfrac{4}{[1 - (x^*)^2 - (y^* + 1)^2]^2} \end{pmatrix}$$

with respect (x, y) coordinates into

$$\bar{G}(x, y) = \begin{pmatrix} \dfrac{[x^2 + (y+2)^2]^2}{16y^2} & 0 \\ 0 & \dfrac{[x^2 + (y+2)^2]^2}{16y^2} \end{pmatrix}$$

According to the general theory, the matrix of the metric in H^2, induced by changing of coordinates and by the metric of the interior of $C(T, 1)$, is computed using the formula

$$G(x, y) = dM^t \cdot \bar{G}(x, y) \cdot dM.$$

After computations, we obtain

$$G(x, y) = \begin{pmatrix} \dfrac{1}{y^2} & 0 \\ 0 & \dfrac{1}{y^2} \end{pmatrix}.$$

We have proved the following

Theorem 6.3.1 *The Poincaré metric of the disk induces, via a change of coordinates expressed by an appropriate inversion, the metric of the superior half-plane*

$$ds^2 = \frac{1}{y^2}(dx^2 + dy^2).$$

Definition 6.3.2 The previous metric is called the Poincaré metric of the superior half plane H^2.

Let us make a short review of what we are expecting. Previously we proved that if A, B, C are three points in this order on an orthogonal arc sS, such that the order on the arc in the interior of the circle $C(T, 1)$ is s, A, B, C, S, then $d(A, B) + d(B, C) = d(A, C)$ where d is the Poincaré distance. Therefore the orthogonal arc sS is a geodesic of the Poincaré distance on the disk. The metric induced in the interior of $C(T, 1)$ is

$$ds^2 = \frac{4}{\left[1 - (x^*)^2 - (y^* + 1)^2\right]^2} \cdot ((dx^*)^2) + (dy^*)^2).$$

Is the orthogonal arc sS a geodesic of this metric?

To underline what we are trying to find out, let ask ourselves again: is the orthogonal arc sS in the same time geodesic with respect the distance and geodesic with respect to the metric induced by the distance?

We know that this metric induces, in the superior half-plane H^2, the coordinates $x = x(x^*, y^*)$, $y = y(x^*, y^*)$ and the Poincaré metric

$$ds^2 = \frac{1}{y^2}(dx^2 + dy^2).$$

The inversion $I(A, 4)$ maps the orthogonal arc sS into an ordinary semicircle having the center on the Ox axis or into a line orthogonal to the Ox axis.

Are these geometric objects geodesics with respect to the Poincaré metric of the half-plane?

If yes, the orthogonal arc sS becomes a geodesic with respect the hyperbolic metric of the translated disk because, under a change of coordinates, geodesics are mapped into geodesics as we have proved.

Theorem 6.3.3 *The semicircles $(x - c)^2 + y^2 = R^2$ and the lines $x = a$ are geodesics with respect the Poincaré metric of the half-plane.*

Proof We have $g_{11} = g_{22} = \frac{1}{y^2}$, $g_{12} = g_{21} = 0$, $g^{11} = g^{22} = y^2$, $g^{12} = g^{21} = 0$. Then considering $x = x^1$, $y = x^2$ the Christoffel symbols are

$$\Gamma_{11,1} = \Gamma_{22,1} = \Gamma_{12,2} = \Gamma_{21,2} = 0, \ \Gamma_{12,1} = \Gamma_{21,1} = \Gamma_{22,2} = -\frac{1}{y^3}, \ \Gamma_{11,2} = \frac{1}{y^3},$$

$$\Gamma^1_{11} = \Gamma^1_{22} = \Gamma^2_{12} = \Gamma^2_{21} = 0, \ \Gamma^1_{12} = \Gamma^1_{21} = \Gamma^2_{22} = -\frac{1}{y}, \ \Gamma^2_{11} = \frac{1}{y}.$$

The geodesic equation written for the first variable is

$$\ddot{x} - \frac{2}{y}\dot{x}\dot{y} = 0.$$

For the second variable y the geodesic equation is

$$\ddot{y} + \frac{1}{y}\dot{x}^2 - \frac{1}{y}\dot{y}^2 = 0.$$

The appropriate parameterization for the semicircle is

$$x = x(s) = c + R \tanh s; \ \ y = y(s) = \frac{R}{\cosh s}$$

instead of

$$x = x(s) = c + R \cos s; \ \ y = y(s) = R \sin s.$$

Why? Because the first formulas lead to a constant speed on the semicircle, the second formulas lead to a variable speed. Indeed,

$$\dot{x}(s) = \frac{R}{\cosh^2 s}, \ \ddot{x}(s) = \frac{-2R \tanh s}{\cosh^2 s}, \ \dot{y}(s) = \frac{-R \sinh s}{\cosh^2 s}, \ \ddot{y}(s) = \frac{R}{\cosh s} - \frac{2R}{\cosh^3 s}.$$

It results that the length of the speed vector $\dot{c}(s) = (\dot{x}(s), \dot{y}(s))$ is

$$||\dot{c}(s)||^2 = \frac{1}{y^2(s)}\left(\dot{x}^2(s) + \dot{y}^2(s)\right), \text{ i.e.}$$

$$||\dot{c}(s)||^2 = \frac{\cosh^2 s}{R^2}\left(\frac{R^2}{\cosh^4 s} + \frac{R^2 \sinh^2 s}{\cosh^4 s}\right) = 1.$$

Continuing, if we replace in the formulas of the geodesic equations, we see that the answer is yes in the case of the first parameterization.

In the case of the second parameterization the same kind of computation leads to

$$||\dot{c}(s)||^2 = \frac{1}{R^2 \sin^2 s}\left(R^2 \sin^2 s + R^2 \cos^2 s\right) = \frac{1}{\sin^2 s},$$

that is a non constant speed. This parameterization is not appropriate for a geodesic.

According to the first parameterization it results that the semicircles are geodesics in the half-plane with respect to the Poincaré metric of the half-plane.

In the same way, the lines orthogonal to the Ox axis, parameterized in the form $x = x(s) = a; \ y = y(s) = e^s$ satisfy the equations of geodesics. The parameterization

$x = x(s) = a;\ y = y(s) = s$ is not appropriate for the same reason: the speed is not constant along the geodesic.   □

Corollary 6.3.4 *The orthogonal arcs sS of the disk are geodesics with respect to the Poincaré metric of the disk.*

To finish, let us recall that the Poincaré metric in H^2 can be deduced from the distance involved in the metric structure of the half-plane. If the ends of the semicircle are denoted by v, V and we have three points on the semicircle such that the order is v, A, B, C, V then, in terms of distance in the H^2 half-plane, we have $d(A, B) + d(B, C) = d(A, C)$, i.e. the semicircle is a geodesic with respect to the distance d. So, in both cases, the geodesics with respect to the corresponding distance are geodesic with respect to the metric, and the two metrics can be derived using the same theorem. In the same time we can deduce one metric from another by a change of coordinates induced by a suitable inversion.

Example 6.3.5 We may observe something interesting not directly related to the models of Non-Euclidean Geometry, our present topic. We know the Minkowski metric

$$ds^2 = dx^2 - dy^2$$

having the constant Minkowski-Gauss curvature $K = 0$ at all points.

In the chapter dedicated to affine universes, we will show that the de Sitter metric

$$ds^2 = dx^2 - \cosh^2 x dy^2$$

has constant Minkowski-Gauss curvature $K = -1$ at all points.

What about positive constant curvature for this kind of metrics? The Poincaré half-plane model can help us to construct one.

If we consider the Minkowski-Poincaré metric

$$ds^2 = \frac{1}{y^2}(dx^2 - dy^2)$$

we have:

$$g_{11} = \frac{1}{y^2}, \quad g_{22} = -\frac{1}{y^2} \quad g_{12} = g_{21} = 0, \quad g^{11} = y^2, \quad g^{22} = -y^2, \quad g^{12} = $$

$$g^{21} = 0, \quad \det(g_{ij}) = -\frac{1}{y^4}.$$

Then, considering $x = x^1$, $y = x^2$, the first kind Christoffel symbols are a little bit different, i.e.

$$\Gamma_{11,1} = \Gamma_{22,1} = \Gamma_{12,2} = \Gamma_{21,2} = 0, \ \Gamma_{12,1} = \Gamma_{21,1} = \Gamma_{11,2} = -\frac{1}{y^3}, \ \Gamma_{22,2} = \frac{1}{y^3}.$$

But the second kind Christoffel symbols are the same as in the Poincaré half-plane model, i.e.

$$\Gamma^1_{11} = \Gamma^1_{22} = \Gamma^2_{12} = \Gamma^2_{21} = 0, \ \Gamma^1_{12} = \Gamma^1_{21} = \Gamma^2_{22} = -\frac{1}{y}, \ \Gamma^2_{11} = \frac{1}{y}.$$

The Minkowski-Gauss curvature of this metric is $K = 1$ at all points of the half plane because $R^1_{212} = -\dfrac{1}{y^2}$ and $R_{1212} = -\dfrac{1}{y^4}$.

The geodesic equations are exactly as in the half-plane case

$$\ddot{x} - \frac{2}{y}\dot{x}\dot{y} = 0$$

$$\ddot{y} + \frac{1}{y}\dot{x}^2 - \frac{1}{y}\dot{y}^2 = 0.$$

It is easy to see that the previous parameterization is not appropriate for this case. But if we write the second equation as

$$\frac{\ddot{y}}{y} + \frac{1}{y^2}\left(\dot{x}^2 - \dot{y}^2\right) = 0$$

the formula

$$\frac{1}{y^2}\left(\dot{x}^2 - \dot{y}^2\right)$$

expresses the speed along geodesic which has to be a constant. Therefore the second equation becomes

$$\frac{\ddot{y}}{y} = k, \ k > 0.$$

Quick exercise: Show that if $y(s) = \cosh(s\sqrt{k}) > 0$, $x(s)$ does not exist.

Consider $y = e^{s\sqrt{k}}$, $s \in \mathbb{R}$. Therefore $x(s) = a$, $a =$ constant. The geodesics of the Minkowski-Poincaré metric

$$ds^2 = \frac{1}{y^2}(dx^2 - dy^2)$$

are, from the Euclidean point of view, lines orthogonal to $y = 0$ axis.

Let us observe a difference between the Geometry of constant positive Gaussian curvature $K = 1$ in the case of the metric

$$ds^2 = dx^2 + \sin^2 x \, dy^2$$

and the Geometry of constant positive Minkowski-Gauss curvature $K = 1$ of this completely different metric, the Minkowski-Poincaré metric

$$ds^2 = \frac{1}{y^2}(dx^2 - dy^2).$$

The geodesics of the first metric (seen related to the sphere) are great circle of the sphere. Therefore, in the Geometry of the first metric, every two lines (geodesics) intersect in exactly two points. In the Geometry of the second metric, every two lines (geodesics) do not intersect. There are no triangles. The geometries are different.

According to the theory we developed, we can directly deduce the following things.

The metric of the Minkowski-Poincaré half-plane is transformed into the Minkowski-Poincaré metric of the disk

$$ds^2 = \frac{4}{[1 - (x^*)^2 - (y^*)^2]^2} \cdot ((dx^*)^2 - (dy^*)^2).$$

The Gaussian curvature of this metric is $K = 1$ at each point, because the Gaussian curvature remains invariant under a change of coordinates, here the change being related to the inversion previously described. The geodesics of the half-plane with respect to the metric

$$ds^2 = \frac{1}{y^2}(dx^2 - dy^2),$$

i.e. the lines orthogonal to $y = 0$ axis, are mapped into orthogonal arcs of circles which pass through the pole of inversion having the tangent in pole exactly the $x = 0$ axis. Therefore, the property of non-intersecting lines is preserved in the disk model of constant positive Minkowski-Gauss curvature.

A question remains after this discussion: Can a Minkowski type metric, that is a non-Riemannian metric, produce the same Geometry as a Riemannian type metric? Even at this moment the answer seems to be no, we will show that all Non-Euclidean Geometry models presented in this book can be represented by a Minkowski type metric of a one-sheet hyperboloid. The construction is related to the hemisphere model we will study below.

6.4   The Exterior Disk Model

Now suppose we are in the two dimensional Euclidean plane E^2; consider the circle $C(O, 1)$ where the center O has the coordinates $(0, 0)$ and the length of the radius is 1. The point $B(x, y)$ belongs to the exterior of our given circle, denoted by $ext C(O, 1)$, i.e. $x^2 + y^2 > 1$. Its inverse $B^*(x^*, y^*)$ with respect to the inversion I, having O as a pole and $\mu = 1$ as a power, has the coordinates

$$x^*(x, y) = \frac{x}{x^2 + y^2}$$

$$y^*(x, y) = \frac{y}{x^2 + y^2}.$$

We obtain these using the condition $OB \cdot OB^* = 1$.

It is easy to see that

$$(x^*)^2 + (y^*)^2 < 1,$$

that is B^* belongs to the interior of the circle $C(O, 1)$.

The mapping $M : extC(O, 1) \to intC(O, 1)$ which describes the change of coordinates $(x, y) \to (x^*, y^*)$ has the differential dM given by the formula

$$dM = \begin{pmatrix} \dfrac{\partial x^*}{\partial x} & \dfrac{\partial x^*}{\partial y} \\ \dfrac{\partial y^*}{\partial x} & \dfrac{\partial y^*}{\partial y} \end{pmatrix}$$

where

$$\frac{\partial x^*}{\partial x} = \frac{-x^2 + y^2}{(x^2 + y^2)^2}; \quad \frac{\partial x^*}{\partial y} = \frac{-2xy}{(x^2 + y^2)^2};$$

$$\frac{\partial y^*}{\partial x} = \frac{-2xy}{(x^2 + y^2)^2}; \quad \frac{\partial y^*}{\partial y} = \frac{x^2 - y^2}{(x^2 + y^2)^2}.$$

We observe that $dM = dM^t$.

The metric of the unit disk in coordinates (x^*, y^*) is

$$ds^2 = \frac{4}{[1 - (x^*)^2 - (y^*)^2]^2} \cdot ((dx^*)^2 + (dy^*)^2).$$

We transform the matrix

$$\bar{G}(x^*, y^*) = \begin{pmatrix} \dfrac{4}{[1 - (x^*)^2 - (y^*)^2]^2} & 0 \\ 0 & \dfrac{4}{[1 - (x^*)^2 - (y^*)^2]^2} \end{pmatrix}$$

with respect to (x, y) coordinates, that is

$$\bar{G}(x, y) = \begin{pmatrix} \dfrac{4(x^2 + y^2)^2}{[1 - (x^2 + y^2)]^2} & 0 \\ 0 & \dfrac{4(x^2 + y^2)^2}{[1 - (x^2 + y^2)]^2} \end{pmatrix}$$

According to the general theory, the matrix of the metric in the exterior of the disk induced

- by changing of coordinates and

- by the metric of the interior of $C(O, 1)$
  is computed using the formula

$$G(x, y) = dM^t \cdot \bar{G}(x, y) \cdot dM.$$

After computations we obtain

$$G(x, y) = \begin{pmatrix} \dfrac{4}{[1 - (x^2 + y^2)]^2} & 0 \\ 0 & \dfrac{4}{[1 - (x^2 + y^2)]^2} \end{pmatrix}$$

that is, we have proved

Theorem 6.4.1 *The Poincaré metric of the disk induces, via a change of coordinates expressed by an appropriate inversion, the metric of the exterior of the disk*

$$ds^2 = \frac{4}{[1 - (x^2 + y^2)]^2}(dx^2 + dy^2).$$

We observe that the metric of the exterior of the disk coincides to the metric of the interior of the disk.

Example 6.4.2 The second possible computation is related to the other formula which describes the change of coefficients of the metric,

$$g_{ij} = \bar{g}_{kl} \frac{\partial \bar{x}^k}{\partial x^i} \frac{\partial \bar{x}^l}{\partial x^j}.$$

Since $\bar{g}_{12} = \bar{g}_{21} = 0$ we have

$$g_{11} = \bar{g}_{kl} \frac{\partial \bar{x}^k}{\partial x^1} \frac{\partial \bar{x}^l}{\partial x^1} = \bar{g}_{11} \frac{\partial \bar{x}^1}{\partial x^1} \frac{\partial \bar{x}^1}{\partial x^1} + \bar{g}_{22} \frac{\partial \bar{x}^2}{\partial x^1} \frac{\partial \bar{x}^2}{\partial x^1}.$$

According to our previous notations $x^* = \bar{x}^1$, $y^* = \bar{x}^2$, $x = x^1$, $y = x^2$, i.e. in (x, y) coordinates we have

$$\bar{g}_{11} = \bar{g}_{22} = \frac{4(x^2 + y^2)^2}{(1 - (x^2 + y^2))^2}.$$

Then

$$g_{11} = \bar{g}_{kl} \frac{\partial \bar{x}^k}{\partial x^1} \frac{\partial \bar{x}^l}{\partial x^1} = \bar{g}_{11} \frac{\partial \bar{x}^1}{\partial x^1} \frac{\partial \bar{x}^1}{\partial x^1} + \bar{g}_{22} \frac{\partial \bar{x}^2}{\partial x^1} \frac{\partial \bar{x}^2}{\partial x^1} =$$

$$= \frac{4(x^2 + y^2)^2}{(1 - (x^2 + y^2))^2} \cdot \left(\frac{-x^2 + y^2}{(x^2 + y^2)^2} \cdot \frac{-x^2 + y^2}{(x^2 + y^2)^2} + \frac{-2xy}{(x^2 + y^2)^2} \cdot \frac{-2xy}{(x^2 + y^2)^2} \right) =$$

$$= \frac{4}{(1 - (x^2 + y^2))^2}$$

In the same way $g_{12} = g_{21} = 0$ and $g_{22} = \frac{4}{[1 - (x^2 + y^2)]^2}$, therefore the metric of the exterior of the disk is the same

$$ds^2 = \frac{4}{[1 - (x^2 + y^2)]^2}(dx^2 + dy^2).$$

Example 6.4.3 Another computation of the metric of the exterior of the disk is suggested from calculus. This is a direct one and it is easier to be applied by the students. The metric of the unit disk in coordinates (x^*, y^*) is

$$ds^2 = \frac{4}{[1 - (x^*)^2 - (y^*)^2]^2} \cdot ((dx^*)^2 + (dy^*)^2)$$

and the formulas which switch from the exterior of the disk to its interior are

$$x^*(x, y) = \frac{x}{x^2 + y^2}; \ \ y^*(x, y) = \frac{y}{x^2 + y^2}.$$

Therefore

$$dx^* = \frac{\partial x^*}{\partial x}dx + \frac{\partial x^*}{\partial y}dy = \frac{-x^2 + y^2}{(x^2 + y^2)^2}dx + \frac{-2xy}{(x^2 + y^2)^2}dy,$$

$$dy^* = \frac{\partial y^*}{\partial x}dx + \frac{\partial y^*}{\partial y}dy = \frac{-2xy}{(x^2 + y^2)^2}dx + \frac{x^2 - y^2}{(x^2 + y^2)^2}dy.$$

We square both formulas, we add and we observe that the terms containing $dxdy$ cancel:

$$(dx^*)^2) + (dy^*)^2 = \frac{1}{(x^2 + y^2)^2}(dx^2 + dy^2)$$

Then, we replace x^* and y^* in $\dfrac{4}{[1 - (x^*)^2 - (y^*)^2]^2}$. We obtain

$$\frac{4}{[1 - (x^*)^2 - (y^*)^2]^2} = \frac{4(x^2 + y^2)^2}{[1 - (x^2 + y^2)]^2}$$

Combining both formulas obtained before, the metric of the exterior of the disk is obtained.

The most important consequence is related to the geodesics of the exterior of the disk. As we saw in the previous example, the geodesic of the interior of the disk are arcs of orthogonal circles to $C(O, 1)$ contained in $int C(O, 1)$. However geodesics are mapped into geodesics under a change of coordinates. If we look at our inversion, an interior arc is mapped into the exterior arc belonging to the same orthogonal circle to $C(O, 1)$. Taking into account that this model is derived from the disk model as we described before we conclude:

Proposition 6.4.4 *The geodesics of the exterior of the disk model are the arcs of circles orthogonal to the circle which determines the disk, the arcs contained outside of the disk.*

6.5 A Hemisphere Model for the Non-Euclidean Geometry

Let us give a concise description of the three models before.

- Disk model
  - set $D^2 := int C(0, 1) = \{(x, y) \in \mathbb{R}^2 \mid x^2 + y^2 < 1\}$
  - metric
  $$ds^2 = \frac{4}{[1 - (x^2 + y^2)]^2}(dx^2 + dy^2)$$

  which comes from the Poincaré hyperbolic distance through Barbilian's Theorem.

- Poincaré half-plane model

  - set $H^2 := \{(x, y) \in \mathbb{R}^2 \mid y > 0\}$
  - it comes from D^2 model using the change of coordinates suggested by an appropriate geometric inversion $(x, y) \in H^2 \longrightarrow (x^*, y^*) \in int C(T, 1)$, $T(0, -1)$,

  $$x^*(x, y) = \frac{4x}{x^2 + (y + 2)^2}; \quad y^*(x, y) = \frac{-2(x^2 + y^2 + 2y)}{x^2 + (y + 2)^2},$$

  which transfer the D^2 metric seen in $C(T, 1)$,

  $$ds^2 = \frac{4}{[1 - (x^*)^2 - (y^* + 1)^2]^2} \cdot ((dx^*)^2 + (dy^*)^2),$$

  into H^2 metric
  $$ds^2 = \frac{1}{y^2}(dx^2 + dy^2).$$

- Exterior disk model
  - set $ext C(0, 1) = \{(x, y) \in \mathbb{R}^2 \mid x^2 + y^2 > 1\}$

- it comes from D^2 model using the change of coordinates suggested by an appropriate geometric inversion which switches the coordinates $(x^*, y^*) \in D^2$ into $(x, y) \in extC(O, 1)$ such that the inverse coordinate transformations is

$$x^*(x, y) = \frac{x}{x^2 + y^2}; \quad y^*(x, y) = \frac{y}{x^2 + y^2}.$$

These ones transfer the D^2 metric

$$ds^2 = \frac{4}{[1 - (x^*)^2 - (y^*)^2]^2} \cdot ((dx^*)^2 + (dy^*)^2)$$

onto $extC(O, 1)$ metric $ds^2 = \dfrac{4}{[1 - (x^2 + y^2)]^2}(dx^2 + dy^2)$.

The Non-Euclidean Geometry in *the hemisphere model* is fixed by an appropriate change of coordinates.

The hemisphere is the set $H_+ = \{(x_1, x_2, x_3) \in S^2 \mid x_1^2 + x_2^2 + x_3^2 = 1, \ x_3 > 0\}$.

Consider the Poincaré half-plane model seen in the form $H_1^2 := \{(1, y_2, y_3) \mid y_3 > 0\}$ endowed with the metric

$$ds^2_{H_1^2} = \frac{1}{y_3^2}\left(dy_2^2 + dy_3^2\right).$$

The stereographic projection π, corresponding to the point $(-1, 0, 0)$, $\pi : H_+ \to H_1^2$ is defined by

$$y_2 = \frac{2x_2}{1 + x_1}, \quad y_3 = \frac{2x_3}{1 + x_1}.$$

In fact, if we consider the line passing through the points $(-1, 0, 0)$ and (x_1, x_2, x_3) $\in H_+$ having the equation

$$\frac{X + 1}{x_1 + 1} = \frac{Y}{x_2} = \frac{Z}{x_3}$$

and the plane $X = 1$, the formulas above defining the stereographic projections obviously appear.

Now, let us use the last technique seen when we found out the metric of the exterior of the disk.

We have

$$dy_2 = \frac{2}{x_1 + 1}dx_2 - \frac{2x_2}{(x_1 + 1)^2}dx_1,$$

$$dy_3 = \frac{2}{x_1 + 1}dx_3 - \frac{2x_3}{(x_1 + 1)^2}dx_1$$

and $x_1^2 + x_2^2 + x_3^2 = 1$ implies both

$$x_2^2 + x_3^2 = 1 - x_1^2$$

and

$$x_1 dx_1 = -x_2 dx_2 - x_3 dx_3.$$

Now, replacing in

$$ds_{H_1^2}^2 = \frac{1}{y_3^2} \left(dy_2^2 + dy_3^2 \right),$$

we obtain

$$ds_{H_+}^2 = \frac{(x_1+1)^2}{4x_3^2} \left[\left(\frac{2}{x_1+1} dx_2 - \frac{2x_2}{(x_1+1)^2} dx_1 \right)^2 + \left(\frac{2}{x_1+1} dx_3 - \frac{2x_3}{(x_1+1)^2} dx_1 \right)^2 \right].$$

After a straightforward computation we obtain the following

Theorem 6.5.1 *The Poincaré half-plane metric induces, via a change of coordinates expressed by an appropriate stereographic projection, the metric*

$$ds_{H_+}^2 = \frac{1}{x_3^2} \left(dx_1^2 + dx_2^2 + dx_3^2 \right)$$

for the hemisphere

$$H_+ = \{(x_1, x_2, x_3) \in S^2 \mid x_1^2 + x_2^2 + x_3^2 = 1, \ x_3 > 0\}.$$

If you imagine the geodesics drawn on H_1^2, their images through the previous stereographic projection are geodesics on H_+.

The hemisphere model of the Non-Euclidean Geometry allows us to discover the hyperboloid model of the Non-Euclidean Geometry. This one is endowed with a Minkowski metric.

6.6   A Minkowski Model for the Non-Euclidean Geometry: The Hyperboloid Model

We start from the hemisphere model expressed by the set

$$H_+ = \{(y_1, y_2, y_3) \in S^2 \mid y_1^2 + y_2^2 + y_3^2 = 1, \ y_3 > 0\}$$

and the metric

$$ds^2_{H_+} = \frac{1}{y^2_3} \left(dy^2_1 + dy^2_2 + dy^2_3 \right).$$

If we consider only one sheet of the two sheets hyperboloid $x^2_1 + x^2_2 - x^2_3 = -1$, we create the set

$$\mathbb{H} := \{(x_1, x_2, x_3) \in E^3 \mid x^2_1 + x^2_2 - x^2_3 = -1, \; x_3 > 0\}.$$

The map

$$y_1 = \frac{x_1}{x_3}, \; y_2 = \frac{x_2}{x_3}, \; y_3 = \frac{1}{x_3}$$

is the stereographic projection α between $\mathbb{H}$ and H_+, the formlas being obtained intersecting the line equation determined by $(0, 0, -1)$ and $(x_1, x_2, x_3) \in \mathbb{H}$,

$$\frac{X}{x_1} = \frac{Y}{x_2} = \frac{Z+1}{x_3 + 1},$$

now with H_+.

Quick exercise: consider $Y = \dfrac{x_2}{x_3}$ and show that $Z = \dfrac{1}{x_3}$ and $X = \dfrac{x_1}{x_3}$.

Quick exercise: if $x^2_1 + x^2_2 - x^2_3 = -1, x_3 > 0$ then $X^2 + Y^2 + Z^2 = 1, \; Z > 0$.
As previously, we consider

$$dy_1 = \frac{1}{x_3} dx_1 - \frac{x_1}{x^2_3} dx_3,$$

$$dy_2 = \frac{1}{x_3} dx_1 - \frac{x_2}{x^2_3} dx_3,$$

$$dy_1 = -\frac{1}{x^2_3} dx_3,$$

together with both

$$x^2_1 + x^2_2 = -1 + x^2_1$$

and

$$x_1 dx_1 + x_2 dx_2 = x_3 dx_3.$$

Replacing in

$$ds^2_{H_+} = \frac{1}{y^2_3} \left(dy^2_1 + dy^2_2 + dy^2_3 \right),$$

it results

$$ds_{\mathbb{H}}^2 = x_3^2 \left[\left(\frac{1}{x_3} dx_1 - \frac{x_1}{x_3^2} dx_3 \right)^2 + \left(\frac{1}{x_3} dx_1 - \frac{x_1}{x_3^2} dx_3 \right)^2 + \left(-\frac{1}{x_3^2} dx_3 \right)^2 \right].$$

Continuing the computations, after convenient replacements, we obtain the induced metric of $\mathbb{H}$,

$$ds_{\mathbb{H}}^2 = dx_1^2 + dx_2^2 - dx_3^2.$$

Therefore, we proved the following result.

Theorem 6.6.1 *The hemisphere metric induces, via a change of coordinates expressed by an appropriate stereographic projection, the metric of the hyperboloid $\mathbb{H}$ model,*

$$ds_{\mathbb{H}}^2 = dx_1^2 + dx_2^2 - dx_3^2.$$

Together with the fact that distances in Non-Euclidean Geometry are described by hyperbolic functions, this last model offers another reason to consider the Non-Euclidean Geometries as hyperbolic geometries. In the next chapters, we use both Euclidean Geometry and the Non-Euclidean geometries to approximate what we can call the physical reality. What reality is? An example we present in Sect. 6.8, due to H. Poincaré, shows how far we can be in our understanding of what we call reality. However, we have to accept the mathematical description, that is the model created, if there are physical evidences. The role of experiment is crucial for validating models.

6.7 The Theoretical Minimum About Non-Euclidean Geometry Models. A Possible Shortcut

Consider the following sets, the following metrics and the following functions which connect the sets. The sets and the metrics are:

1. The Poincaré disk model

$$D^2 := int\, C(0, 1) = \{(x^*, y^*) \in \mathbb{R}^2 \mid (x^*)^2 + (y^*)^2 < 1\}$$

$$ds_{D^2}^2 = \frac{4}{[1 - ((x^*)^2 + (y^*)^2)]^2}((dx^*)^2 + (y^*)^2).$$

1'. The translated disk model

$$D_T^2 =: \{(x^*, y^*) \in \mathbb{R}^2 \mid (x^*)^2 + (y^* + 1)^2 < 1\}$$

$$ds_{D_T^2}^2 = \frac{4}{[1 - (x^*)^2 - (y^* + 1)^2]^2} \cdot ((dx^*)^2 + (dy^*)^2).$$

Obviously, 1 and 1' are representing the same model.

2. The Poincaré half-plane model

$$H^2 := \{(x, y) \in \mathbb{R}^2 \mid y > 0\}$$

$$ds_{H^2}^2 = \frac{1}{y^2}(dx^2 + dy^2).$$

2′. The spatial Poincaré half-plane model

$$H_1^2 := \{(1, y_2, y_3) \mid y_3 > 0\}$$

$$ds_{H_1^2}^2 = \frac{1}{y_3^2}\left(dy_2^2 + dy_3^2\right).$$

Also in this case, 2 and 2′ are obviously representing the same model.

3. The exterior disk model

$$\mathbb{E} := extC(0, 1) = \{(x, y) \in \mathbb{R}^2 \mid x^2 + y^2 > 1\}$$

$$ds_{\mathbb{E}}^2 = \frac{4}{[1 - (x^2 + y^2)]^2}(dx^2 + dy^2)$$

4. The hemisphere model

$$H_+ = \{(y_1, y_2, y_3) \in S^2 \mid y_1^2 + y_2^2 + y_3^2 = 1, \ y_3 > 0\}$$

$$ds_{H_+}^2 := \frac{1}{y_3^2}\left(dy_1^2 + dy_2^2 + dy_3^2\right).$$

5. The hyperboloid model

$$\mathbb{H} := \{(x_1, x_2, x_3) \in E^3 \mid x_1^2 + x_2^2 - x_3^2 = -1, \ x_3 > 0\}$$

$$ds_{\mathbb{H}}^2 = dx_1^2 + dx_2^2 - dx_3^2.$$

The connection between the models are described by the functions:

$$\beta : H^2 \longrightarrow D_T^2, \ x^*(x, y) = \frac{4x}{x^2 + (y + 2)^2}; \ y^*(x, y) = \frac{-2(x^2 + y^2 + 2y)}{x^2 + (y + 2)^2}.$$

$$\gamma : \mathbb{E} \longrightarrow D^2, \ x^*(x, y) = \frac{x}{x^2 + y^2}; \ y^*(x, y) = \frac{y}{x^2 + y^2}.$$

$$\pi : H_+ \to H_1^2, \quad y_2 = \frac{2x_2}{1 + x_1}, \quad y_3 = \frac{2x_3}{1 + x_1}.$$

$$h : \mathbb{H} \to H_+, \quad y_1 = \frac{x_1}{x_3}, y_2 = \frac{x_2}{x_3}, y_3 = \frac{1}{x_3}.$$

An important remark is in order at this point:

Our way to describe the models starts from obtaining all the geometric properties of the construction of the Poincaré disk model.

It was a continuous "struggle" passing through geometric transformations, projective Geometry, algebraic invariants, hyperbolic distances until the Differential Geometry helped us to obtain the Poincaré metric of the disk. And not only for the interior of the disk, we also did it for the Poincaré half-plane. Then, a transfer process of metrics was described: supposing to know the Poincaré disk model D^2 and the function γ, we find the metric of the exterior disk model $\mathbb{E}$.

Supposing to know the Poincaré disk model D^2, we obtain the metric of the translated disk D_T^2 and now, using the function β, we find the metric of the of the Poincaré half-plane model H^2.

Knowing the Poincaré half-plane model H^2, we obtain the metric of spatial half-plane H_1^2 and now, using the function π, we find the metric of the hemisphere model H_+.

Knowing the hemisphere model H_+ and the function h, we find the Minkowski metric of the hyperboloid model $\mathbb{H}$.

A Possible Shortcut

For somebody who is not interested in the historic-geometrical description of the Poincaré disk model, as we tried to present until now, there is a shortcut.

Consider the superior half-plane H^2 and a half circle having its center at the point $O_1(x, 0)$. Denote by A and B the points (x, y), $(x + dx, y + dy)$ respectively, which belong to the half-circle.

If ds is $\angle AO_1B$, then the length of the arc AB is yds and can be approximated by the length of the segment AB, that is $\sqrt{dx^2 + dy^2}$. Therefore the Poincaré metric

$$ds^2 = \frac{1}{y^2}(dx^2 + dy^2)$$

is obtained.

Now, the person has to look only at the previous Poincaré half-plane metric denoted now by

$$ds_{H^2}^2 = \frac{1}{y^2}(dx^2 + dy^2)$$

and to establish the geodesics equations

$$\ddot{x} - \frac{2}{y}\dot{x}\dot{y} = 0, \quad \ddot{y} + \frac{1}{y}\dot{x}^2 - \frac{1}{y}\dot{y}^2 = 0.$$

Then, to find their solutions

$$x(s) = c + R \tanh s, \quad y(s) = \frac{R}{\cosh s}$$

who represent a parameterization for a semicircle having the center on the Ox axis.

Other solutions are the lines orthogonal to the Ox axis parameterized in the form $x(s) = a, \ y(s) = e^s$. The geodesic becomes the "lines" of H^2.

Then, the person has to observe that from a given point of H^2 with respect to a given line, we can construct at least two non-secant lines, etc. A model for the Non-Euclidean Geometry occurs in H^2. Transferring the metrics using the functions above, the person can obtain the other models. However such a person does not really understand the substance of these models.

6.8 A Physical Interpretation

On Internet at the address https://archive.org/details/lascienceetlhypo00poin, it can be found Henri Poincaré' famous book *Science et Hypothèse* [13]. Pages 83–87 offer us a beautiful physical example of Universe related to the non-Euclidean Geometry in the disk model. The example is given in the interior of a sphere. Consider the interior of a sphere $S(O, R)$ where O is the center and R is the radius. This interior is the Universe for some intelligent inhabitants.

For Poincaré, who thought to this special Universe, in the interior of the sphere, both the Euclidean Geometry and a special temperature law are acting.

The temperature is maximum at the center, decreases to 0 on the surface of the sphere in which this Universe is included. The law of temperature variation is: if M is a point such that $OM = r$ then, the temperature at M is proportional to $R^2 - r^2$. Poincaré allows temperature to contract or to dilate the length of the creatures according to their position after a rule we describe as: the length of a ruler is proportional to its absolute temperature. So, a ruler having a side in O and the other side in M, such that the Euclidean length is $|OM| = r$, has in fact a length proportional to $R^2 - r^2$.

The last Poincaré axiom is about how light travels in this Universe: the index of refraction of this Universe is inversely proportional to $R^2 - r^2$. We can suppose it as $\dfrac{4}{R^2 - r^2}$.

Having all these facts in mind, let us understand how the inhabitants will perceive their Universe. First at all, it is enough to understand the Geometry of a disk containing the center of the sphere. For Poincaré, this disk is Euclidean and it has the form of an open R-disk. For the inhabitants, their length is smaller and smaller when they try to reach the border of this slice of Universe. They become shorter and shorter, their legs become shorter, their steps become shorter. These things happen because the temperature acts by contracting the dimensions when they step to the border. The finite Euclidean Universe for Poincaré seems to be infinite for the small creatures.

The rule establishes by Poicaré for distance will be understood by the inhabitants as

$$d^n(O, M) = \frac{1}{2} \cdot \ln \frac{R - x}{R - 0} : \frac{-R - x}{-R - 0},$$

that is, when x approaches R, d^n approaches infinity. Of course, here OM is the x-axis. One inhabitant, mister "H", will observe that it is possible to describe this distance for two arbitrary points A, B in the form

$$d^n(A, B) := \ln \frac{\max_{P \in K} g_{AB}(P)}{\min_{P \in K} g_{AB}(P)}, \quad g_{AB} = \frac{|PA|}{|PB|}$$

where K is the boundary, that is the circle of radius R, and $|PA|$ is the Euclidean distance between P in K and A in the interior of the disk.

The intelligent inhabitants will understand that light is moving on the "straight lines" of the Geometry of their Universe. Since the law of propagation of light depends on the index of refraction, they will deduce the metric of their Universe as

$$ds^2 = \frac{4}{(R^2 - (x^2 + y^2))^2}(dx^2 + dy^2).$$

The straight lines (the geodesics), induced by the trajectories of ray lights, are diameters or arcs of circles bi-orthogonal to the border as we explained above.

There are two "parallel lines" to a given "line" through a given point. The sum of angles of a "triangle" is less than two right angles. Now they conclude they live in a non-Euclidean Universe.

Finally, the inhabitants have two ideas about their slice of Universe, ideas which can be extended to the entire interior of the sphere:

(i) the Universe is infinite
(ii) the Universe is governed by the laws of hyperbolic Geometry and is curved. In each slice the Gaussian curvature is a negative constant, $K(x, y) = -\dfrac{1}{R^2}$.

But this is not true, their Universe is a finite interior of a R-sphere and the underlying Geometry is Euclidean, not hyperbolic!

Poincaré established that the inhabitants of his physical model are perfectly right to use hyperbolic Geometry as the foundation of their Physics because it is convenient, but there is a nonsense to speak about the philosophical abstract truth or about an approximation of any truth, because intelligent inhabitants point of view is in collision with the way and laws their Universe were established.

Poincaré opinion is that the reality is not described by the most "realistic" Geometry "*la géométrie la plus vrai*", but by the most comfortable for description of the physical laws *(la géométrie la plus commode)*. Therefore, Poincaré believed that the Geometry of physical space is a conventional one.

NOTE: *A possible sequel which can be written after Einstein field's equation were established.*

A very intelligent inhabitant, "A.E." succeeded to determine the equations related to the physical structure of any Universe,

$$R_{ij} - \frac{1}{2} \cdot R \cdot g_{ij} = \frac{8\pi G}{c^4} \cdot T_{ij},$$

where R_{ij} is the Ricci tensor, $R^i_j = g^{is} R_{sj}$, $R = R^i_i$ is called the curvature scalar, T_{ij} is a tensor related to the matter and energy contained in the Universe, $k = \dfrac{8\pi G}{c^4} \neq 0$ is a constant. Let us remember that in the 2-dimensional case $R_{ij} = K \cdot g_{ij}$ where K is the Gaussian curvature, and R is computed with the formula $R = R^1_1 + R^2_2$.
   "A.E." computed $R_{ij} - \frac{1}{2} \cdot R \cdot g_{ij}$ in the case of the metric

$$ds^2 = \frac{4}{[R^2 - (x^2 + y^2)]^2}(dx^2 + dy^2),$$

which is determined from the Lagrangian attached to the trajectory of a light ray. Let us remember the Gaussian curvature of this metric: $K = -\dfrac{1}{R^2}$. The result is

$$R_{ij} - \frac{1}{2} \cdot R \cdot g_{ij} = K \cdot g_{ij} - \frac{1}{2} \cdot (R^1_1 + R^2_2) \cdot g_{ij} = -\frac{1}{R^2} \cdot g_{ij} - \frac{1}{2} \cdot (-\frac{1}{R^2} - \frac{1}{R^2}) \cdot g_{ij} = 0.$$

Therefore $T_{ij} = 0$. "A.E." was shocked: the Universe in which he is living has no matter. So, "A.E." modifies the equation related to the physical structure by adding a term in the left hand side, $\Lambda \cdot g_{ij}$, where Λ was called "the cosmological constant"; the new "A.E." equation for a physical Universe is

$$R_{ij} - \frac{1}{2} \cdot R \cdot g_{ij} + \Lambda \cdot g_{ij} = \frac{8\pi G}{c^4} \cdot T_{ij}.$$

These new equations highlight a Poincaré disk as an infinite non-Euclidean Universe with matter inside it. By adding a term at the equations of structure with the aim to offer a chance to have matter inside the Universe, "A.E." seems to strength Poincaré conclusion about the fact that the Geometry of a physical space is a conventional one.

In fact, Einstein did not add the term with this aim, Einstein added the term to preserve a static structure for a Universe in which the gravity attracts together all masses. Here, the role of Λ is only to follow the Poincaré style of thinking.

We do not comment here other ideas relative to the Geometry of a physical space but, in conclusion, the arguments presented above can constitute a sort of *big picture* for Non-Euclidean Geometry models.

Chapter 7
Gravity in Newtonian Mechanics

Per Aspera ad Astra.

Newtonian mechanics is a branch of Physics which studies the way in which the bodies are changing in time their position in space. The space in which the objects are at rest (or they change their position) is the Euclidean 3-dimensional space E^3. All objects, regardless of size, can be identified as points with a given mass in the previous space. So, the Euclidean frame of coordinates $Oxyz$ becomes the absolute place where all is happening. Newtonian Mechanics accepts an universal time in which all changes in position take place. Forces are seen as vectors. For a given point M in space, the vector $\vec{X} = \overrightarrow{OM}$ is called a position vector. If the point evolves in time, we write this as

$$\vec{X}(t) = (x(t), y(t), z(t)).$$

The velocity vector is

$$\dot{\vec{X}} = (\dot{x}(t), \dot{y}(t), \dot{z}(t))$$

and the acceleration vector is

$$\ddot{\vec{X}} = (\ddot{x}(t), \ddot{y}(t), \ddot{z}(t)).$$

Of course, we make the assumption that the coordinates functions are indefinitely differentiable on their domain of definition which differs from a model to another. The foundations of Newtonian Mechanics are based on three fundamental principles, the so called Newton's Laws of motion. They were introduced by Isaac Newton in "Philosophiae Naturalis Principia Mathematica", book published in 1687. The Principle of Inertia, or the first law, asserts: "A physical body preserves its state of

W. Boskoff and S. Capozziello, *A Mathematical Journey to Relativity*, UNITEXT for Physics, https://doi.org/10.1007/978-3-030-47894-0_7

rest or will continue moving at its current velocity conserving its direction, until a force causes a change in its state of moving or rest. The physical body will change the velocity and the direction according to this force." A particular case is related to the rectilinear uniform motion, when the body is moving on a straight line at constant speed. The frames where this principle is available are called inertial frames. These frames are at rest or they move rectilinear at constant speed. This fundamental principle was first enunciated by Galilei. We can say that this principle tells us where, according to Newton, the two others fundamental principles make sense: in inertial frames. In the same time, it tells us that it is impossible to make a distinction between the state "at rest" and the state "rectilinear motion at constant speed." Imagine you are in the bowl of a ship and you have no possibility to observe outside. You slept and you waked up. You can not distinguish between the two states without an observation, a possible comparison. You will play table tennis alike in both states, the object fall down in same way in both states, etc. The two states are equivalent for you in the given conditions. Newton introduces a concept, the quantity of motion of a body as the product between the mass m and its velocity $\vec{v}$. This quantity of motion is known today as momentum and it is denoted by $\vec{p}$, therefore $\vec{p} := m \vec{v}$. The second law asserts: "The force who acts on a body is the variation in time of the quantity of motion." Its differential form is $\vec{F} = \dfrac{d \vec{p}}{dt}$. If m does not depend on time, then

$$\vec{F} = \frac{d \vec{p}}{dt} = m \frac{d \vec{v}}{dt} = m \vec{a},$$

that is the force who acts on a body is proportional to the body acceleration through its mass. Newton's third law: "When a body acts on a second body by the force $\vec{F}$, the second body simultaneously reacts on the first body by the force $- \vec{F}$."

    This chapter is devoted to gravity. We try to outline the basic facts about gravity, we prove the vacuum field equation and the general gravitational field equation. The artifact we use to express these laws is the gravitational potential. Later, in the chapter devoted to General Relativity, the same gravitational potential is involved, in general, in metric components, and, specifically, in the coefficients of the Schwarzschild metric. The step towards General Relativity is made when the tidal acceleration equations are written in a geometric form corresponding to a space endowed with a metric. However, our journey to Relativity has to wait because we need some other tools until the moment we derive Einstein's field equations via the Einstein–Hilbert action. We study Lagrangians and metrics induced by Lagrangians, where Euler–Lagrange equations become the geodesic equations of these metrics. Finally, we will connect these results to Non-Euclidean Geometry models. Kepler's laws are derived. Later, in the same General Relativity chapter, we understand how the conic curve, found as the trajectory of a planet, is still the geodesic trajectory approximation of the same planet in a given metric. An excellent discussion on Newtonian Mechanics, in gravitational perspective, can be found in the book [14].

7.1   Gravity. The Vacuum Field Equation

Let us start to study of the gravity. Later on, in the book, gravity will be studied following Einstein's ideas. Now, we concentrate on gravity as a force trying to understand it from the Classical Mechanics point of view.

In the Euclidean 3-dimensional space E^3, let us consider two bodies of masses M and m, $M > m$, located at the points $X_1(x_1, y_1, z_1)$ and $X(x, y, z)$. The position vectors $\overrightarrow{OX_1}$ and $\overrightarrow{OX}$, where $O(0, 0, 0)$ is the origin, are simply denoted by $X_1 = (x_1, y_1, z_1)$ and $X = (x, y, z)$. Let us define

$$\vec{r} := \vec{X} - \vec{X_1} = (x - x_1, y - y_1, z - z_1).$$

The length of $\vec{r}$ is

$$r := \sqrt{(x - x_1)^2 + (y - y_1)^2 + (z - z_1)^2}$$

and the unit vector pointing the point X_1 from the point X is

$$\vec{u} = -\frac{\vec{X} - \vec{X_1}}{r} = -\frac{\vec{r}}{r} = -\left(\frac{x - x_1}{r}, \frac{y - y_1}{r}, \frac{z - z_1}{r}\right).$$

Newton stated that the *gravitational force* induced by the body of mass M which acts on the body of mass m has the intensity $F = G\dfrac{mM}{r^2}$, where $G = 6.67 \cdot 10^{-11}\dfrac{(m)^3}{(kg) \cdot (s)^2}$ is the *gravitational constant*. It can be described by the gravitational force vector

$$\vec{F} = \frac{GmM}{r^2}\,\vec{u} = -\frac{GmM}{r^2}\frac{\vec{r}}{r} = -\frac{GmM}{r^2}\left(\frac{x - x_1}{r}, \frac{y - y_1}{r}, \frac{z - z_1}{r}\right).$$

Before continuing, let us write the previous formula in the form

$$\vec{F} = m\frac{GM}{r^2}\,\vec{u},$$

where $\vec{u}$ is a unitary vector. The mass m of the body gravitationally attracted seems to be like a "gravitational charge", if we compare $F = G\dfrac{mM}{r^2}$ with the similar formula which describes the intensity of an electric force, $F = k\dfrac{q_1 q_2}{r^2}$. Therefore we can think at m to be a *gravitational mass* denoted by m_g.

In the special case, when we consider a body gravitational attracted by the Earth, M is the mass of the Earth, r is the radius of the Earth and G the gravitational constant, it results

$$F = m_g \cdot A,$$

where $A = \dfrac{GM}{r^2}$ is a constant acceleration denoted by g, where $g = 9.81 \dfrac{\text{(m)}}{\text{(s)}^2}$.

In Newton's second law of motion, the mass m seems to be a constant which makes possible to compare the intensity of the force and the magnitude of acceleration, $F = ma$. This is an *inertial mass*, denoted by m_i, because the first Newton's law establishes the frames were the all three laws are true: the inertial frames. Therefore $F = m_i a$. In the case when F is the gravitational force exerted by the Earth on the body of mass m_i, $F = m_i g$. It results

$$\frac{m_g}{m_i} = \frac{gr^2}{GM} = k.$$

The constant k is not equal to 1 by definition, but, if we measure the weight, the space and the time with some other scaled units, the ratio $\dfrac{m_g}{m_i}$ results 1.

So we can accept that the gravitational mass is the same as the inertial mass, and we can denote by m the value $m_g = m_i$. This is the Equivalence Principle as formulated by Galileo.[1] We will see that it assumes a fundamental role in the formulation of General Relativity.

Let us return to the formula

$$\vec{F} = m \frac{GM}{r^2} \vec{u}$$

seen as $\vec{F} = m \, \vec{A}$. We can define the *gravitational acceleration* as the vector

$$\vec{A} = \frac{GM}{r^2} \vec{u} .$$

This gravitational acceleration is also called a *gravitational field* induced by the body of mass M. This definition suggests how the gravity acts. In coordinates we have

$$\vec{A} = -\frac{GM}{r^2} \left(\frac{x - x_1}{r}, \frac{y - y_1}{r}, \frac{z - z_1}{r} \right).$$

We define the *gravitational potential* of the field $\vec{A}$ to be the function

---

[1] It is worth noticing that this is a peculiarity of gravitational force. For example, for the Coulomb force involving electric charges q, it is $m_i \neq q$. This means that Equivalence Principle is proper of gravity.

$$\Phi(x, y, z) = -\frac{GM}{r}.$$

This definition makes sense at all points of the Euclidean 3-dimensional space except (x_1, y_1, z_1) where the gravitational source is located. It is easy to observe that

$$\frac{\partial \Phi}{\partial x} = \frac{GM}{r^2}\frac{\partial r}{\partial x} = \frac{GM}{r^2}\left(\frac{x - x_1}{r}\right).$$

If we define the *gradient* of the gravitational potential Φ by $\nabla \Phi := \left(\dfrac{\partial \Phi}{\partial x}, \dfrac{\partial \Phi}{\partial y}, \dfrac{\partial \Phi}{\partial z}\right)$, using the previous computation, we can prove

$$\nabla \Phi = \frac{GM}{r^2}\left(\frac{x - x_1}{r}, \frac{y - y_1}{r}, \frac{z - z_1}{r}\right) = -\vec{A}.$$

The *Laplace operator*, or simply, the Laplacian, denoted by ∇^2, is defined as

$$\nabla^2 := \frac{\partial^2}{\partial x^2} + \frac{\partial^2}{\partial y^2} + \frac{\partial^2}{\partial z^2}.$$

The Laplacian of the gravitational potential is

$$\nabla^2 \Phi = \frac{\partial^2 \Phi}{\partial x^2} + \frac{\partial^2 \Phi}{\partial y^2} + \frac{\partial^2 \Phi}{\partial z^2}$$

and can be computed. As we know

$$\frac{\partial \Phi}{\partial x} = \frac{GM}{r^2}\frac{x - x_0}{r},$$

therefore

$$\frac{\partial^2 \Phi}{\partial x^2} = GM \cdot \frac{r^3 - 3r^2\dfrac{\partial r}{\partial x}}{r^6} = GM\left(\frac{1}{r^3} - 3\frac{(x - x_1)^2}{r^5}\right),$$

i.e.

$$\nabla^2 \Phi = GM\left(\frac{3}{r^3} - 3\frac{r^2}{r^5}\right) = 0.$$

Therefore we showed that for all the points $(x, y, z) \neq (x_1, y_1, z_1)$ the gravitational potential

$$\Phi(x, y, z) = -\frac{GM}{r}.$$

satisfies $\nabla^2 \Phi = 0$. Having in mind that the gravitational source is located at (x_1, y_1, z_1), we have proved that in the remaining "empty space", i.e. in vacuum, the Newtonian equation of the gravitational field, expressed with respect to its gravitational potential, is

$$\nabla^2 \Phi = 0.$$

The previous formula is known as *Newton's vacuum field equation*.

What is happening at a source point? We remember our previous construction with the gravitational potential

$$\Phi(x, y, z) = -\frac{1}{r}$$

where

$$r := \sqrt{x^2 + y^2 + z^2}$$

and

$$\nabla^2 \Phi(x, y, z) = 0$$

for all $(x, y, z) \neq (0, 0, 0)$.

Let us introduce the gravitational potential

$$\Phi_b(x, y, z) = -\frac{1}{\bar{r}_b}$$

where

$$\bar{r}_b := \sqrt{(x - b)^2 + y^2 + z^2},$$

that is the source is now $(b, 0, 0)$. The corresponding gravitational field is

$$\vec{A}_b(x, y, z) = -\nabla \Phi_b(x, y, z) = -\frac{1}{\bar{r}_b^2} \left(\frac{x - b}{\bar{r}_b}, \frac{y}{\bar{r}_b}, \frac{z}{\bar{r}_b} \right).$$

After easy computations

$$\frac{\partial \vec{A}_b}{\partial x}(0, 0, 0) = \left(\frac{2}{b^3}, 0, 0 \right)$$

$$\frac{\partial \vec{A}_b}{\partial y}(0, 0, 0) = \left(0, -\frac{1}{b^3}, 0 \right)$$

$$\frac{\partial \vec{A}_b}{\partial z}(0, 0, 0) = \left(0, 0, -\frac{1}{b^3} \right).$$

Now, we observe that the Hessian of the gravitational potential $d^2\Phi_b$ is the matrix with components $\dfrac{\partial \vec{A_b}}{\partial x_j}$, where $x_i \in \{x, y, z\}$, satisfying the relation

$$d^2\Phi_b(0, 0, 0) = \begin{pmatrix} \dfrac{\partial \vec{A_b}}{\partial x}(0, 0, 0) \\[2mm] \dfrac{\partial \vec{A_b}}{\partial y}(0, 0, 0) \\[2mm] \dfrac{\partial \vec{A_b}}{\partial z}(0, 0, 0) \end{pmatrix} = \begin{pmatrix} \dfrac{2}{b^3} & 0 & 0 \\[2mm] 0 & -\dfrac{1}{b^3} & 0 \\[2mm] 0 & 0 & -\dfrac{1}{b^3} \end{pmatrix}.$$

On the other hand, it can be seen as the matrix with the components $\dfrac{\partial^2 \Phi_b}{\partial x_i \partial x_j}$, that is

$$d^2\Phi_b(0, 0, 0) = \begin{pmatrix} \dfrac{\partial^2 \Phi_b}{\partial x^2} & \dfrac{\partial^2 \Phi_b}{\partial x \partial y} & \dfrac{\partial^2 \Phi_b}{\partial x \partial z} \\[2mm] \dfrac{\partial^2 \Phi_b}{\partial y \partial x} & \dfrac{\partial^2 \Phi_b}{\partial y^2} & \dfrac{\partial^2 \Phi_b}{\partial y \partial z} \\[2mm] \dfrac{\partial^2 \Phi_b}{\partial z \partial x} & \dfrac{\partial^2 \Phi_b}{\partial z \partial y} & \dfrac{\partial^2 \Phi_b}{\partial z^2} \end{pmatrix}.$$

In fact the first line of the previous matrix is $\dfrac{\partial \vec{A_b}}{\partial x} = \left(\dfrac{\partial^2 \Phi_b}{\partial x^2}, \dfrac{\partial^2 \Phi_b}{\partial x \partial y}, \dfrac{\partial^2 \Phi_b}{\partial x \partial z} \right)$, etc. Combining the previous results, the trace of Hessian matrix is the Laplacian of the gravitational potential, i.e.

$$Tr\left(d^2\Phi_b\right)(0, 0, 0) = \nabla^2\Phi_b(0, 0, 0) = \dfrac{2}{b^3} - \dfrac{1}{b^3} - \dfrac{1}{b^3} = 0$$

for all points $(x, y, z) \neq (b, 0, 0)$. When $b \to 0$, the gravitational potential $\nabla\Phi_b$ approaches the gravitational potential $\nabla\Phi$, therefore $\nabla\Phi_b^2(0, 0, 0) = 0 \to \nabla\Phi^2(0, 0, 0)$. It means $\nabla\Phi^2(0, 0, 0) = 0$. We may conclude that the vacuum equation becomes

$$\nabla\Phi^2 = 0$$

everywhere, not only for all points without the source.

Let us now suppose that there are many gravitational sources, and we label the gravitational potentials. For each point (x_j, y_j, z_j), one can define

$$r_j := \sqrt{(x - x_j)^2 + (y - y_j)^2 + (z - z_j)^2}$$

and the gravitational potentials

$$\Phi_j(x, y, z) = -\frac{GM_j}{r_j}.$$

The total gravitational potential determined by the N sources is

$$\Phi(x, y, z) = \sum_1^N \Phi_j(x, y, z) = -\sum_1^N \frac{GM_j}{r_j}.$$

Theorem 7.1.1 *For $(x, y, z) \neq (x_j, y_j, z_j)$, $j \in \{1, 2, \ldots, N\}$, the total gravitational potential satisfies the gravitational field equation in vacuum*

$$\nabla^2 \Phi = 0.$$

Proof The linearity of Φ allows to work as previously, for all $j \in \{1, 2, \ldots, N\}$ having

$$\frac{\partial \Phi_j}{\partial x} = \frac{GM_j}{r_j^2} \frac{x - x_j}{r_j}.$$

Therefore

$$\frac{\partial^2 \Phi_j}{\partial x^2} = GM_j \cdot \frac{r_j^3 - 3r_j^2 \frac{\partial r_j}{\partial x}}{r_j^6} = GM_j \left(\frac{1}{r_j^3} - 3\frac{(x - x_j)^2}{r_j^5} \right),$$

i.e.

$$\nabla^2 \Phi = G \sum_1^N M_j \left(\frac{3}{r_j^3} - 3\frac{r_j^2}{r_j^5} \right) = 0.$$

The equation $\nabla^2 \Phi = 0$ is also known as the *Laplace equation for gravity*. □

In a similar way it can be proved.

Corollary 7.1.2 *For multiple sources, the equation $\nabla \Phi^2 = 0$ holds everywhere.*

7.2 Divergence of a Vector Field in an Euclidean 3D-Space

Let us consider an *incompressible fluid flow* described by the vector $\vec{F} := \rho \vec{V}$, where $\rho := \rho(x, y, z)$ is the *density of the incompressible fluid* at (x, y, z) and $\vec{V} = \vec{V}(x, y, z)$ is the speed vector at each point of a given region D of the Euclidean space.

If we are looking at the fact that F is measured in $\dfrac{(kg)}{(m)^2 \cdot (s)}$, we see in fact how much matter flows through a unit surface area in a unit time.

Consider a small parallelepiped centered at $(x, y, z) \in D$ and with sides of lengths $\Delta x, \Delta y, \Delta z$ parallel to the axis of coordinates. The vector flow $\overrightarrow{F}$ has three components, $\overrightarrow{F} = (F_x, F_y, F_z)$. We can suppose the parallelepiped small enough to have the flow $\overrightarrow{F}$ constant over each face, that is at each point of a face, $\overrightarrow{F}$ has the same three given components. We are interested in expressing the *net outflow* through this parallelepiped, i.e. the algebraic sum of all outward flow vectors through the six faces.

The flow through the face of area $\Delta y \Delta z$ at the point $\left(x - \dfrac{\Delta x}{2}, y, z\right)$ is

$$F_x\left(x - \frac{\Delta x}{2}, y, z\right) \Delta y \Delta z.$$

Suppose this is an inflow. In the same way, the flow through the face of area $\Delta y \Delta z$ at the point $\left(x + \dfrac{\Delta x}{2}, y, z\right)$ is

$$F_x\left(x + \frac{\Delta x}{2}, y, z\right) \Delta y \Delta z$$

and this one is an outflow. Therefore the total outflow through these two parallel faces is

$$F_x\left(x + \frac{\Delta x}{2}, y, z\right) \Delta y \Delta z - F_x\left(x - \frac{\Delta x}{2}, y, z\right) \Delta y \Delta z \approx \frac{\partial F_x}{\partial x}(x, y, z) \Delta x \Delta y \Delta z,$$

where the last approximation was made taking into consideration the small dimensions of the parallelepiped.

Considering the contribution of the other two pairs of parallel faces, the total outflow through the parallelepiped faces becomes

$$\left(\frac{\partial F_x}{\partial x}(x, y, z) + \frac{\partial F_y}{\partial y}(x, y, z) + \frac{\partial F_z}{\partial z}(x, y, z)\right) \Delta x \Delta y \Delta z.$$

The *divergence* of $\overrightarrow{F}$ is defined by

$$\mathrm{div}\ \overrightarrow{F} := \frac{\partial F_x}{\partial x}(x, y, z) + \frac{\partial F_y}{\partial y}(x, y, z) + \frac{\partial F_z}{\partial z}(x, y, z)$$

and a physical interpretation of it as total outflow over the parallelepiped is that presented above.

We can conclude: On the entire region D, the total outflow over D is

$$\mathcal{F}(D) := \int_D \text{div } \vec{F} \, d^3x = \text{div } \vec{F} \, (\vec{u_\eta}) \cdot \text{vol } D$$

where d^3x is the volume element $dx\,dy\,dz$ and the last equality is a consequence of a mean value theorem for the given triple integral. A consequence of the last formula is

$$\lim_{D \to \vec{u}} \frac{\mathcal{F}(D)}{\text{vol } D} = \text{div } \vec{F} \, (\vec{u}).$$

7.3 Covariant Divergence

We have discussed about a flow of an incompressible fluid in an Euclidean space. How this discussion changes if we are talking about an incompressible fluid in a region where the parallelism is not the Euclidean one? The problem appears when we consider the difference

$$F_x\left(x + \frac{\Delta x}{2}, y, z\right) \Delta y \Delta z - F_x\left(x - \frac{\Delta x}{2}, y, z\right) \Delta y \Delta z$$

because it means that we have moved by parallel transport the vector $\left(-F_x\left(x - \frac{\Delta x}{2}\right), 0, 0\right)$ to the other face at the point $\left(x + \frac{\Delta x}{2}, y, z\right)$.

Therefore we parallel transport the contravariant vector $\left(-F_x\left(x - \frac{\Delta x}{2}\right), 0, 0\right)$ along the infinitesimal vector $A^1 = (\Delta x, 0, 0)$.

Since, in general, $\Gamma^k_{ij} \neq 0$, the parallel transport along $A^1 = (\Delta x, 0, 0)$ for a contravariant vector $V = (V^1, 0, 0)$ leads to a vector whose first component is

$$V^1\left(x - \frac{\Delta x}{2}, y, z\right) + \Delta V^1,$$

where

$$\Delta V^1 = -\Gamma^1_{ij} V^j \Delta x^i = -\Gamma^1_{1j} V^j \Delta x = -\Gamma^1_{11} V^1 \Delta x.$$

The difference

$$\left[V^1\left(x + \frac{\Delta x}{2}, y, z\right) - V^1\left(x - \frac{\Delta x}{2}, y, z\right) + \Gamma^1_{11} V^1 \Delta x\right] \Delta y \Delta z$$

is

$$\left(\frac{\partial V^1}{\partial x} + \Gamma^1_{11} V^1\right) \Delta x \Delta y \Delta z,$$

i.e. the covariant derivative with respect to the first variable denoted by

$$V^1_{;1} \Delta x \Delta y \Delta z.$$

We have three pairs of opposite faces corresponding to the three directions, therefore the net outflow is

$$(V^1_{;1} + V^2_{;2} + V^3_{;3}) \Delta x \Delta y \Delta z$$

for a parallelepiped in a region where the Euclidean parallel transport is replaced by the general parallel transport.

The quantity $V^s_s := V^1_{;1} + V^2_{;2} + V^3_{;3}$ is the covariant divergence of a contravariant vector (V^1, V^2, V^3).

In our case, we obtain

$$\left(F_x \left(x + \frac{\Delta x}{2}, y, z \right) - F_x \left(x - \frac{\Delta x}{2}, y, z \right) + F_x \Gamma^1_{11} \Delta x \right) \Delta y \Delta z \approx$$

$$\approx \left(\frac{\partial F_x}{\partial x} + F_x \Gamma^1_{11} \right) \Delta x \Delta y \Delta z = F_{x;1} \Delta x \Delta y \Delta z.$$

For the entire parallelepiped we have the total net outflow

$$(F_{x;1} + F_{y;2} + F_{z;3}) \Delta x \Delta y \Delta z$$

expressed with respect the covariant derivative.

Definition 7.3.1 The quantity $(F_{x;1} + F_{y;2} + F_{z;3})$ expressed with respect to the covariant derivatives of components is called a *covariant divergence* of the field F.

7.4 The General Newtonian Gravitational Field Equations

If a gravitational source of mass M is placed at (x_1, y_1, z_1) and no other gravitational source exists, we have deduced the vacuum fields equation

$$\nabla^2 \Phi(x, y, z) = 0.$$

If there are many gravity sources (x_j, y_j, z_j), $j \in \{1, 2, \ldots, N\}$, we have defined

$$r_j := \sqrt{(x - x_j)^2 + (y - y_j)^2 + (z - z_j)^2}$$

and the corresponding gravitational potentials

$$\Phi_j(x, y, z) = -\frac{GM_j}{r_j}.$$

The total gravitational potential, determined by the N sources, was

$$\Phi(x, y, z) = \sum_1^N \Phi_j(x, y, z) = -\sum_1^N \frac{GM_j}{r_j}.$$

We have proved that the vacuum field equation, in this case, is

$$\nabla^2 \Phi(x, y, z) = 0,$$

and it makes sense for all (x, y, z) of the space.

Now suppose that in a bounded region D of the Euclidean space E^3 there is a continuous distribution of matter and point sources. This continuous distribution of matter is defined by a density function $\rho = \rho(x, y, z)$ measured in $\frac{(\text{kg})}{(\text{m})^3}$. Outside D we have $\rho \equiv 0$.

How it looks like the gravitational field equation in this case? Let us prove the following

Theorem 7.4.1 (General Gravitational Field Equation) *If D is a region of the space where it exists a continuous distribution of matter defined by the density function ρ, then*

$$\nabla^2 \Phi(x, y, z) = 4\pi G\rho(x, y, z)$$

everywhere in D.

Proof Outside D, where $\rho = 0$, the theorem reduces to the vacuum field equation. It remains to prove the statement for all the points of D. We cover D with parallelepipeds. To do this, we consider points on Ox axis and parallel planes to yOz through these points. In the same way, we take into account parallel planes to xOz through points on Oy and parallel planes to xOy through points on Oz. We obtain parallelepipeds with the faces parallel to the planes determined by the axes of coordinates. Some of parallelepipeds are completely inside D, some are completely outside D and some of them contain parts inside and outside.

Now we can index the points and we can denote the centers of parallelepipeds which cover D as being (x_i, y_j, z_k) and the corresponding dimensions of sides as $\Delta x_i, \Delta y_j, \Delta z_k$.

We can suppose the mass of such parallelepiped is $\rho(x_i, y_j, z_k)\Delta x_i \Delta y_j \Delta z_k$.

The corresponding gravitational potential at a point $(x, y, z) \in E^3$ is

$$\Phi(x, y, z) \approx \sum \Phi_j(x, y, z),$$

that is

$$\Phi(x, y, z) = -\sum_{i=1}^{N}\sum_{j=1}^{M}\sum_{k=1}^{P} \frac{G\rho(x_i, y_j, z_k)}{\sqrt{(x - x_i)^2 + (y - y_j)^2 + (z - z_k)^2}} \Delta x_i \Delta y_j \Delta z_k.$$

We can improve the approximation of the gravitational potential formula considering more points on the axes and, at limit, we obtain

$$\Phi(x, y, z) = -G\int_D \rho(u, v, w) \frac{1}{\sqrt{(x - u)^2 + (y - v)^2 + (z - w)^2}} d^3u,$$

where d^3u is the volume element $du\,dv\,dw$. If $(x, y, z) \notin D$, the integral has sense. We are able to show that the integral has sense even for points $(x, y, z) \in D$. Consider a change of coordinates in E^3 defined by

$$u = x + r \sin x^2 \cos x^1$$

$$v = y + r \sin x^2 \sin x^1$$

$$w = z + r \cos x^2.$$

We observe

$$r(x, y, z) = \sqrt{(x - u)^2 + (y - v)^2 + (z - w)^2}.$$

Then, according to our knowledge in calculus, the volume element for spherical coordinates is changing after the formula

$$du\,dv\,dw = r^2 \sin x^2 dr dx^2 dx^1$$

and the integral becomes

$$\Phi(r, x^1, x^2) = -G\int_{D^*} \rho\frac{1}{r}r^2 \sin x^2 dr dx^2 dx^1 = -G\int_{D^*} \rho r \sin x^2 dr dx^2 dx^1,$$

where D^* is the transformed of D with respect to the previous change of coordinates. The last integral is not singular, therefore the definition of the gravitational potential makes sense in D, too.

If we apply the Laplace operator

$$\frac{\partial^2}{\partial x^2} + \frac{\partial^2}{\partial y^2} + \frac{\partial^2}{\partial z^2}$$

to

$$\Phi(x, y, z) = -G\int_D \rho(u, v, w) \frac{1}{r(x, y, z)} d^3u,$$

we obtain

$$\nabla^2 \Phi(x, y, z) = \int_D \nabla^2 \left(-\frac{G\rho(u, v, w)}{r(x, y, z)} \right) d^3 u.$$

The gradient operator $\nabla := \left(\dfrac{\partial}{\partial x}, \dfrac{\partial}{\partial y}, \dfrac{\partial}{\partial z} \right)$ leads to the Laplace operator via a formal dot product:

$$\nabla^2 := \nabla \cdot \nabla = \left(\frac{\partial}{\partial x}, \frac{\partial}{\partial y}, \frac{\partial}{\partial z} \right) \cdot \left(\frac{\partial}{\partial x}, \frac{\partial}{\partial y}, \frac{\partial}{\partial z} \right) = \frac{\partial^2}{\partial x^2} + \frac{\partial^2}{\partial y^2} + \frac{\partial^2}{\partial z^2}.$$

It implies

$$\nabla^2 \Phi = \nabla \cdot \nabla \Phi = -\nabla \cdot \vec{A} = -\left(\frac{\partial A_x}{\partial x} + \frac{\partial A_y}{\partial y} + \frac{\partial A_z}{\partial z} \right) = -div \, \vec{A},$$

that is

$$\nabla^2 \Phi(x, y, z) = -div \, \vec{A} \, (x, y, z).$$

Now, if $(x, y, z) \notin D$, we have proved $\nabla^2 \left(-\dfrac{G\rho}{r} \right) = 0$, therefore $\nabla^2 \Phi(x, y, z) = 0$. In the same time we have proved that $div \, \vec{A} \, (x, y, z) = 0$ when $(x, y, z) \notin D$.

If $(x, y, z) \in D$, let us make some considerations.

We define the gravitational field $\vec{A} = (A_x, A_y, A_z)$ attached to the potential Φ, $\vec{A} := -\nabla \Phi$. It remains to evaluate $-div \, \vec{A} \, (x, y, z)$ when $(x, y, z) \in D$. To do this, we consider a sphere $S(r)$ centered at (x, y, z) with a small radius r such that the mass density ρ can be considered constant in all its interior, interior here denoted by $B(r)$. Therefore we suppose $\rho(u, v, w) = \rho(x, y, z)$ for all $(u, v, w) \in B(r)$. Let us decompose D in $B(r) \cup (D - B(r))$. We have

$$\vec{A} = \vec{A}_{B(r)} + \vec{A}_{D-B(r)}$$

and, since $(x, y, z) \notin D - B(r)$, using the previous case result, it follows

$$div \, \vec{A}_{D-B(r)} \, (x, y, z) = 0,$$

i.e.

$$div \, \vec{A} \, (x, y, z) = div \, \vec{A}_{B(r)} \, (x, y, z) + div \, \vec{A}_{D-B(r)} \, (x, y, z) = div \, \vec{A}_{B(r)} \, (x, y, z).$$

Now, the problem reduces to the evaluation of $div \, \vec{A}_{B(r)} \, (x, y, z)$.

Let us observe that the gravitational field $\vec{A}_{B(r)}$ at every $(\bar{x}, \bar{y}, \bar{z}) \in B(r)$ is

$$\vec{A}_{B(r)}\,(\bar{x}, \bar{y}, \bar{z}) = -\frac{G \cdot M_{B(r)}}{\bar{r}^2}\,\vec{n} = -\frac{G \cdot \rho \cdot vol\,B(r)}{\bar{r}^2}\,\vec{n},$$

where $\bar{r}$ is the length of the vector who points from (x, y, z) to $(\bar{x}, \bar{y}, \bar{z})$ and $\vec{n}$ is its unit vector. On the entire surface of $S(r)$, the gravitational field becomes the constant magnitude vector field

$$\vec{A}_{B(r)}\,(x, y, z) = -\frac{G \cdot \rho \cdot vol\,B(r)}{r^2}\,\vec{n}\,.$$

The total outflow over $B(r)$ is

$$\mathcal{F}(B(r)) = -\frac{G \cdot \rho \cdot vol\,B(r)}{r^2} \cdot 4\pi r^2 = -4\pi G \cdot \rho \cdot vol\,B(r).$$

Therefore

$$\lim_{r \to 0} \frac{\mathcal{F}(B(r))}{vol\,B(r)} = \operatorname{div}\vec{A}_{B(r)}\,(x, y, z) = -4\pi G \cdot \rho,$$

that is

$$\nabla^2 \Phi = 4\pi G \cdot \rho.$$

Since the ρ chosen is $\rho = \rho(x, y, z)$ and all computations are done at the point (x, y, z), the proof is complete. The previous equation is also known as the *Poisson equation for gravity*. □

7.5 Tidal Acceleration Equations

We met before the gravitational potential

$$\Phi_b(x, y, z) = -\frac{1}{\bar{r}_b}$$

determined by a source at $(b, 0, 0)$, $b > 0$. The denominator is

$$\bar{r}_b := \sqrt{(x - b)^2 + y^2 + z^2}$$

and the corresponding gravitational field is

$$\vec{A}_b\,(x, y, z) = -\nabla \Phi_b(x, y, z) = -\frac{1}{\bar{r}_b^2}\left(\frac{x - b}{\bar{r}_b}, \frac{y}{\bar{r}_b}, \frac{z}{\bar{r}_b}\right).$$

We have observed $\vec{A}_b\,(0, 0, 0) = \left(\frac{1}{b^2}, 0, 0\right).$

Definition 7.5.1 The tidal acceleration $\vec{T}$ (x, y, z), generated by the gravitational field $\vec{A_b}$ (x, y, z) at $(0, 0, 0)$, is defined by the formula

$$\vec{T} \ (x, y, z) := \vec{A_b} \ (x, y, z) - \vec{A_b} \ (0, 0, 0).$$

We may use a Taylor approximation to compute the tidal acceleration at some points of the axes as follows

$$\vec{T} \ (a, 0, 0) := \vec{A_b} \ (a, 0, 0) - \vec{A_b} \ (0, 0, 0) \approx a \frac{\partial \vec{A_b}}{\partial x} (0, 0, 0) = \left(\frac{2a}{b^3}, 0, 0 \right).$$

In the same way

$$\vec{T} \ (0, a, 0) := \vec{A_b} \ (0, a, 0) - \vec{A_b} \ (0, 0, 0) \approx a \frac{\partial \vec{A_b}}{\partial y} (0, 0, 0) = \left(0, -\frac{a}{b^3}, 0 \right)$$

and

$$\vec{T} \ (0, 0, a) := \vec{A_b} \ (0, 0, a) - \vec{A_b} \ (0, 0, 0) \approx a \frac{\partial \vec{A_b}}{\partial z} (0, 0, 0) = \left(0, 0, -\frac{a}{b^3} \right).$$

The effect of translation due to a tidal acceleration is called a *tidal effect*.
We can better see the tidal effect, if we consider slices in Oxy and Oxz planes.
We focus on Oxy plane and let us consider the unit vector $(\cos u, \sin u)$.
If we compute $\vec{T}$ $(a \cos u, a \sin u)$, we describe the tidal effect at all points of the circle centered in O having a as radius. Therefore

$$\vec{T} \ (a \cos u, a \sin u) := \vec{A_b} \ (a \cos u, a \sin u) - \vec{A_b} \ (0, 0) \approx$$

$$\approx a \cos u \frac{\partial \vec{A_b}}{\partial x} (0, 0) + a \sin u \frac{\partial \vec{A_b}}{\partial y} (0, 0),$$

the approximation being given by the directional derivative of $\vec{A_b}$ in the direction $(\cos u, \sin u)$. It results

$$\vec{T} \ (a \cos u, a \sin u) := \left(\frac{2a \cos u}{b^3}, -\frac{a \sin u}{b^3} \right).$$

This is the image of the tidal effect around $(0, 0, 0)$ in Oxy plane. There is a similar image in Oxz plane. In fact, if you rotate the Oxy plane around Ox axis, you have the big picture of the tidal effect at all the points of a sphere surface.

Now, you can imagine the Moon at $(b, 0, 0)$ and the Earth as a sphere centered at $(0, 0, 0)$ having radius a and the oceans tides appears when you rotate the sphere. This is the animated picture of the tidal effect.

The tidal effect appears and it can be studied as previously.

If we wish to highlight the equations of the tidal effect, we need to consider free falling particles in the gravitational field created by the source which start from the points of a given curve $c(q) = (x(q), y(q), z(q))$.

Here q is not a time parameter, it is only a geometric parameter which allows us to describe the image of the curve c.

Denote by $\bar{x}(t, q) = (x^1(t, q), x^2(t, q), x^3(t, q))$ the system of free falling particles $q \in [a, b]$. The particle $\bar{x}(t, q_0)$ starts from $c(q_0)$ and has a time evolution. We can prove the following.

Theorem 7.5.2 *The tidal acceleration equations are*

$$\frac{d^2}{dt^2} \frac{\partial \bar{x}}{\partial q} = -d^2 \Phi_{\bar{x}} \frac{\partial \bar{x}}{\partial q},$$

where $d^2 \Phi_{\bar{x}} = \left(\dfrac{\partial^2 \Phi(\bar{x})}{\partial x^i \partial x^k} \right)_{i,k}$ *is the Hessian matrix attached to the gravitational potential* Φ.

Proof For each particle $\bar{x}(t, q)$, Newton's second law leads to the formulas

$$\frac{d^2 x^k}{dt^2}(t, q) = -\frac{\partial \Phi}{\partial x^k}(\bar{x}(t, q)) , \, k \in \{1, 2, 3\},$$

because the particle experiences the gravitational acceleration due to the source. If we consider a nearby point $c(q + \Delta q)$, the same considerations leads to the equations

$$\frac{d^2 x^k}{dt^2}(t, q + \Delta q) = -\frac{\partial \Phi}{\partial x^k}(\bar{x}(t, q + \Delta q)).$$

If we subtract the first equation from the second, we divide by Δq and consider the limit as Δq approaches 0, we obtain

$$\lim_{\Delta q \to 0} \frac{\dfrac{d^2 x^k}{dt^2}(t, q + \Delta q) - \dfrac{d^2 x^k}{dt^2}(t, q)}{\Delta q} = -\lim_{\Delta q \to 0} \frac{\dfrac{\partial \Phi}{\partial x^k}(\bar{x}(t, q + \Delta q)) - \dfrac{\partial \Phi}{\partial x^k}(\bar{x}(t, q))}{\Delta q},$$

that is

$$\frac{d^2}{dt^2} \frac{\partial x^k}{\partial q}(t, q) = -\frac{\partial^2 \Phi}{\partial q \partial x^k}(\bar{x}(t, q)) = -\sum_{i=1}^{3} \frac{\partial^2 \Phi}{\partial x^i \partial x^k} \frac{\partial x^i}{\partial q}(\bar{x}(t, q))$$

for each $k \in \{1, 2, 3\}$. The last equality is obtained from the chain rule. We highlighted the vector $\dfrac{\partial \bar{x}}{\partial q} = \left(\dfrac{\partial \bar{x}^1}{\partial q}, \dfrac{\partial \bar{x}^2}{\partial q}, \dfrac{\partial \bar{x}^3}{\partial q} \right)$ which satisfies the tidal acceleration equations

$$\frac{d^2}{dt^2} \frac{\partial \bar{x}}{\partial q} = -d^2 \Phi_{\bar{x}} \frac{\partial \bar{x}}{\partial q},$$

where the Hessian matrix $d^2 \Phi_{\bar{x}} = \left(\dfrac{\partial^2 \Phi(\bar{x})}{\partial x^i \partial x^k} \right)_{i,k}$ encapsulates in its trace the vacuum field equation $\nabla^2 \Phi = 0$.   □

Definition 7.5.3 $\dfrac{\partial \bar{x}}{\partial q} = \left(\dfrac{\partial \bar{x}^1}{\partial q}, \dfrac{\partial \bar{x}^2}{\partial q}, \dfrac{\partial \bar{x}^3}{\partial q} \right)$ is called a *tidal vector*.

The tidal vector measures, as we saw, the variation of nearby trajectories due to the tidal acceleration. Therefore, the tidal vector and the tidal acceleration equations naturally appear when objects experience a gravitational field.

7.6 Geometric Separation of Geodesics

Suppose we work in a space of coordinates denoted by $(x^0, x^1, \ldots, x^n)$ endowed with a metric $ds^2 = g_{ij} dx^i dx^j$. For each coordinate x^k, we imagine two parameters, now denoted (τ, q), such that $x^k = x^k(\tau, q)$ and we define a difference between the two parameters denoting by $\dfrac{dx^k}{d\tau}$ the derivative of the coordinate function x^k with respect to the first parameter and by $\dfrac{\partial x^k}{\partial q}$ with respect to the second parameter.

The geodesic equations are written with respect to the first variable τ which may be thought as a time parameter; the covariant derivative of vectors will be denoted by ∇ as we did when we studied surfaces.

A remark: We refer here to the covariant derivative, denoted by $\dfrac{\nabla}{d\tau}$, which is obviously different from the gradient denoted by ∇. In the same way below, there is no connection between the second iteration of the covariant derivative $\dfrac{\nabla^2}{d\tau^2}$ and the Laplacian ∇^2.

According to our previous notations, we have the covariant derivative formula

$$\frac{\nabla}{d\tau} \frac{dx^k}{d\tau} = \frac{d^2 x^k}{d\tau^2} + \Gamma^k_{ij} \frac{dx^i}{d\tau} \frac{dx^j}{d\tau}$$

which implies, for a given geodesic $c(\tau, q) = (x^0(\tau, q), x^1(\tau, q), \ldots, x^n(\tau, q))$, $q = q_0$, that

$$\frac{\nabla}{d\tau} \frac{dx^k}{d\tau} = 0, \; k \in \{0, 1, \ldots, n\},$$

i.e.

$$\frac{d^2 x^k}{d\tau^2} + \Gamma^k_{ij} \frac{dx^i}{d\tau} \frac{dx^j}{d\tau} = 0, \; k \in \{0, 1, \ldots, n\}.$$

Now, for each q from an interval which contains q_0, it is possible to consider the corresponding geodesic $c(\tau, q)$. A family of geodesics, starting from the points of the curve $c(0, q) = (x^0(0, q), x^1(0, q), \ldots, x^n(0, q))$, having each one the initial vector $\left(\dfrac{dx^0}{d\tau}(0, q), \dfrac{dx^1}{d\tau}(0, q), \ldots, \dfrac{dx^n}{d\tau}(0, q) \right)$, is immediately defined. As in the previous case, the vector

$$\frac{\partial x}{\partial q} := \left(\frac{\partial x^0}{\partial q}, \frac{\partial x^1}{\partial q}, \ldots, \frac{\partial x^n}{\partial q} \right)$$

is called the tidal vector and measures, according to its orientation, the *rate of separation of geodesics*.

Let us see which are the equivalent of tidal acceleration equations. They are

$$\frac{d^2}{dt^2} \frac{\partial \bar{x}}{\partial q} = -d^2 \Phi_{\bar{x}} \frac{\partial \bar{x}}{\partial q},$$

where $d^2 \Phi_{\bar{x}}$ is the Hessian matrix $\left(\dfrac{\partial^2 \Phi(\bar{x})}{\partial x_i \partial x_k} \right)_{i,k}$.

We can prove:

Theorem 7.6.1 *The tidal vector above satisfies the equations*

$$\frac{\nabla^2}{d\tau^2} \frac{\partial x^h}{\partial q} = -R^h_{ijk} \frac{dx^i}{d\tau} \frac{\partial x^j}{\partial q} \frac{dx^k}{d\tau},$$

where R^i_{jkl} is the Riemann mixed tensor defined by the metric g_{ij}.

Proof We start from the covariant derivative $\dfrac{\nabla}{d\tau}$ of $\dfrac{\partial x^h}{\partial q}$:

$$\frac{\nabla}{d\tau} \frac{\partial x^h}{\partial q} = \frac{d}{d\tau} \left(\frac{\partial x^h}{\partial q} \right) + \Gamma^h_{ij} \frac{dx^i}{d\tau} \frac{\partial x^j}{\partial q}.$$

Then

$$\frac{\nabla^2}{d\tau^2}\frac{\partial x^h}{\partial q} = \frac{\nabla}{d\tau}\left(\frac{\nabla}{d\tau}\frac{\partial x^h}{\partial q}\right) = \frac{d}{d\tau}\left(\frac{\nabla}{d\tau}\frac{\partial x^h}{\partial q}\right) + \Gamma^h_{mk}\left(\frac{\nabla}{d\tau}\frac{\partial x^m}{\partial q}\right)\frac{dx^k}{d\tau} =$$

$$= \frac{d}{d\tau}\left[\frac{d}{d\tau}\left(\frac{\partial x^h}{\partial q}\right) + \Gamma^h_{ij}\frac{dx^i}{d\tau}\frac{\partial x^j}{\partial q}\right] + \Gamma^h_{mk}\left[\frac{d}{d\tau}\left(\frac{\partial x^m}{\partial q}\right) + \Gamma^m_{ij}\frac{dx^i}{d\tau}\frac{\partial x^j}{\partial q}\right]\frac{dx^k}{d\tau} =$$

$$= \frac{\partial}{\partial q}\frac{d^2 x^h}{d\tau^2} + \frac{\partial\Gamma^h_{ij}}{\partial x^k}\frac{dx^k}{d\tau}\frac{dx^i}{d\tau}\frac{\partial x^j}{\partial q} +$$

$$+ \Gamma^h_{ij}\frac{d^2 x^i}{d\tau^2}\frac{\partial x^j}{\partial q} + \Gamma^h_{ij}\frac{dx^i}{d\tau}\frac{\partial^2 x^j}{\partial\tau\partial q} + \Gamma^h_{mk}\frac{\partial^2 x^m}{\partial\tau\partial q}\frac{dx^k}{d\tau} + \Gamma^h_{mk}\Gamma^m_{ij}\frac{dx^i}{d\tau}\frac{\partial x^j}{\partial q}\frac{dx^k}{d\tau}.$$

Now, we replace $\dfrac{d^2 x^k}{d\tau^2}$ by $-\Gamma^k_{ij}\dfrac{dx^i}{d\tau}\dfrac{dx^j}{d\tau}$ and in some terms we replace the dummy indexes in a convenient way. It results

$$\frac{\nabla^2}{d\tau^2}\frac{\partial x^h}{\partial q} = \frac{\partial}{\partial q}\left(-\Gamma^h_{ik}\frac{dx^i}{d\tau}\frac{dx^k}{d\tau}\right) + \frac{\partial\Gamma^h_{mj}}{\partial x^k}\frac{dx^i}{d\tau}\frac{\partial x^j}{\partial q}\frac{dx^k}{d\tau} + \Gamma^h_{mj}\left(-\Gamma^m_{ik}\frac{dx^i}{d\tau}\frac{dx^k}{d\tau}\right)\frac{\partial x^j}{\partial q} +$$

$$+ \Gamma^h_{ij}\frac{dx^i}{d\tau}\frac{\partial^2 x^j}{\partial\tau\partial q} + \Gamma^h_{mk}\frac{\partial^2 x^m}{\partial\tau\partial q}\frac{dx^k}{d\tau} + \Gamma^h_{mk}\Gamma^m_{ij}\frac{dx^i}{d\tau}\frac{\partial x^j}{\partial q}\frac{dx^k}{d\tau} =$$

$$= -\frac{\partial\Gamma^h_{ik}}{\partial x^j}\frac{\partial x^j}{\partial q}\frac{dx^i}{d\tau}\frac{dx^k}{d\tau} - 2\Gamma^h_{ik}\frac{\partial^2 x^i}{\partial\tau\partial q}\frac{dx^k}{d\tau} + \frac{\partial\Gamma^h_{ij}}{\partial x^k}\frac{dx^i}{d\tau}\frac{\partial x^j}{\partial q}\frac{dx^k}{d\tau} - \Gamma^h_{mj}\Gamma^m_{ik}\frac{dx^i}{d\tau}\frac{dx^k}{d\tau}\frac{\partial x^j}{\partial q} +$$

$$+ 2\Gamma^h_{ij}\frac{dx^i}{d\tau}\frac{\partial^2 x^j}{\partial\tau\partial q} + \Gamma^h_{mk}\Gamma^m_{ij}\frac{dx^i}{d\tau}\frac{\partial x^j}{\partial q}\frac{dx^k}{d\tau},$$

that is

$$\frac{\nabla^2}{d\tau^2}\frac{\partial x^h}{\partial q} = \left(\frac{\partial\Gamma^h_{ij}}{\partial x^k} - \frac{\partial\Gamma^h_{ik}}{\partial x^j} + \Gamma^h_{mk}\Gamma^m_{ij} - \Gamma^h_{mj}\Gamma^m_{ik}\right)\frac{dx^i}{d\tau}\frac{\partial x^j}{\partial q}\frac{dx^k}{d\tau}.$$

Therefore, we have obtained the tidal acceleration equations

$$\frac{\nabla^2}{d\tau^2}\frac{\partial x^h}{\partial q} = R^h_{ikj}\frac{dx^i}{d\tau}\frac{\partial x^j}{\partial q}\frac{dx^k}{d\tau} = -R^h_{ijk}\frac{dx^i}{d\tau}\frac{\partial x^j}{\partial q}\frac{dx^k}{d\tau}.$$

□

If we denote by

$$K^h_j := R^h_{ijk} \frac{dx^i}{d\tau} \frac{dx^k}{d\tau}$$

the previous equality becomes

$$\frac{\nabla^2}{d\tau^2} \frac{\partial x^h}{\partial q} = -K^h_j \frac{\partial x^j}{\partial q}.$$

These last formulas are the geometric equivalent of the classical tidal acceleration equation. A further comment is necessary. The trace of the matrix K^h_j is K^h_h, that is R^h_{ihk}. The tidal acceleration equations can hide an equality as $R_{ik} = 0$ which can become the geometric equivalent of the classical vacuum field equation $\nabla^2 \Phi = 0$. Later, in the book, we will see how this becomes possible.

7.7 Kepler's Laws

According to the previous considerations, we intend now to obtain the three Kepler laws regarding the motion of planets around the Sun. It is necessary to understand how Newtonian Mechanics together with Euclidean Geometry describe these laws and, for this reason, let us prepare the geometric framework we need.

An *ellipse* of foci $F_1(f, 0)$ and $F_2(-f, 0)$, $f > 0$ is the locus of points P in the Euclidean plane such that $|PF_1| + |PF_2| = 2a$, where a is a positive constant, $a > f$. The equation of the ellipse can be found after we transform the condition $|PF_1| + |PF_2| = 2a$ into the equation

$$\sqrt{(x - f)^2 + y^2} + \sqrt{(x + f)^2 + y^2} = 2a.$$

The result is

$$\frac{x^2}{a^2} + \frac{y^2}{b^2} = 1$$

where $b^2 = a^2 - f^2$.

The line $F_1 F_2$ is called the *major axis* and the points where the ellipse cuts the major axis have the coordinates $(a, 0)$ and $(-a, 0)$.

The middle of the interval $F_1 F_2$ is called the center of the ellipse. In this case, the *center of ellipse* is the origin $O(0, 0)$.

The minor axis is perpendicular to the major axis at $O(0, 0)$. The minor axis intersects the ellipse at the points $(0, b)$ and $(0, -b)$.

The *eccentricity of the ellipse* is, by definition, $e := \dfrac{f}{a} = \dfrac{\sqrt{a^2 - b^2}}{a} = \sqrt{1 - \dfrac{b^2}{a^2}}.$

The area enclosed by the previous ellipse can be computed using the function $y(x) = b\sqrt{1 - \dfrac{x^2}{a^2}}$ which describes the arc of the ellipse $\{(x, y),\ x \in (-a, a),\ y > 0\}$. If we use the change of variable $x = a \sin t$ the enclosed area is

$$\mathbb{A} = 2 \int_{-a}^{a} y(x)dx = 2\frac{b}{a} \int_{-a}^{a} \sqrt{a^2 - x^2}dx = \pi a b.$$

If the ellipse has its center at (x_0, y_0) and the axes parallel to the axes of the system, i.e. the foci are $(x_0 + f, y_0)$ and $(x_0 - f, y_0)$, the equation is

$$\frac{(x - x_0)^2}{a^2} + \frac{(y - y_0)^2}{b^2} = 1.$$

In fact, the previous ellipse is parallel shifted with respect to the axis such that the old center $O(0, 0)$ becomes $O_1(x_0, y_0)$.

Consider an ellipse of eccentricity $0 < e < 1$ with a focus at $O(0, 0)$. Its major axis intersects the ellipse at the points $V\left(\dfrac{k}{1+e}, 0\right)$, $k > 0$, and $V'\left(-\dfrac{k}{1-e}, 0\right)$. The length of the major semi-axis is $a = \dfrac{k}{1 - e^2}$, the center of the ellipse is $\left(-\dfrac{ke}{1 - e^2}, 0\right)$ and the length of the minor semi-axis is $b = \dfrac{k}{\sqrt{1 - e^2}}$. The equation of this ellipse is

$$\frac{\left(x + \dfrac{ke}{1 - e^2}\right)^2}{\dfrac{k^2}{(1 - e^2)^2}} + \frac{y^2}{\dfrac{k^2}{1 - e^2}} = 1.$$

Problem 7.7.1 Find the locus of points $M(x, y)$ such that

$$r = r(\theta) = \frac{k}{1 + e \cos \theta},$$

where $r = |OM| = \sqrt{x^2 + y^2}$ and θ is the counterclockwise angle $\angle VOM$, $V \in Ox$.

Hint. The geometric meaning of $r + er \cos \theta = k$, $k > 0$, leads to the equation $\sqrt{x^2 + y^2} + ex = k$, i.e. $\sqrt{x^2 + y^2} = k - ex$. If $e = 1$, we obtain a parabola. If $e \neq 1$, after squaring, the previous equation can be written in the form

$$\frac{\left(x + \dfrac{ke}{1 - e^2}\right)^2}{\dfrac{k^2}{(1 - e^2)^2}} + \frac{y^2}{\dfrac{k^2}{1 - e^2}} = 1.$$

Let us observe that, for $0 < e < 1$, we have an ellipse equation. For $e > 1$, the equation is

$$\frac{(x - x_0)^2}{a^2} - \frac{(y - y_0)^2}{b^2} = 1,$$

i.e. we deal with a hyperbola. $\square$

We are ready to study the motion of planets under the action of gravitational force.

Consider the position of the Sun as $O(0, 0, 0)$. The motion of the Earth around the Sun depends on time, i.e. the position of the Earth is given by the vector $\overrightarrow{X}(t) = (x(t), y(t), z(t))$. Denote the length of this vector by

$$r(t) = \sqrt{x^2(t) + y^2(t) + z^2(t)}.$$

The Earth is attracted by the Sun via the gravitational force

$$\overrightarrow{F}(t) = -\frac{GmM}{r^3(t)} \overrightarrow{X}(t),$$

where M is the mass of the Sun, m is the mass of the Earth and G is the gravitational constant. The equation of motion of the Earth around the Sun, established by the Newton's second law, is

$$m\overset{..}{\overrightarrow{X}}(t) = -\frac{GmM}{r^3(t)} \overrightarrow{X}(t),$$

which can be written as

$$\overset{..}{\overrightarrow{X}}(t) = -\frac{GM}{r^3(t)} \overrightarrow{X}(t),$$

due to the validity of Galileo's Equivalence Principle. Let us denote $\mu = GM$ and $\overrightarrow{V} = \overset{.}{\overrightarrow{X}}$.

Theorem 7.7.2 *The motion of the Earth is planar, that is the entire trajectory is included in a plane which contains the Sun.*

Proof If we consider the derivative of the cross product between $\overrightarrow{X}(t)$ and $\overrightarrow{V}(t)$, successively we have

$$\frac{d}{dt}\left(\overrightarrow{X} \times \overrightarrow{V}\right) = \overset{.}{\overrightarrow{X}} \times \overrightarrow{V} + \overrightarrow{X} \times \overset{.}{\overrightarrow{V}} = \overrightarrow{V} \times \overrightarrow{V} + \overrightarrow{X} \times \overset{..}{\overrightarrow{X}} = \overrightarrow{0},$$

that is $\overrightarrow{X}(t) \times \overrightarrow{V}(t) = \overrightarrow{J}$, where $\overrightarrow{J}$ does not depend on t. Therefore, the vector $\overrightarrow{J}$ of length J is a constant vector, more precisely, it is the normal vector to the plane in which the motion of the Earth around the Sun happens. $\square$

Let us consider $z = 0$ the equation of the plane of motion, that is the position of the Earth is given by the vector $\vec{X}(t) = (x(t), y(t), 0)$. In the plane of motion, we consider polar coordinates $x = r \cos \theta$, $y = r \sin \theta$, with $r = r(t) = \sqrt{\dot{x}^2(t) + \dot{y}^2(t)}$; $\theta = \theta(t)$. We can prove:

Theorem 7.7.3 *If* $\vec{X}(t) = (r(t) \cos \theta(t), r(t) \sin \theta(t), 0)$ *and* $\vec{V}(t) = \dot{\vec{X}}(t)$ *it results*

(i) $\vec{J} = (0, 0, r^2 \dot{\theta})$

(ii) $r^2 \dot{\theta} = J$.

Proof We cancel t to write easier the next computations. Then

$$\vec{V} = \dot{\vec{X}} = (\dot{r} \cos \theta - r\dot{\theta} \sin \theta, \ \dot{r} \sin \theta + r\dot{\theta} \cos \theta, 0)$$

and

$$\vec{J} = \vec{X} \times \vec{V} = (0, 0, r^2 \dot{\theta}).$$

Since $\vec{J}$ is a constant vector, the last component does not depend on time, therefore it is a positive constant equal to its length J. So, both assertions are proved. □

Theorem 7.7.4 *The equation of motion for* $\vec{X}(t)$ *is transformed into the equation*

$$r(t)\ddot{r}(t) = \frac{J^2}{r^2(t)} - \frac{\mu}{r(t)}.$$

Proof We started from the equation of motion

$$\ddot{\vec{X}}(t) = -\frac{\mu}{r^3(t)} \vec{X}(t)$$

and, using it, we obtained that the motion is planar. In the plane of motion, the polar coordinates allow us to describe the normal vector $\vec{J}$ and to obtain the relation $r^2 \dot{\theta} = J$.

The derivative with respect to t of the relation $r^2 = \langle \vec{X}, \vec{X} \rangle$ leads to $r\dot{r} = \langle \vec{X}, \vec{V} \rangle$. Then, we have

$$(\dot{r})^2 + r\ddot{r} = \langle \vec{V}, \vec{V} \rangle + \langle \vec{X}, \vec{V} \rangle,$$

i.e.

$$(\dot{r})^2 + r\ddot{r} = \langle \vec{V}, \vec{V} \rangle + \langle \vec{X}, -\frac{\mu}{r^3} \vec{X} \rangle,$$

that is

$$(\dot{r})^2 + r\ddot{r} = |\vec{V}|^2 - \frac{\mu}{r}.$$

To compute $|\overrightarrow{V}|^2$, we start from the identity

$$\left(\overrightarrow{X}, \overrightarrow{V}\right)^2 + |\overrightarrow{X} \times \overrightarrow{V}|^2 = |\overrightarrow{X}|^2 |\overrightarrow{V}|^2.$$

If we replace, in the previous identity, $\left(\overrightarrow{X}, \overrightarrow{V}\right)$ with $r\dot{r}$, $|\overrightarrow{X} \times \overrightarrow{V}|^2$ with J^2, i.e. $(r^2\dot{\theta})^2$, and $|\overrightarrow{X}|^2$ with r^2, it results

$$(r\dot{r})^2 + (r^2\dot{\theta})^2 = r^2 |\overrightarrow{V}|^2,$$

thus

$$(\dot{r})^2 + r^2(\dot{\theta})^2 = |\overrightarrow{V}|^2.$$

Using $\dot{\theta} = \dfrac{J}{r^2}$, we obtain

$$|\overrightarrow{V}|^2 = (\dot{r})^2 + \frac{J^2}{r^2}.$$

It results

$$(\dot{r})^2 + r\ddot{r} = (\dot{r})^2 + \frac{J^2}{r^2} - \frac{\mu}{r},$$

which complete the proof. $\qquad\qquad\qquad\qquad\qquad\qquad\qquad\qquad\qquad\quad$ □

Theorem 7.7.5 *If* $r = \dfrac{1}{u}$ *and* $u = u(\theta)$, *the equation*

$$r\ddot{r} = \frac{J^2}{r^2} - \frac{\mu}{r}$$

becomes

$$\frac{d^2u}{d\theta^2} + u = \frac{\mu}{J^2}.$$

Proof We first show that $\dot{r} = -J\dfrac{du}{d\theta}$. To obtain this, let us observe that, successively, we have

$$\dot{r} = -\frac{\dot{u}}{u^2} = -\frac{1}{u^2}\frac{du}{dt} = -\frac{1}{u^2}\frac{du}{d\theta}\frac{d\theta}{dt} = -r^2\frac{du}{d\theta}\dot{\theta} = -J\frac{du}{d\theta}.$$

Then, $\ddot{r} = -J\dfrac{d^2u}{d\theta^2}\dot{\theta}$, i.e.

$$\ddot{r} = -J^2\frac{1}{r^2}\frac{d^2u}{d\theta^2}.$$

Taking into account $r = \dfrac{1}{u}$ and replacing into

$$r\ddot{r} = \frac{J^2}{r^2} - \frac{\mu}{r},$$

we obtain the desired equation

$$\frac{d^2 u}{d\theta^2} + u = \frac{\mu}{J^2}.$$

$\square$

The general solution is $u(\theta) = A\cos(\theta - \theta_0) + \dfrac{\mu}{J^2}$, where A is an arbitrary constant and θ_0 is the initial value, called phase, who leads to the starting point of the trajectory.

If we are interested only in the shape of the solution, we may consider

$$u(\theta) = A\cos\theta + \frac{\mu}{J^2}.$$

The solution in r is

$$r(\theta) = \frac{\dfrac{J^2}{\mu}}{1 + \dfrac{J^2 A}{\mu}\cos\theta}.$$

If A is in such a way that

$$0 < e := \frac{J^2 A}{\mu} < 1$$

the trajectory is an ellipse. Therefore we have proved.

Theorem 7.7.6 (Kepler's first law) *In the case of the pair {Sun, Earth}, the gravity makes Earth to move around the Sun after an elliptical orbit having the Sun as one of the foci.*

This is the Kepler first law. It generally describes how a planet moves around a star.

Let see again the big picture of the motion of the Earth around Sun. We have the Sun at the origin of the coordinate system and the Earth position given by the vector $\vec{X}(t) = (x(t), y(t), z(t))$. The gravitational force acting between the two bodies leads to the equation of motion

$$\ddot{\vec{X}}(t) = -\frac{GM}{r^3(t)}\vec{X}(t).$$

The motion is planar. Using polar coordinates, we can transform the equation into the new equation

$$r\ddot{r} = \frac{J^2}{r^2} - \frac{\mu}{r},$$

and finally into the equation

$$\frac{d^2u}{d\theta^2} + u = \frac{\mu}{J^2}$$

which can be solved. The solution, in polar coordinates, is the ellipse

$$r(\theta) = \frac{\dfrac{J^2}{\mu}}{1 + \dfrac{J^2 A}{\mu}\cos\theta}.$$

Defining $k := \dfrac{J^2}{\mu}$ and $e := \dfrac{J^2 A}{\mu}$, in Cartesian coordinates, the equation is

$$\frac{\left(x + \dfrac{ke}{1 - e^2}\right)^2}{\dfrac{k^2}{(1-e^2)^2}} + \frac{y^2}{\dfrac{k^2}{1-e^2}} = 1,$$

where the semi-axes are $a := \dfrac{k}{(1-e^2)}$ and $b := \dfrac{k}{\sqrt{(1-e^2)}}$. The *perihelion* of the trajectory, that is the closest position to the Sun, is at the point $V\left(\dfrac{k}{1+e}, 0\right)$. The *aphelion*, that is the furthest position from the Sun, is located at the point $V'\left(-\dfrac{k}{1-e}, 0\right)$.

If we look at comets, the trajectories can be elliptic, hyperbolic and parabolic. The case $e = 1$ is a possible case, but it is difficult for an astronomer to say that a comet has a parabolic orbit. It is more probable to have a hyperbolic orbit with $e > 1$ but very close to 1. We prefer to remain at the case {planet, Sun} where the trajectories are always ellipses. Now we are able to prove the Kepler second law.

Theorem 7.7.7 (Kepler's second law) *Areas swept out by $\overrightarrow{OX}$ in equal time intervals are equal.*

Proof Consider two close positions of $\overrightarrow{OX}$, that is $\overrightarrow{OX'}$ and $\overrightarrow{OX''}$. The angle between this two positions is $d\theta$. The infinitesimal area swept by $\overrightarrow{OX}$ is $dA = \dfrac{1}{2}r^2 d\theta$. It results $\dfrac{dA}{dt} = \dfrac{1}{2}r^2\dot{\theta}$, i.e.

$$\frac{dA}{dt} = \frac{1}{2}J,$$

which ends the proof.   □

Let us continue with the Kepler third law. The time necessary to have a complete revolution around the Sun is called the *orbital period of a planet*. It is denoted by T.

Theorem 7.7.8 (Kepler's third law) *The ratio between the square of the orbital period and the cube of the major semi-axis is a constant, that is* $\dfrac{T^2}{a^3} = \dfrac{4\pi^2}{GM}$.

Proof Let us observe that $\overrightarrow{OX}$ sweeps the area of the ellipse during a revolution. Thus

$$\pi ab = T\frac{1}{2}r^2\dot{\theta}.$$

It results $\dfrac{T}{a} = 2\pi\dfrac{b}{J}$, that is $\dfrac{T^2}{a^2} = 4\pi^2\dfrac{b^2}{J^2}$.

According to previous formulas for semi-axes we have $b^2 = \dfrac{k^2}{1-e^2} = k\dfrac{k}{1-e^2} = ka$, therefore

$$\frac{T^2}{a^2} = 4\pi^2\frac{ka}{J^2}.$$

Taking into account that $k := \dfrac{J^2}{\mu}$ we finally obtain

$$\frac{T^2}{a^3} = 4\pi^2\frac{1}{\mu}.$$

                                                                                         □

The third law is called the *Harmony law* because, if we consider two different planets moving around the Sun, the same constant is the ratio between $\dfrac{T_1^2}{a_1^3}$ and $\dfrac{T_2^2}{a_2^3}$.

7.8 Circular Motion, Centripetal Force and Dark Matter Problem

Before continuing, let us discuss a little bit about circular motion and observe the differences with respect to the elliptical motion presented above. Circular motion means a movement of an object along the circumference of a circle. A boy rotating a tide up ball with a chord, a car moving at constant speed on a circular track, or even a satellite on its orbit around the Earth can be mathematically modeled as circular motions. So, the trajectory is a circle of radius R, the object in circular motion can be imagined as a point (with a mass, say m) moving at constant speed v. The speed vector $\overrightarrow{v}$ is tangent at each point of the circle. To maintain the point on this trajectory, the force vector (that is the acceleration vector, too) has to be imagined as an arrow oriented from the point to the center. Of course, the magnitude of the force has to be the same for all the possible positions, because there are not differences between these vectors except the possible directions. This force is called a *centripetal force*. The corresponding acceleration is called *centripetal acceleration*.

Let us consider two tangent vectors corresponding to two close points on the circumference separated by a $d\theta$ angle. Denote by dx the length of the arc determined by the two points and observe that between the two tangent vectors there is the same angle $d\theta$. We have $dx = R d\theta$ and $v = \dfrac{dx}{dt}$. If dv is the vector which connects their ends, we may approximate $d\theta = \dfrac{dv}{v}$. It results

$$dx = R d\theta = R \frac{dv}{v} = v dt,$$

that is

$$a := \frac{dv}{dt} = \frac{v^2}{R}.$$

This is the formula of the centripetal acceleration which allows to write the formula of the centripetal force:

$$F_c := m \frac{v^2}{R}.$$

How it can be imagined the rotation of the Earth around the Sun using this force? The mathematical answer is

$$F_c = \frac{mv^2}{R} = \frac{GMm}{R^2} = F,$$

i.e.

$$v^2 = \frac{GM}{R},$$

thanks to the Equivalence Principle by which m can be simplified in both sides of the equation. This is important because if the radius R is increasing the orbital speed has to decrease.

Now, since the period of revolution around the circular trajectory is $T = \dfrac{2\pi R}{v}$, we obtain

$$T^2 = 4\pi^2 R^2 \frac{1}{v^2} = \frac{4\pi^2 R^3}{GM},$$

that is T^2 is proportional to R^3, or

$$\frac{T^2}{R^3} = \frac{4\pi^2}{GM}.$$

It is a sort of approximation of the third Kepler law. The centripetal force is often used in approximations of trajectories in Astronomy. An interesting application of the centripetal force is the possible existence of *dark matter* or, according to Fritz Zwicky the *missing matter* [15]. The formula

$$v^2 = \frac{GM}{R},$$

which asserts that if R is increasing, the speed v decreases (if M remains constant), is crucial.

In a galaxy, there are billions of stars. We may think that these stars are in an imaginary sphere having as a center, the center of the galaxy. Some stars are closer to the center of the galaxy, some of them are far. Some other stars are out of the edge of the galaxy, or more precisely, they are in the area where, if we increase the radius of the galaxy, we add few stars. For stars in the zone with a lot of stars, if we increase the radius we have more stars, i.e. more mass. Here, the fact that the observed speed of stars rotating around the center is the same it is not a problem. The speed v can be kept constant, if the mass M increases when R increases. But for distant stars, when we increase the radius, we do not add more mass inside. However the measured speed v is the same and it is more or less constant also very far from the galactic center (more than 10 kpc). According to this situation, we have to suppose the existence of a sort of (sub-luminous) matter that cannot be detected by the standard electromagnetic emission. However, the amount of such a matter increases with the increasing of the distance from the center. The problem is known as the dark matter problem and can be solved in two alternative ways: Either one suppose the existence of exotic matter interacting only gravitationally, or one assumes deviation from the Kepler laws at large distances. More details can be found in [18, 34, 35].

Later in the book, we will study the trajectory of planets in a given metric. Specifically, we will study the trajectory of planets both in the Schwarzschild metric

$$ds^2 = c^2 \left(1 - \frac{2GM}{c^2 r}\right) dt^2 - \frac{1}{1 - \frac{2GM}{c^2 r}} dr^2 - r^2 d\varphi^2 - r^2 \sin^2 \varphi d\theta^2$$

and in the Einstein metric

$$ds^2 = c^2 \left(1 - \frac{2GM}{c^2 r}\right) dt^2 - \left(1 + \frac{2GM}{c^2 r}\right) dr^2 - r^2 d\varphi^2 - r^2 \sin^2 \varphi d\theta^2.$$

The planet equation of motion in both metric is

$$\frac{d^2u}{d\theta^2} + u = \frac{\mu}{J^2} + \frac{3\mu}{c^2}u^2$$

where c is the constant speed of light in vacuum and $\mu = GM$ as previously defined.

Why we study the equation of motion in a metric and how close is the solution of this new equation to the above classical solution? These topics will be discussed in the General Relativity chapter of this book.

7.9 The Mechanical Lagrangian

In a system of coordinates (t, x), let $(t, x(t))$ be the trajectory of a particle of mass m moving under the influence of a force derived from a time independent *potential* V. Since V depends only on the position, we denote this by $V := V(x)$.

Newton's equation of motion is

$$m\ddot{x}(t) = F(x),$$

where the force acting on the particle is $F(x) = -\dfrac{dV}{dx}$.

Given some initial conditions, the trajectory $(t, x(t))$ is comprised between the initial point $(t_1, x(t_1))$ and the final point $(t_2, x(t_2))$.

Let us underline that this trajectory is the expression of the force acting on the particle under some initial conditions. Therefore, there is an unique trajectory determined by the force and the initial conditions.

Now let us consider all the paths connecting $(t_1, x(t_1))$ and $(t_2, x(t_2))$. They can be thought as $y(t) + \eta(t)$, with $y(t_1) = x(t_1)$, $y(t_2) = x(t_2)$, $\eta(t_1) = \eta(t_2) = 0$.

Having all these paths, what new theory do we need to imagine in order to discover the original path described by the Newton's equation of motion?

To answer this question, we need some technical details (see also [24]).

Let us insist on this first part when we have described what we want to do. We have used V such that $F = -\dfrac{dV}{dx}$. We defined V as an independent potential and we suggested its connection with the force F, $dV = -F dx$.

Is this definition connected to the facts seen in our previous sections when we have studied the gravitational force and the gravitational potential? The answer is yes, but we need to point out a major difference between this V and the gravitational potential Φ.

Consider a body of mass M at the origin O of a line whose current coordinate is denoted by x. Suppose that at point $N(x)$, a body of mass m exists. The gravitational force in this case has the intensity $F = \dfrac{GMm}{x^2}$. The work done by the body of mass M to move the body of mass m from x to $x - dx$ is $-F dx$. There is an energy transferred to do this work. Its variation ΔE is $-Fdx$. By definition, the *gravitational potential energy* $P_E(r)$ is related to the work done to move the body of mass m from the infinity to the point having coordinate r, that is

$$P_E(r) = \int_\infty^r Fdx = \int_\infty^r \frac{GMm}{x^2}dx = -\frac{GMm}{r}.$$

The potential energy can be denoted by P_E. If one looks at the formula obtained and takes into account the formula of the gravitational potential $-\dfrac{GM}{r}$, we can understand both the explanations above and the relation

$$P_E = m\Phi.$$

Therefore, another definition for the gravitational potential appears: the work (energy transferred) per unit mass necessary to move a body from infinity to the point having the coordinate r. Indeed,

$$\Phi(r) = \frac{1}{m}\int_\infty^r Fdx = \frac{1}{m}\int_\infty^r \frac{GMm}{x^2}dx = -\frac{GM}{r}.$$

In the case when we consider the constant gravitational field determined by the constant acceleration g between the origin O and a point H at the coordinate h, the potential energy is expressed by the formula $P_E = mgh$. The explanation is related to the difference of formal integrals

$$P_E := P_E(h) - P_E(0) = \int_\infty^h gmdx - \int_\infty^0 gmdx = \int_0^h gmdx = gmh$$

which describes the amount of energy necessary to move the body at h to 0.

In the same way, we can define the kinetic energy. Let us start from $F = ma = m\dfrac{dv}{dt}$ written in its discrete form, $F = m\dfrac{\Delta v}{\Delta t}$. If we multiply by Δr, we obtain $F\Delta r = m\dfrac{\Delta v}{\Delta t}\Delta r = m\dfrac{\Delta r}{\Delta t}\Delta v = mv\Delta v$, which can be written in the differential way as

$$Fdr = mvdv.$$

Now, the amount of energy necessary to bring a body initially at rest to the speed v is

$$T(v) = \int_0^v Fdx = \int_o^v mxdx = m\frac{v^2}{2}.$$

Since v can be seen as $\dot{x}(t)$, we may consider the *kinetic energy of the mechanical system* defined by the formula $T = T(\dot{x}) := \dfrac{1}{2}m(\dot{x}(t))^2$. Another possible notation is K_E. Here, with *mechanical system* we intend a system of elements that interact on mechanical principles. A material point and a force which acts on it is a possible example. Two materials points which interact through the gravitational force offer another example. In this perspective, the next exercise has important consequences in Newtonian mechanics.

Exercise 7.9.1 Consider a mechanical system whose kinetic energy is $T(\dot{x}) := \frac{1}{2}m(\dot{x}(t))^2$ and its potential energy is V (such that the force which acts is $F(x) = -\frac{dV}{dx}$). Show that the total energy of the system, $T + V$, is a constant.

Hint. If we derive with respect t the total energy, we obtain

$$\frac{d}{dt}(T+V) = \left(m\dot{x}(t)\ddot{x}(t) + \frac{dV}{dx}\frac{dx}{dt} \right) = (m\ddot{x}(t) - F)\,\dot{x}(t) = 0,$$

that is $T + V$ is a constant.

We define the *mechanical Lagrangian* of the system by

$$L = L(x, \dot{x}) := T - V = \frac{1}{2}m(\dot{x}(t))^2 - V(x).$$

In this section, where there is no possibility of confusion, we simply use the definition "Lagrangian" instead of mechanical Lagrangian. Later in the book, we will see that exist general Lagrangians which come from Geometry, therefore we have to well understand the nature of the Lagrangian we are considering.

Let us observe that, even if x and $\dot{x}$ depends on t, this Lagrangian is only implicitly a function of time.

In this formalism, it makes sense to consider a functional called *action*,

$$S[y] = \int_{t_1}^{t_2} \left[\frac{1}{2}m(\dot{y}(t))^2 - V(y) \right] dt$$

which exists for any path $y(t)$, not only for the "physical right on" which is $x(t)$.

Now consider the action corresponding to $y(t) + \eta(t)$,

$$S[y + \eta] = \int_{t_1}^{t_2} \left[\frac{1}{2}m(\dot{y}(t) + \dot{\eta}(t))^2 - V(y(t) + \eta(t)) \right] dt.$$

We have, after expanding V in Taylor series with respect to $y(t)$,

$$S[y + \eta] = S[y] + \int_{t_1}^{t_2} \left[m\dot{y}(t)\dot{\eta}(t) - \frac{dV}{dy}(y(t))\eta(t) \right] dt + \mathbb{O}(\eta^2),$$

where $\mathbb{O}(\eta^2)$ are terms of order $\eta^2 := \eta^2(t)$ or higher. We can write

$$S[y + \eta] = S[y] + \delta S + \mathbb{O}(\eta^2),$$

where

$$\delta S = \int_{t_1}^{t_2} \left[m\dot{y}(t)\dot{\eta}(t) - \frac{dV}{dy}(y(t))\eta(t) \right] dt,$$

is called the *first order variation of the action S*. Since $\eta(t_1) = \eta(t_2) = 0$, we obtain

$$\delta S = \int_{t_1}^{t_2} \left[m\dot{y}(t)\dot{\eta}(t) - \frac{dV}{dy}(y(t)\eta(t)) \right] dt =$$

$$= \int_{t_1}^{t_2} \left[m\frac{d(\dot{y}(t)\eta(t))}{dt} - m\ddot{y}(t)\eta(t) - \frac{dV}{dy}(y(t)\eta(t)) \right] dt =$$

$$= m\dot{y}(t_2)\eta(t_2) - m\dot{y}(t_1)\eta(t_1) - \int_{t_1}^{t_2} \left[m\ddot{y}(t) + \frac{dV}{dy}(y(t)) \right] \eta(t)dt =$$

$$= - \int_{t_1}^{t_2} \left[m\ddot{y}(t) + \frac{dV}{dy}(y(t)) \right] \eta(t)dt.$$

Therefore, $\delta S \equiv 0$ means

$$\int_{t_1}^{t_2} \left[m\ddot{y}(t) + \frac{dV}{dy}(y(t)) \right] \eta(t)dt = 0$$

for every η, and it happens if and only if $m\ddot{y}(t) + \frac{dV}{dy}(y(t)) = 0$, i.e. for $y(t) = x(t)$.
We have proved:

Theorem 7.9.2 *The first order variation of the action S vanishes, i.e.*

$$\delta S = \int_{t_1}^{t_2} \left[m\dot{y}(t)\dot{\eta}(t) - \frac{dV}{dy}(y(t))\eta(t) \right] dt = 0,$$

if and only if $y(t)$ satisfies Newton's equation of motion

$$m\ddot{x}(t) - F(x) = 0.$$

So, the answer is: *The "physical right path" happens when the first order variation δS vanishes*. Therefore the right path is described by the condition $\delta S \equiv 0$. This is known as the *Hamilton's stationary action principle*.

7.10   Geometry Induced by a Lagrangian

Now, let us consider another problem.

Can we find an equation, satisfied by a general function $L(x, \dot{x})$, not only by the mechanical Lagrangian $L = T - V$ as before, such that the function $x = x(t)$, which connects the given points $(t_1, x(t_1))$; $(t_2, x(t_2))$ where the functional

$$S[x] = \int_{t_1}^{t_2} L(x(t), \dot{x}(t))dt .$$

is extremized?

Let us explain first what is the mathematical meaning of the words "extremizes the functional S." Consider all the perturbation of $x(t)$, say

$$y_\lambda(t) = x(t) + \lambda \eta(t), \ \lambda \in \mathbb{R}$$

which preserves the endpoints $(t_1, x(t_1))$; $(t_2, x(t_2))$, that is $\eta(t_1) = \eta(t_2) = 0$ and construct the action

$$S_\lambda[y_\lambda] = \int_{t_1}^{t_2} L_\lambda(y_\lambda(t), \dot{y}_\lambda(t))dt = \int_{t_1}^{t_2} L_\lambda(x(t) + \lambda \eta(t), \dot{x}(t) + \lambda \dot{\eta}(t))dt.$$

Extremizing the functional $S[x]$ means or $S_\lambda[y_\lambda] \geq S[x]$ for any $\lambda \in \mathbb{R}$ or $S_\lambda[y_\lambda] \leq S[x]$ for any $\lambda \in \mathbb{R}$, where the equality works if and only if $\lambda = 0$.

Therefore, *extremizing the functional $S[x]$ implies the condition* $\left.\dfrac{dS_\lambda}{d\lambda}\right|_{\lambda=0} \equiv 0.$

Since

$$\frac{dL_\lambda}{d\lambda} = \frac{\partial L_\lambda}{\partial y_\lambda}\frac{\partial y_\lambda}{\partial \lambda} + \frac{\partial L_\lambda}{\partial \dot{y}_\lambda}\frac{\partial \dot{y}_\lambda}{\partial \lambda} = \frac{\partial L_\lambda}{\partial y_\lambda}\eta(t) + \frac{\partial L_\lambda}{\partial \dot{y}_\lambda}\dot{\eta}(t),$$

it results

$$\left.\frac{dL_\lambda}{d\lambda}\right|_{\lambda=0} = \frac{\partial L}{\partial x}\eta(t) + \frac{\partial L}{\partial \dot{x}}\dot{\eta}(t),$$

therefore the condition $\left.\dfrac{dS_\lambda}{d\lambda}\right|_{\lambda=0} \equiv 0$ is written as

$$\left.\frac{dS_\lambda}{d\lambda}\right|_{\lambda=0} = \int_{t_1}^{t_2}\left[\frac{\partial L}{\partial x}\eta(t) + \frac{\partial L}{\partial \dot{x}}\dot{\eta}(t)\right]dt \equiv 0.$$

Definition 7.10.1 The curve $x = x(t)$ which extremizes the functional

$$S[x] = \int_{t_1}^{t_2} L(x(t), \dot{x}(t))dt$$

is called a stationary point of the functional $S[x]$.

Theorem 7.10.2 (Euler–Lagrange equation) *The curve $x = x(t)$ which connects the given points $(t_1, x(t_1))$, $(t_2, x(t_2))$ satisfies the Euler–Lagrange equation*

$$\frac{d}{dt}\left(\frac{\partial L}{\partial \dot{x}}\right) - \frac{\partial L}{\partial x} = 0$$

if and only if it is a stationary point of the functional

$$S[x] = \int_{t_1}^{t_2} L(x(t), \dot{x}(t))dt.$$

Proof Using the integration by parts

$$\int_{t_1}^{t_2} \frac{\partial L}{\partial x}\eta(t)dt + \int_{t_1}^{t_2} \frac{\partial L}{\partial \dot{x}}\dot{\eta}(t)dt =$$

$$= \int_{t_1}^{t_2} \frac{\partial L}{\partial x}\eta(t)dt + \frac{\partial L}{\partial \dot{x}}\eta(t_2) - \frac{\partial L}{\partial \dot{x}}\eta(t_1) - \int_{t_1}^{t_2} \frac{d}{dt}\left(\frac{\partial L}{\partial \dot{x}}\right)\eta(t)dt =$$

$$= \int_{t_1}^{t_2} \left[\frac{\partial L}{\partial x} - \frac{d}{dt}\left(\frac{\partial L}{\partial \dot{x}}\right)\right]\eta(t)dt.$$

The condition $\left.\dfrac{dS_\lambda}{d\lambda}\right|_{\lambda=0} \equiv 0$ means

$$\int_{t_1}^{t_2} \left[\frac{\partial L}{\partial x} - \frac{d}{dt}\left(\frac{\partial L}{\partial \dot{x}}\right)\right]\eta(t)dt = 0\,,$$

for every function η. We obtain

$$\frac{\partial L}{\partial x} - \frac{d}{dt}\left(\frac{\partial L}{\partial \dot{x}}\right) = 0.$$

$\square$

Another proof can be considered for the Euler–Lagrange equation. As previously, let us consider the action

$$S[y] = \int_{t_1}^{t_2} L(y(t), \dot{y}(t))dt\,.$$

Now consider the action corresponding to $y(t) + \eta(t)$,

$$S[y + \eta] = \int_{t_1}^{t_2} L(y(t) + \eta(t), \dot{y}(t) + \dot{\eta}(t))dt,$$

where $\eta(t_1) = \eta(t_2) = 0$. After expanding L in Taylor series with respect the variables y and $\dot{y}$ we obtain

$$L(y(t) + \eta(t), \dot{y}(t) + \dot{\eta}(t)) = L(y(t), \dot{y}(t)) + \frac{\partial L}{\partial y}\eta + \frac{\partial L}{\partial \dot{y}}\dot{\eta} + \mathbb{O}(\eta^2) + \mathbb{O}(\dot{\eta}^2).$$

The first order variation of the action S is

$$\delta S = \int_{t_1}^{t_2} \left[\frac{\partial L}{\partial y}\eta(t) + \frac{\partial L}{\partial \dot{y}}\dot{\eta}(t)\right] dt.$$

Using the integration by parts and the conditions $\eta(t_1) = \eta(t_2) = 0$, it results successively

$$\delta S = \int_{t_1}^{t_2} \frac{\partial L}{\partial y}\eta(t)dt + \int_{t_1}^{t_2} \frac{\partial L}{\partial \dot{y}}\dot{\eta}(t)dt =$$

$$= \int_{t_1}^{t_2} \frac{\partial L}{\partial y}\eta(t)dt + \frac{\partial L}{\partial \dot{y}}\eta(t_2) - \frac{\partial L}{\partial \dot{y}}\eta(t_1) - \int_{t_1}^{t_2} \frac{d}{dt}\left(\frac{\partial L}{\partial \dot{y}}\right)\eta(t)dt =$$

$$= \int_{t_1}^{t_2} \left[\frac{\partial L}{\partial y} - \frac{d}{dt}\left(\frac{\partial L}{\partial \dot{y}}\right)\right]\eta(t)dt$$

The first order variation of action vanishes if the last integral vanishes, i.e. $\delta S \equiv 0$ iff

$$\int_{t_1}^{t_2} \left[\frac{\partial L}{\partial y} - \frac{d}{dt}\left(\frac{\partial L}{\partial \dot{y}}\right)\right]\eta(t)dt = 0$$

for any function η. This means

$$\frac{\partial L}{\partial y} - \frac{d}{dt}\left(\frac{\partial L}{\partial \dot{y}}\right) = 0. \ \square$$

Both proofs reported before hold even if the Lagrangian is $L(t, y(t), \dot{y}(t))$ instead of $L(y(t), \dot{y}(t))$. In the particular case, when the Lagrangian does not depend explicitly on t, the Euler–Lagrange equation reduces to the Beltrami identity. The following theorem holds.

Theorem 7.10.3 (Beltrami's identity) *If the Lagrangian does not depend explicitly on t, then a constant C exists such that*

$$L(y, \dot{y}) - \dot{y}\frac{\partial L(y, \dot{y})}{\partial \dot{y}} = C.$$

Proof The total derivative of $L(t, y(t), \dot{y}(t))$ is

$$\frac{dL}{dt} = \frac{\partial L}{\partial t} + \frac{\partial L}{\partial y}\dot{y} + \frac{\partial L}{\partial \dot{y}}\ddot{y},$$

i.e.

$$\frac{\partial L}{\partial y}\dot{y} = \frac{dL}{dt} - \frac{\partial L}{\partial t} - \frac{\partial L}{\partial \dot{y}}\ddot{y}.$$

If $\dfrac{\partial L}{\partial t} = 0$, the previous equality becomes

$$\frac{\partial L}{\partial y}\dot{y} = \frac{dL}{dt} - \frac{\partial L}{\partial \dot{y}}\ddot{y}.$$

Multiplying the Euler–Lagrange equation by $\dot{y}$, we obtain

$$\dot{y}\frac{\partial L}{\partial y} = \dot{y}\frac{d}{dt}\left(\frac{\partial L}{\partial \dot{y}}\right),$$

therefore, after combining the last two equalities we have

$$\frac{dL}{dt} - \dot{y}\frac{d}{dt}\left(\frac{\partial L}{\partial \dot{y}}\right) - \frac{\partial L}{\partial \dot{y}}\ddot{y} = 0,$$

that is

$$\frac{d}{dt}\left(L - \dot{y}\frac{\partial L}{\partial \dot{y}}\right) = 0,$$

which is equivalent to the statement. □

Let us consider now a important problem solved first using the equilibrium of the forces involved, afterwards using the Euler–Lagrange equation. We are talking about the problem of hanging rope.

Problem 7.10.4 The catenary problem: Suppose that a rope is hanged with its ends at the same height above the floor and its mass on the unit length is ρ. Find the function which describes the shape of the rope.

Solution I: Consider a frame of coordinates such that the two given points are $A(-a, b)$, $B(a, b)$ and the shape is described by the points of the curve $(x, y(x))$. The statement conditions allow us to consider a minimum point at $O(0, 0)$, a symmetry with respect to Oy-axis and Ox-axis as a tangent to the curve at O. Consider a point $M(x, y(x))$ on the arc OB and the tangent at M. Let us denote by s the length of the arc OM, that is

$$s(x) = \int_0^x \sqrt{1 + (\dot{y}(t))^2}\,dt.$$

From Leibniz integral rule,

$$s(x) = \int_0^x \sqrt{1 + (\dot{y}(t))^2}dt = F(x) - F(0)$$

where $\dfrac{dF}{dt} = \sqrt{1 + (\dot{y}(t))^2}$. Therefore

$$\frac{ds}{dx} = \sqrt{1 + (\dot{y}(x))^2}.$$

There are three forces at equilibrium which act on the given arc. The tension $(-T_0, 0)$ at O, the weight of the arc $(0, -\rho g s)$, where g is the acceleration due to gravity, and the tension of magnitude T at M, $(T \cos \theta, T \sin \theta)$, this one acting along the tangent to $(x, y(x))$ at M. Therefore

$$(-T_0, 0) + (0, -\rho g s) + (T \cos \theta, T \sin \theta) = (0, 0).$$

The equilibrium conditions are

$$\begin{cases} T \cos \theta = T_0 \\ T \sin \theta = \rho g s. \end{cases}$$

It results

$$\dot{y}(x) = \frac{dy}{dx} = \tan \theta = \frac{\rho g}{T_0} s,$$

i.e.

$$\ddot{y}(x) = \frac{\rho g}{T_0}\frac{ds}{dx} = \frac{\rho g}{T_0}\sqrt{1 + (\dot{y}(x))^2}.$$

If we denote $k := \dfrac{\rho g}{T_0}$ and $u = \dot{y}(x)$, it remains to solve the equation

$$\frac{\dot{u}(x)}{\sqrt{1 + (u(x))^2}} = k$$

which leads to

$$\int \frac{du}{\sqrt{1 + u^2}} = k \int dx.$$

Since $u(0) = \dot{y}(0) = 0$, the equality

$$u + \sqrt{1 + u^2} = e^{bx+l}$$

implies $l = 0$ and

$$u(x) = \sinh 2kx,$$

i.e.

$$y(x) = \frac{T_0}{2\rho g} \cosh\left(\frac{2\rho g}{T_0}x\right) - \frac{T_0}{2\rho g}. \quad \square$$

Solution II: The rope has a given length

$$\mathbf{l}_{a,b} = \int_{-a}^{a} \sqrt{1 + (\dot{y}(x))^2}dx,$$

and we can think at a Lagrangian induced by the potential energy of the rope combined with the constraint of finite length for the rope,

$$\mathcal{L} = \rho g y(x)\sqrt{1 + (\dot{y}(x))^2} + \alpha\left(\sqrt{1 + (\dot{y}(x))^2} - \mathbf{l}_{a,b}\right),$$

where α is a constant. Without the length constraint, the potential energy is smaller and smaller while the rope is longer and longer. Finally, we can try to derive the curve starting from the Lagrangian

$$\mathcal{L} = (\rho g y + \alpha)\sqrt{1 + \dot{y}^2} + \beta,$$

where β is a constant. Since $\dfrac{\partial \mathcal{L}}{\partial t} = 0$, we can use Beltrami's identity. Therefore it exists a constant C such that

$$\mathcal{L} - \dot{y}\frac{\partial \mathcal{L}}{\partial \dot{y}} = C$$

which means

$$(\rho g y + \alpha)\sqrt{1 + \dot{y}^2} - \dot{y}(\rho g y + \alpha)\frac{\dot{y}}{\sqrt{1 + \dot{y}^2}} = C,$$

i.e.

$$(\rho g y + \alpha)\frac{1}{\sqrt{1 + \dot{y}^2}} = C.$$

It remains to solve

$$\dot{y}^2 = \frac{(\rho g y + \alpha)^2}{C^2} - 1.$$

The substitution $Cu = \rho g y + \alpha$ leads to

$$\frac{C}{\rho g}\dot{u} = \sqrt{u^2 - 1}$$

i.e.

$$\frac{du}{\sqrt{u^2 - 1}} = \frac{\rho g}{C} dx$$

with the solution $u(x) = \cosh\left(\frac{\rho g}{C} x + \gamma\right)$, where γ is a constant. Therefore

$$y(x) = \frac{C}{\rho g} \cosh\left(\frac{\rho g}{C} x + \gamma\right) - \alpha.$$

The constants are determined from the symmetry condition with respect to Oy-axis, that is $\gamma = 0$, $y(0) = 0$ that is $\alpha = \dfrac{C}{\rho g}$ and C from

$$\mathbf{l}_{a,b} = \int_{-a}^{a} \sqrt{1 + (\dot{y}(x))^2} dx. \ \square$$

Let us return at the first proof we offered for the Euler–Lagrange equation. That proof can be used to obtain the general Euler–Lagrange equations.

For the Lagrangian $L = L(x^0, x^1, \ldots, x^n, \dot{x}^0, \dot{x}^1, \ldots, \dot{x}^n)$, we obtain

$$\frac{\partial L}{\partial x^k} - \frac{d}{dt}\left(\frac{\partial L}{\partial \dot{x}^k}\right) = 0, \quad k = 0, 1, \ldots, n.$$

It is easy to see that we have to act as before on each pair of variables x^k, $\dot{x}^k$, $k = 0, 1, \ldots, n$. We are looking for a system of equations satisfied by the previous Lagrangian, such that a curve $x = x(t) = (x^0(t), x^1(t), \ldots, x^n(t))$, which connects the given points $(t_1, x^0(t_1), x^1(t_1), \ldots, x^n(t_1))$, $(t_2, x^0(t_2), x^1(t_2), \ldots, x^n(t_2))$, extremizes the functional

$$S[x] = \int_{t_1}^{t_2} L(x^0(t), \dot{x}^0(t), x^1(t), \dot{x}^1(t), \ldots, x^n(t), \dot{x}^n(t)) dt.$$

As previously, a perturbation of $x(t)$ which preserves the endpoints is

$$y_\lambda(t) = (y^0_\lambda(t), y^1_\lambda(t), \ldots, y^n_\lambda(t)) = (x^0(t) + \lambda\eta_0(t), x^1(t) + \lambda\eta_1(t), \ldots, x^n(t) + \lambda\eta_n(t)),$$

$\lambda \in \mathbb{R}$ with $\eta_k(t_1) = \eta_k(t_2) = 0$, $k = 0, 1, \ldots, n$. Consider

$$S_\lambda[y_\lambda] = \int_{t_1}^{t_2} L_\lambda(y^0_\lambda(t), \dot{y}^0_\lambda(t), \ldots, y^n_\lambda(t), \dot{y}^n_\lambda(t)) dt =$$

$$= \int_{t_1}^{t_2} L_\lambda(x^0(t) + \lambda\eta_0(t), \dot{x}^0(t) + \lambda\dot{\eta}_0(t), \ldots, x^n(t) + \lambda\eta_n(t), \dot{x}^n(t) + \lambda\dot{\eta}_n(t)) dt.$$

Extremizing the functional S implies the condition $\left.\dfrac{dS_\lambda}{d\lambda}\right|_{\lambda=0} \equiv 0$.

Or,

$$\frac{dL_\lambda}{d\lambda} = \sum_{k=0}^{n} \left[\frac{\partial L_\lambda}{\partial y_\lambda^k} \frac{\partial y_\lambda^k}{\partial \lambda} + \frac{\partial L_\lambda}{\partial \dot{y}_\lambda^k} \frac{\partial \dot{y}_\lambda^k}{\partial \lambda} \right] = \sum_{k=0}^{n} \left[\frac{\partial L_\lambda}{\partial y_\lambda^k} \eta_k(t) + \frac{\partial L_\lambda}{\partial \dot{y}_\lambda^k} \dot{\eta}_k(t) \right],$$

therefore

$$\left.\frac{dL_\lambda}{d\lambda}\right|_{\lambda=0} = \sum_{k=0}^{n} \left[\frac{\partial L}{\partial x^k} \eta_k(t) + \frac{\partial L}{\partial \dot{x}^k} \dot{\eta}_k(t) \right].$$

The condition $\left.\dfrac{dS_\lambda}{d\lambda}\right|_{\lambda=0} \equiv 0$ becomes

$$\left.\frac{dS_\lambda}{d\lambda}\right|_{\lambda=0} = \sum_{k=0}^{n} \int_{t_1}^{t_2} \left[\frac{\partial L}{\partial x^k} \eta_k(t) + \frac{\partial L}{\partial \dot{x}^k} \dot{\eta}_k(t) \right] dt \equiv 0.$$

Definition 7.10.5 The curve $x = x(t) = (x^0(t), x^1(t), \ldots, x^n(t))$ which extremizes the functional

$$S[x] = \int_{t_1}^{t_2} L(x^0(t), \dot{x}^0(t), x^1(t), \dot{x}^1(t), \ldots, x^n(t), \dot{x}^n(t)) dt$$

is called a stationary point of the functional.

Theorem 7.10.6 (Euler–Lagrange equations) *The curve* $x = x(t) = (x^0(t),$ $x^1(t), \ldots, x^n(t))$ *which connects the given points* $(t_1, x^0(t_1), x^1(t_1), \ldots, x^n(t_1))$, $(t_2, x^0(t_2), x^1(t_2), \ldots, x^n(t_2))$ *satisfies the Euler–Lagrange equations*

$$\frac{d}{dt}\left(\frac{\partial L}{\partial \dot{x}^k} \right) - \frac{\partial L}{\partial x^k} = 0, \ k = 0, 1, \ldots, n$$

if and only if $x = x(t) = (x^0(t), x^1(t), \ldots, x^n(t))$ *is a stationary point of the functional*

$$S[x] = \int_{t_1}^{t_2} L(x^0(t), \dot{x}^0(t), x^1(t), \dot{x}^1(t), \ldots, x^n(t), \dot{x}^n(t)) dt.$$

Proof Using the integration by parts, it is

$$\sum_{k=0}^{n} \int_{t_1}^{t_2} \frac{\partial L}{\partial x^k} \eta_k(t) dt + \sum_{k=0}^{n} \left[\frac{\partial L}{\partial \dot{x}^k} \eta_k(t_2) - \frac{\partial L}{\partial \dot{x}^k} \eta_k(t_1) \right] - \sum_{k=0}^{n} \int_{t_1}^{t_2} \frac{d}{dt}\left(\frac{\partial L}{\partial \dot{x}^k} \right) \eta_k(t) dt =$$

$$= \sum_{k=0}^{n} \int_{t_1}^{t_2} \left[\frac{\partial L}{\partial x^k} - \frac{d}{dt}\left(\frac{\partial L}{\partial \dot{x}^k} \right) \right] \eta_k(t) dt.$$

Therefore the condition $\dfrac{dS_\lambda}{d\lambda}\Big|_{\lambda=0} \equiv 0$ reduces to

$$\sum_{k=0}^{n} \int_{t_1}^{t_2} \left[\frac{\partial L}{\partial x^k} - \frac{d}{dt}\left(\frac{\partial L}{\partial \dot{x}^k} \right) \right] \eta_k(t)dt = 0$$

for every function η_k. We obtain

$$\frac{\partial L}{\partial x^k} - \frac{d}{dt}\left(\frac{\partial L}{\partial \dot{x}^k} \right) = 0, \; k = 0, 1, \ldots, n.$$

☐

These equations are called the *Euler–Lagrange equations*.

They represent an equivalent way to express Newton's equations of motion in several variables for the Lagrangian $L = T - V$. However, they are more general than the Newton equations because accelerations are not required in an explicit form. See [36] for a general discussion.

Example 7.10.7 Consider a curve in the Euclidean plane, $c(t) = (t, x(t))$, $t \in [a, b] \subset \mathbb{R}$. We know, from standard calculus textbooks, that its length between the points $c(a)$ and $c(b)$ is given by the formula

$$l_a^b = \int_a^b \|\dot{c}(t)\|dt = \int_a^b \sqrt{1 + \dot{x}^2(t)}dt.$$

For the Lagrangian $L(x, \dot{x}) = \sqrt{1 + \dot{x}^2}$, extremizing the functional

$$S[x] = \int_a^b \sqrt{1 + \dot{x}^2}dt ,$$

means to find out a curve connecting the points $A(a, x(a))$, $B(b, x(b))$ such that it has minimum length. Any other curve has a longer length. Such a curve is a line and its minimum length is the length of the segment $[AB]$.

Let us see what happens if we use the Euler–Lagrange equation. We have $\dfrac{\partial L}{\partial x} = 0$ and $\dfrac{\partial L}{\partial \dot{x}} = \dfrac{\dot{x}}{\sqrt{1 + \dot{x}^2}}$. Therefore the Euler–Lagrange equation is $\dfrac{d}{dt}\left(\dfrac{\dot{x}}{\sqrt{1 + \dot{x}^2}} \right) = 0$.

It results $\dfrac{\dot{x}}{\sqrt{1 + \dot{x}^2}} = k$=constant, i.e. $\dot{x} = \dfrac{k}{\sqrt{1 - k^2}} := m$, and finally $x = mt + n$, that is a line equation in the Euclidean plane. The reader has to try to understand why $\sqrt{1 - k^2}$ exists.

Let us observe that the Euclidean metric is obtained from the previous Lagrangian, that is

$$ds^2 = L^2 dt^2 = \left(\sqrt{(\dot{t})^2 + \dot{x}^2}\right)^2 dt^2 = dt^2 + dx^2.$$

We may conclude that this is another proof for the fact that Euclidean lines are the geodesics of the Euclidean metric.

Example 7.10.8 Using the rule $ds^2 = L^2 dt^2$, the Poincaré metric of the half-plane written as

$$ds^2 = \frac{1}{(x^2)^2}\left[(dx^1)^2 + (dx^2)^2\right]$$

allows us to highlight a Lagrangian. This is

$$L(x^1, x^2, \dot{x}^1, \dot{x}^2) := \sqrt{\frac{1}{(x^2)^2}\left[(\dot{x}^1)^2 + (\dot{x}^2)^2\right]}.$$

Let us write some modified equations in which L^2 is involved, in the form

$$\frac{d}{dt}\left(\frac{\partial L^2}{\partial \dot{x}^i}\right) - \frac{\partial L^2}{\partial x^i} = 0, \quad i \in \{1, 2\}.$$

Denote $x := x^1$, $y := x^2$. The first one becomes

$$\frac{d}{dt}\left(\frac{\partial L^2}{\partial \dot{x}}\right) - \frac{\partial L^2}{\partial x} = 0,$$

that is

$$\ddot{x} - \frac{2}{y}\dot{x}\dot{y} = 0.$$

The second one becomes

$$\frac{d}{dt}\left(\frac{\partial L^2}{\partial \dot{y}}\right) - \frac{\partial L^2}{\partial y} = 0,$$

that is

$$\ddot{y} + \frac{1}{y}\dot{x}^2 - \frac{1}{y}\dot{y}^2 = 0.$$

Therefore, we observe that we have obtained the equations of the geodesics of the Poincaré half-plane. The solutions are

$$x = x(t) = c + R\tanh t, \quad y = y(t) = \frac{R}{\cosh t}$$

and

$$x(t) = a, \ y(t) = e^t,$$

therefore the curves $c_1(t) = \left(c + R \tanh t, \ \dfrac{R}{\cosh t}\right)$ and $c_2(t) = (a, \ e^t)$ are the
stationary points of the functional

$$S[c] = \int_{t_1}^{t_2} \frac{1}{y^2} \left[\dot{x}^2 + \dot{y}^2\right] dt.$$

If we look back at the first example and we work with L^2 instead L, we obtain the
same segment line as a geodesic.

These facts involving the extremization of a functional and the examples, rise
some fundamental questions.

- *Is there Geometry involved?*
- *Are the Euler–Lagrange equations, the geodesic equations for a given metric in
  which the Lagrangian is involved?*
- *Why L^2 appeared?*

The next theorem answers at all these questions.

Theorem 7.10.9 *Consider the Lagrangian $L = \sqrt{g_{ij}\dot{x}^i\dot{x}^j}$ where $g_{ij} = g_{ji}$ and g_{ij}
depends only on the variables $(x^0, x^1, \dots, x^n)$. Then the Euler–Lagrange equations*

$$\frac{\partial L}{\partial x^k} - \frac{d}{dt}\left(\frac{\partial L}{\partial \dot{x}^k}\right) = 0, \ k = 0, 1, \dots, n,$$

are the geodesic equations of the metric $ds^2 = L^2 dt^2$.

Proof First, we prove that Euler–Lagrange equations have an equivalent form written
with respect to L^2,

$$\frac{\partial L^2}{\partial x^k} - \frac{d}{dt}\left(\frac{\partial L^2}{\partial \dot{x}^k}\right) = -2\frac{dL}{dt}\frac{\partial L}{\partial \dot{x}^k}.$$

Let us start from the Euler–Lagrange equations and multiply by $2L$. We have

$$2L\frac{\partial L}{\partial x^k} - 2L\frac{d}{dt}\left(\frac{\partial L}{\partial \dot{x}^k}\right) = 0,$$

that is

$$\frac{\partial L^2}{\partial x^k} - 2L\frac{d}{dt}\left(\frac{\partial L}{\partial \dot{x}^k}\right) = 0.$$

Next, we compute $\dfrac{d}{dt}\left(\dfrac{\partial L^2}{\partial \dot{x}^k}\right)$. We obtain

$$\frac{d}{dt}\left(\frac{\partial L^2}{\partial \dot{x}^k}\right) = \frac{d}{dt}\left(2L\frac{\partial L}{\partial \dot{x}^k}\right) = 2\frac{dL}{dt}\cdot\frac{\partial L}{\partial \dot{x}^k} + 2L\frac{d}{dt}\left(\frac{\partial L}{\partial \dot{x}^k}\right),$$

therefore

$$2L\frac{d}{dt}\left(\frac{\partial L}{\partial \dot{x}^k}\right) = \frac{d}{dt}\left(\frac{\partial L^2}{\partial \dot{x}^k}\right) - 2\frac{dL}{dt}\cdot\frac{\partial L}{\partial \dot{x}^k}.$$

So, the transformed equations

$$\frac{\partial L^2}{\partial x^k} - \frac{d}{dt}\left(\frac{\partial L^2}{\partial \dot{x}^k}\right) = -2\frac{dL}{dt}\cdot\frac{\partial L}{\partial \dot{x}^k}$$

are obtained.

Second, using $L^2 = g_{ij}\dot{x}^i\dot{x}^j$, where $g_{ij} = g_{ij}(x^0, x^1, \ldots, x^n)$, we have

$$\left(\frac{\partial L^2}{\partial x^k}\right) = \frac{\partial g_{ij}}{\partial x^k}\dot{x}^i\dot{x}^j.$$

Third: we prove the relation

$$\frac{d}{dt}\left(\frac{\partial L^2}{\partial \dot{x}^k}\right) = 2g_{ks}\ddot{x}^s + 2\frac{\partial g_{ks}}{\partial x^m}\dot{x}^m\dot{x}^s.$$

This is not difficult. Successively

$$\frac{d}{dt}\left(\frac{\partial L^2}{\partial \dot{x}^k}\right) = \frac{d}{dt}\left(\frac{\partial}{\partial \dot{x}^k}\left[g_{ij}\dot{x}^i\dot{x}^j\right]\right) = \frac{d}{dt}\left(g_{ij}\frac{\partial \dot{x}^i}{\partial \dot{x}^k}\dot{x}^j + g_{ij}\dot{x}^i\frac{\partial \dot{x}^j}{\partial \dot{x}^k}\right) =$$

$$= \frac{d}{dt}\left(g_{kj}\dot{x}^j + g_{ik}\dot{x}^i\right) = \frac{d}{dt}\left(2g_{ks}\dot{x}^s\right),$$

then

$$\frac{d}{dt}\left(\frac{\partial L^2}{\partial \dot{x}^k}\right) = 2g_{ks}\ddot{x}^s + 2\frac{\partial g_{ks}}{\partial x^m}\frac{dx^m}{dt}\dot{x}^s = 2g_{ks}\ddot{x}^s + 2\frac{\partial g_{ks}}{\partial x^m}\dot{x}^m\dot{x}^s.$$

The forth relation to be proved is

$$\frac{dL}{dt}\cdot\frac{\partial L}{\partial \dot{x}^k} = \frac{\ddot{S}}{\dot{S}}g_{ks}\dot{x}^s$$

where

$$S = \int_{t_0}^{t} L\,d\tau = \int_{t_0}^{t}\sqrt{g_{ij}\dot{x}^i\dot{x}^j}\,d\tau,\ \ \dot{S} = L,\ \ \ddot{S} = \frac{dL}{dt}.$$

Step by step, we have

$$\frac{dL}{dt} \cdot \frac{\partial L}{\partial \dot{x}^k} = \frac{dL}{dt} \cdot \frac{\partial}{\partial \dot{x}^k}\left[\sqrt{g_{ij}\dot{x}^i\dot{x}^j}\right] = \frac{dL}{dt} \cdot \left[\frac{1}{2\sqrt{g_{ij}\dot{x}^i\dot{x}^j}}\frac{\partial}{\partial \dot{x}^k}\left[g_{ij}\dot{x}^i\dot{x}^j\right]\right] =$$

$$= \frac{dL}{dt} \cdot \left[\frac{1}{2L}(2g_{ks}\dot{x}^s)\right] = \frac{\ddot{S}}{\dot{S}}\, g_{ks}\dot{x}^s.$$

Now, replacing in the modified Euler–Lagrange equations

$$-\frac{\partial L^2}{\partial x^k} + \frac{d}{dt}\left(\frac{\partial L^2}{\partial \dot{x}^k}\right) = 2\frac{dL}{dt} \cdot \frac{\partial L}{\partial \dot{x}^k}$$

we obtain

$$-\frac{\partial g_{ij}}{\partial x^k}\dot{x}^i\dot{x}^j + 2g_{ks}\ddot{x}^s + 2\frac{\partial g_{ks}}{\partial x^m}\dot{x}^m\dot{x}^s = 2\frac{\ddot{S}}{\dot{S}}\, g_{ks}\dot{x}^s.$$

Manipulating the dummy indexes, the previous relation can be written in the form

$$2g_{ks}\ddot{x}^s + \left(\frac{\partial g_{ks}}{\partial x^m} + \frac{\partial g_{km}}{\partial x^s}\right)\dot{x}^m\dot{x}^s - \frac{\partial g_{ms}}{\partial x^k}\dot{x}^m\dot{x}^s = 2\frac{\ddot{S}}{\dot{S}}\, g_{ks}\dot{x}^s.$$

The Christoffel symbols appear if we put together the last two terms of the left member,

$$g_{ks}\ddot{x}^s + \frac{1}{2}\left(\frac{\partial g_{ks}}{\partial x^m} + \frac{\partial g_{km}}{\partial x^s} - \frac{\partial g_{ms}}{\partial x^k}\right)\dot{x}^m\dot{x}^s = \frac{\ddot{S}}{\dot{S}}\, g_{ks}\dot{x}^s,$$

therefore, after multiplying by g^{ik}, we have

$$\ddot{x}^i + \Gamma^i_{ms}\dot{x}^m\dot{x}^s = \frac{\ddot{S}}{\dot{S}}\dot{x}^i,\ i \in \{0, 1, \ldots, n\}.$$

Still we have not the desired geodesic equations, but we are close. It remains to consider the parameter t in such a way to have a curve which is canonically parameterized.

So, we choose t such that $L = \dfrac{dS}{dt} = \dot{S} = 1$. It results $\dfrac{dL}{dt} = \ddot{S} = 0$, i.e.

$$\ddot{x}^i + \Gamma^i_{ms}\dot{x}^m\dot{x}^s = 0,\ i \in \{0, 1, \ldots, n\}.$$

□

We can see a new feature of Lagrangians: they are important because they induce metrics whose geodesics are described by the Euler–Lagrange equations.

Finally, we can see a possible switch between the traditional mechanical point of view for several models in Physics to the geometric point of view. Somehow the forces, the energies, some other functions involved in describing "the reality" can be replaced by geometric objects from Differential Geometry. The trajectories created by forces are now geodesics of spaces with metrics induced by Lagrangians. As we will see below, this point of view is fundamental in General Relativity.

Chapter 8
Special Relativity

Numerus omnium aptantur.

Pythagoras

In XVIIth century, Newton considered light as a collection of particles, now called photons according to Quantum Mechanics, traveling through space. Reflection and refraction of light were explained in a satisfactory way interpreting light rays as trajectory of photons.

James Clark Maxwell results on Electrodynamics, in the middle of the XIXth century, offered another view: the light is an electromagnetic wave.

Maxwell's equations of Electromagnetism are not simple at all, and, putting them in accordance with Newton's theory, points out the necessity of considering a medium in which the electromagnetic waves travel through space. This hypothetical medium was called "ether".

Ernst Mach did not agree with the idea of ether and observed the necessity of the revision of all fundamental concepts of Physics. Michelson–Morley experiment, who initially was designed to reveal such an ether, had a result completely different with respect the expectations and hard to interpret in view of Classical Mechanics. Albert Einstein explained the result of the experiment in a theory, the Special Relativity, where he revised, in a fundamental way, the ideas of space and time. After this achievement, no place remained for ether. For a comprehensive exposition of Special Relativity, see [37].

W. Boskoff and S. Capozziello, *A Mathematical Journey to Relativity*,
UNITEXT for Physics, https://doi.org/10.1007/978-3-030-47894-0_8

8.1 Principles of Special Relativity

Let us first discuss about Michelson–Morley experiment.

Suppose we have a platform of a railway train wagon, an open one, on an existing straight railway line. During the Michelson–Morley experiment, the platform is considered at rest or it moves at constant speed v.

On this platform, let us imagine two perpendicular lines which intersect at I, one, say d_1, coincident to the sense of motion, the other one, say d_2, perpendicular to the sense of motion. On d_1, called the longitudinal direction, in this order, there exist: a source of light denoted by S_L, an interferometer placed in I and a mirror denoted by M_1, such that the distance between I and M_1 is l.

The interferometer is a device able to split a light-ray in the two perpendicular directions d_1 and d_2, but also to receive two light-rays from perpendicular directions and to send them separately to another given direction.

On the line d_2, which corresponds to the transversal direction, there is another mirror denoted by M_2, such that the distance between I and M_2 is the same l and a receiver-device R_L such that the interferometer I is between M_2 and R_L.

The receiver-device is able to capture the light rays coming from the interferometer and to decide which one reached first the device.

The experiment is like this: when the platform is at rest or it is moving at constant speed v in the $S_L I$ longitudinal direction, a light-ray is sent by the source S_L to the interferometer I. The interferometer splits the light-ray in two light-rays. The first one is sent to the mirror M_1, it is reflected by the mirror and it is returned to the interferometer which sends it to R_L. The second one is directed to M_2, it is reflected and sent to the interferometer which sends it to R_L. Which one reaches first R_L?

This is something as: we are interested in identifying the influence of the speed v on the splitted light-rays. There is, or there is not, a difference between what is happening when the platform is at rest comparing with the case when the platform is moving at constant speed v?

Let us observe something obvious: if the platform is at rest, both light-rays reach at same time R_L.

Now, let us try to use Classical Mechanics to describe what is happening when the platform is moving at constant speed v. First at all, let us observe that it is enough to establish only the time necessary to cover the routes $I M_1 I$ and $I M_2 I$ and to compare them.

Denote by c the speed of light. The time to cover the longitudinal route $I M_1 I$ is

$$t_1 = \frac{l}{c - v} + \frac{l}{c + v} = \frac{2lc}{c^2 - v^2},$$

because $c - v$ and $c + v$ are in Newtonian mechanics the speeds for the directions $I M_1$, $M_1 I$ respectively. To be sure that the reader understands why the speeds are like this, let us focus on the first direction case. Moving at constant speed v in the sense $I M_1$, the photon is slowed down by the air, that is by the medium in which it is traveling, with the speed $-v$. Therefore, according to mechanics rules, the speed of the photon traveling in IM_1 direction is $c - v$.

For the transversal direction, denote by t' the time necessary for the light-ray to reach the mirror M_2. During this time, the platform, therefore the mirror, travels in the longitudinal direction a $t'v$ space. The Pythagoras theorem in the rectangle triangle formed is $(t'c)^2 = l^2 + (t'v)^2$, that is

$$t' = \frac{l}{\sqrt{c^2 - v^2}}.$$

It is obvious that the time necessary to the transversal ray to reach again the interferometer I is $t_2 := 2t'$, so we have

$$t_2 = \frac{2l}{\sqrt{c^2 - v^2}}.$$

Therefore

$$\frac{t_2}{t_1} = \sqrt{1 - \frac{v^2}{c^2}},$$

which implies

$$t_2 < t_1,$$

i.e. the transversal light-ray reaches earlier R_L compared to the longitudinal light-ray.

The mathematical model made with respect to the rules of Classical Mechanics has a prediction, let us repeat it: the first light-ray arriving in R_L is the transversal one.

If we make the experiment the result is: the transversal and the longitudinal light-rays reach R_L at the same time. If we repeat it, the same results holds. There is not a difference between what happens when the platform is at rest, comparing with the case when the platform is moving at constant speed v.

As we explained in the introduction, the error is in the model: it is related to the fact we thought that v could affect the speed of light. It seems that $c - v$ and $c + v$ are not correctly thought, therefore we have not to consider Classical Mechanics when we try to understand this experiment. Another rule has to be applied when we "add" velocities.

This experiment can be also seen making a parallel between the platform moving in Earth atmosphere at constant speed v and the Earth moving through the ether at constant speed v. After we establish a new theory to explain the experimental result, the main consequence is the fact that there is no ether.[1]

---

[1] In modern physics, it has been realized that "ether" is the "physical vacuum" that is a maximally symmetric configuration of spacetime where no physical field is present. This means that matter-energy density is extremely low. In this "vacuum", electromagnetic waves propagate at the speed of light.

The consequences of Einstein's postulates give the chance to understand how the light propagates in the context of a new physical theory, the Special Relativity, which changes the rules of Classical Mechanics when we are dealing with bodies moving at very large speeds.

Part of these results were also obtained by Henry Poincaré in his effort to explain the Michelson–Morley experiment.

Essentially, Einstein formulated the Special Relativity starting from two main postulates:

1. The laws of Physics are the same in all inertial reference frames.

2. The speed of light in vacuum, denoted by $c \approx 2,99 \cdot 10^8$ **m/s, is the same for all the observers and it is the maximal speed reached by a moving object.**

Einstein used the word *observer* with the meaning of *reference frame* from which a set of objects or events are measured. Since the measurement are generally made with respect to the center O of the frame, this special point is often called the "O observer" or we may refer to a frame with "the observer placed at O". We know that the laws of mechanics are the same in all inertial frames. The first postulate asks for the same form of electromagnetic laws in any inertial reference frames, as the mechanics laws have. And in general, all laws of Physics must have the same form in all reference frames (this result will be fully achieved in General Relativity).

The second postulates plays a key role in Special Relativity being involved in the way in which we derive the Lorentz transformations.

The framework of Newton's laws of mechanics is the 3-dimensional Euclidean space. Each object is described by a point or by a collection of points of it. Time is given by a universal clock and allows us to see the evolution of objects.

In Special Relativity, we have to work in a 4-dimensional space, but not in an usual one. Three of the dimensions are the standard dimensions used in mechanics. We can denote them with the letters as x^1, x^2, x^3. The fourth dimension is related to time.

Definition 8.1.1 A frame of coordinates (t, x^1, x^2, x^3) is called a spacetime.

The Geometry of a spacetime is in fact what we are trying to develop, and this is made according to some given physical postulates we have to accept.

Definition 8.1.2 Each point of such a spacetime is called an event.

Definition 8.1.3 A curve of the spacetime is called a world line and represents a successions of events.

Example 8.1.4 Suppose we work in a two dimensional slice of the previous frame, with the coordinates (t, x^3). Consider a world line starting from the origin $O(0, 0)$. Suppose the next point is $A(1, x_0^3)$. Then the object remain t_0 seconds at rest with respect our perspective. This means that the world line has to be continued with the segment AB, where B has the coordinates $B(1 + t_0, x_0^3)$. Next, suppose the object advances in the direction $-v_1$. The line followed has the equation $x^3 - x_0^3 = -v_1(t - (1 + t_0))$, etc.

Example 8.1.5 From the origin $O(0, 0)$ an object is moving t_1 seconds in the direction $-v$. It reaches the point $M(t_1, -vt_1)$. Negative speed means only the direction of evolution in time.

Example 8.1.6 A photon is released from the origin O. There are two possible directions, c and $-c$. If it is released in the direction c, its trajectory will be the line $x^3 = ct$. Or, it can be released in the direction $-c$. Its trajectory in this case is $x^3 = -ct$. In this case, after $t_0 > 0$ seconds, the photon reaches the point $L(t_0, -ct_0)$.

In order to advance into the theory, we have to consider two local frames of coordinates, one moving at constant speed v, denoted by S, and another one considered at rest, denoted by R. The letters are chosen from the words "speed" and "rest." Two observers are placed at the origins of each system denoted by $\bar{O}$, respectively O. The first local frame S is considered described by the coordinates $(\tau = \bar{x}^0, \bar{x}^1, \bar{x}^2, \bar{x}^3)$, while the frame R is described by the coordinates $(t = x^0, x^1, x^2, x^3)$.

Now, the reference frames of the two observers have to adapt to the second postulate of the Special Relativity. To be easier in our reasonings, let us suppose the bidimensional case when the frame S consists of the coordinates $(\tau = \bar{x}^0, \bar{x}^3)$ and it is moving, at constant speed v, in the same plane as the one determined by R, here denoted as $(t = x^0, x^3)$.

First at all, how can we express the fact that S is moving at constant speed v with respect to R? The simple mathematical answer is: the axis $\bar{O}\tau$ in R has the equation $x^3 = vt$.

Even if the light can be seen as an electromagnetic wave and we check the conservation of the form of Maxwell's equations by the Lorentz transformations, in order to develop Special Relativity, we can consider the light-rays as trajectories of photons.

What can we say about the world line of a photon in these inertial reference frames? With respect to the observers in each frame, two world lines are highlighted: a photon moving at constant speed c with a trajectory $x^3 = ct$ in R and $\bar{x}^3 = c\tau$ in S, while, for a photon moving at speed $-c$, we have the lines $x^3 = -ct$ in R and $\bar{x}^3 = -c\tau$ in S.

The two world lines of photons at O form the *light cone* of the frame R. A similar definition holds in S.

Therefore, if we use a same diagram for both frames, the second postulate has the following mathematical expression:

1. The lines $x^3 = ct$ in R and $\bar{x}^3 = c\tau$ in S have the same image;
2. The lines $x^3 = -ct$ in R and $\bar{x}^3 = -c\tau$ in S have the same image.

In other words, the two light cones are coincident.

Since we deal with inertial frames, as a rule, objects moving at constant speed in S move at constant speed in R, and vice versa. So, a straight line representing a world line of an object moving at constant speed in S, it is seen as a straight line representing the world line of the same object moving at (another) constant speed in R and vice versa. Transforming lines into lines, the change of coordinates between the two frames is described by a linear map; we denote it by L_v and we call it a *Lorentz transformation* corresponding to the speed v.

Theorem 8.1.7 *In the context described before, the matrix of the Lorentz transformation corresponding to the speed v has the form*

$$L_v = \frac{1}{\sqrt{1 - v^2/c^2}} \begin{pmatrix} 1 & v/c^2 \\ v & 1 \end{pmatrix}.$$

Proof A linear map $L_v : S \to R$ has the form

$$L_v = \begin{pmatrix} a & b \\ d & e \end{pmatrix}.$$

Since $\bar{O}\tau$ axis in R has the equation $x^3 = vt$ we have

$$\begin{pmatrix} a & b \\ d & e \end{pmatrix} \begin{pmatrix} 1 \\ 0 \end{pmatrix} = \begin{pmatrix} t \\ vt \end{pmatrix},$$

that is $d = va$. In mathematical language, the second postulate is:
The eigenvectors of L_v are $\begin{pmatrix} 1 \\ c \end{pmatrix}$ and $\begin{pmatrix} 1 \\ -c \end{pmatrix}$, that is

$$L_v \cdot \begin{pmatrix} 1 \\ c \end{pmatrix} = \lambda_1 \begin{pmatrix} 1 \\ c \end{pmatrix}$$

and

$$L_v \cdot \begin{pmatrix} 1 \\ -c \end{pmatrix} = \lambda_2 \begin{pmatrix} 1 \\ -c \end{pmatrix}.$$

To preserve the sense of movement of photons, it is necessary to impose two inequalities for the eigenvalues $\lambda_1 > 0$, $\lambda_2 > 0$.

Replacing L_v, it results the equations

$$\begin{cases} a\,c + b\,c^2 = a\,v + e\,c \\ \\ -a\,c + b\,c^2 = a\,v - e\,c \end{cases}$$

that is

$$L_v = a \begin{pmatrix} 1 & v/c^2 \\ v & 1 \end{pmatrix}.$$

To determine a, we need to observe who is the inverse of the considered Lorentz transformation.

L_v^{-1} has to act from R to S, such that $L_v\, L_v^{-1} = L_v^{-1}\, L_v = I_2$. It is standard to think at $L_v^{-1} := L_{-v}$, that is to see S at rest and R moving at constant speed $-v$. This leads to

$$I_2 = a^2 \begin{pmatrix} 1 - v^2/c^2 & 0 \\ 0 & 1 - v^2/c^2 \end{pmatrix},$$

i.e. $a^2 = \dfrac{1}{1 - v^2/c^2}$.

To determine the right sign of a, we use the Cayley Theorem. It is a simple matrix exercise: For a 2×2 real matrix B, it is

$$B^2 - 2\,Tr\,B \cdot B + \det B \cdot I_2 = O_2.$$

In our case, $Tr\,L_v = 2a = \lambda_1 + \lambda_2 > 0$.

The Lorentz transformation, in final form, is

$$L_v = \frac{1}{\sqrt{1 - v^2/c^2}} \begin{pmatrix} 1 & v/c^2 \\ v & 1 \end{pmatrix}.$$

$\square$

We can write how the transformation looks like in four dimensions:

$$\begin{cases} t = \dfrac{\tau + \bar{x}^3\, v/c^2}{\sqrt{1 - v^2/c^2}} \\ x^1 = \bar{x}^1 \\ x^2 = \bar{x}^2 \\ x^3 = \dfrac{\tau v + \bar{x}^3}{\sqrt{1 - v^2/c^2}}. \end{cases}$$

Exercise 8.1.8 *Express in four dimensions the corresponding inverse of the Lorentz transformation L_v.*

Solution. According to the proof, the inverse transformation is $L_{-v} : R \to S$. In four dimensions, we have

$$\begin{cases} \tau = \dfrac{t - x^3\, v/c^2}{\sqrt{1 - v^2/c^2}} \\ \bar{x}^1 = x^1 \\ \bar{x}^2 = x^2 \\ \bar{x}^3 = \dfrac{-t\, v + x^3}{\sqrt{1 - v^2/c^2}}. \end{cases}$$

Let us observe that, for a small velocity v with respect to c, the ratios v/c^2 and v^2/c^2 are small enough. We can consider the influence of these terms almost zero, that is the Lorentz transformations become the usual way, in Classical Mechanics to pass from the inertial reference frame S to the inertial reference frame R, that is

$$\begin{cases} t = \tau \\ x^1 = \bar{x}^1 \\ x^2 = \bar{x}^2 \\ x^3 = \tau\, v + \bar{x}^3. \end{cases}$$

These formulas are called *Galilean transformations* for Classical Mechanics.

Consider three inertial reference frames, S', S and R, such that S' is moving at constant speed w with respect to S and S is moving at constant speed v with respect to R.

The two corresponding Lorentz transformations are $L_w = \dfrac{1}{\sqrt{1 - w^2/c^2}}$ $\begin{pmatrix} 1 & w/c^2 \\ w & 1 \end{pmatrix}$ and $L_v = \dfrac{1}{\sqrt{1 - v^2/c^2}} \begin{pmatrix} 1 & v/c^2 \\ v & 1 \end{pmatrix}$.

The natural question is: which is the speed of S' with respect to R?

The answer is: We have to describe the linear map between S' and R via S, that is $L_v \cdot L_w$.

Theorem 8.1.9 $L_v \cdot L_w = L_{v \oplus w}$, where $v \oplus w = \dfrac{v + w}{1 + vw/c^2}$.

Proof After multiplying, we have

$$L_v \cdot L_w = \frac{1}{\sqrt{1 - v^2/c^2}} \frac{1}{\sqrt{1 - w^2/c^2}} \begin{pmatrix} 1 & v/c^2 \\ v & 1 \end{pmatrix} \cdot \begin{pmatrix} 1 & w/c^2 \\ w & 1 \end{pmatrix} =$$

$$= \frac{1 + vw/c^2}{\sqrt{(1 - v^2/c^2)(1 - w^2/c^2)}} \begin{pmatrix} 1 & \dfrac{v + w}{1 + vw/c^2} \cdot \dfrac{1}{c^2} \\ \dfrac{v + w}{1 + vw/c^2} & 1 \end{pmatrix} =$$

$$= \frac{1}{\sqrt{1 - \left(\dfrac{v + w}{1 + vw/c^2}\right)^2 \cdot \dfrac{1}{c^2}}} \begin{pmatrix} 1 & \dfrac{v + w}{1 + vw/c^2} \cdot \dfrac{1}{c^2} \\ \dfrac{v + w}{1 + vw/c^2} & 1 \end{pmatrix} = L_{v \oplus w},$$

where

$$v \oplus w = \frac{v + w}{1 + vw/c^2}.$$

$\square$

Definition 8.1.10 The last formula is called the *relativistic velocities addition*.

The relativistic velocities-addition formula, in the case of small velocities, reduces to the standard sum of velocities of Classical Mechanics.

Exercise 8.1.11 *Show that the set $K = (-c, c)$ endowed with the operation*

$$v \oplus w = \frac{v + w}{1 + vw/c^2}$$

is an abelian group.

Exercise 8.1.12 *Show that the set of Lorentz transformations*

$$L := \{L_v \in M_{2\times 2}(\mathbb{R})|\ v \in (-c, c)\}$$

endowed with the usual product of matrices is an Abelian group.

8.2 Lorentz Transformations in Geometric Coordinates and Consequences

In Physics, systems of coordinates are thought with axes whose coordinates are related to the physical units as second, meter, etc. The systems of coordinates corresponding to the physical units can be called systems of *physical coordinates*. In the previous sections, we worked in physical coordinates. The units of measure in Physics were thought before to understand how deeply is the Geometry involved in the description of the physical phenomena. If we choose an appropriate "length" (e.g. the meter) and an appropriate "time duration" (e.g. the second), the speed of light can be $c = 1$. We call these new units *geometric units*. All formulas become simpler and the geometric images are more intuitive.

Definition 8.2.1 The coordinates corresponding to geometric units are called geometric coordinates.

If we adapt the second postulate conditions, seen on the same diagram, we have:
1. The lines $x^3 = t$ in R and $\bar{x}^3 = \tau$ in S have the same image
2. The lines $x^3 = -t$ in R and $\bar{x}^3 = -\tau$ in S have the same image,
in geometric coordinates, it is easier to understand how it looks like the frame S seen in R: since $O = \bar{O}$, the axis $O\bar{x}^3$ and $O\tau$ are symmetric with respect the line $x^3 = t$.

Before obtaining the Lorentz transformations in geometric coordinates, let us consider the concept of simultaneity.

8.2.1 *The Relativity of Simultaneity*

Two events, E_1 and E_2, are called *simultaneous* in S, if they happen at the same moment of time τ_0 in S, that is they are $E_1(\tau_0, \tau_0)$ and $E_2(\tau_0, -\tau_0)$. The same, two events, U_1 and U_2, are called simultaneous in R if they happen at the same moment of time t_0 in R, i.e. they are $U_1(t_0, t_0)$ and $U_2(t_0, -t_0)$.

On the same diagram, it is easy to see that U_1 and U_2 are simultaneous in R, but $U_1(t_0, t_0)$ and $V_2\left(t_0\dfrac{1-v}{1+v}, -t_0\dfrac{1-v}{1+v}\right)$ are simultaneous in S.

Let us explain the result from the mathematical point of view.

It is not very difficult to show that, in geometric coordinates, if $O\tau$ has the equation $x^3 = vt$, then $O\bar{x}^3$ has the equation $x^3 = \dfrac{1}{v}t$. Therefore the line $y - t_0 = \dfrac{1}{v}(t - t_0)$ intersects $x^3 = -t$, if $t = t_0\dfrac{1-v}{1+v}$.

For the observer in R, the events $U_1(t_0, t_0)$ and $U_2(t_0, -t_0)$ happen simultaneously. The observer in S cannot agree: for him $U_1(t_0, t_0)$ and $V_2\left(t_0\dfrac{1-v}{1+v}, -t_0\dfrac{1-v}{1+v}\right)$ happen simultaneously. Therefore it exists the Relativity of the *simultaneity*.

8.2.2 The Lorentz Transformations in Geometric Coordinates

In geometric coordinates, we choose the Lorentz transformation as the linear map $L_v : S \to R$,

$$L_v = \begin{pmatrix} a & b \\ d & e \end{pmatrix}.$$

Since $\bar{O}\tau$ axis in R has the equation $x^3 = vt$, we have

$$\begin{pmatrix} a & b \\ d & e \end{pmatrix}\begin{pmatrix} 1 \\ 0 \end{pmatrix} = \begin{pmatrix} t \\ vt \end{pmatrix},$$

that is $d = va$. In mathematical language, the second postulate is:

The eigenvectors of L_v are $\begin{pmatrix} 1 \\ 1 \end{pmatrix}$ and $\begin{pmatrix} 1 \\ -1 \end{pmatrix}$, that is

$$L_v \cdot \begin{pmatrix} 1 \\ 1 \end{pmatrix} = \lambda_1 \begin{pmatrix} 1 \\ 1 \end{pmatrix}$$

and

$$L_v \cdot \begin{pmatrix} 1 \\ -1 \end{pmatrix} = \lambda_2 \begin{pmatrix} 1 \\ -1 \end{pmatrix}.$$

To preserve the direction of movement of photons, it is necessary to impose $\lambda_1 > 0$, $\lambda_2 > 0$.

Replacing L_v, the following equations result

$$\begin{cases} a + b = a\,v + e \\ \\ -a + b = a\,v - e \end{cases}$$

that is

$$L_v = a \begin{pmatrix} 1 & v \\ v & 1 \end{pmatrix}.$$

In the same way, as in the physical coordinates case, the inverse of the Lorentz transformation L_v in geometric coordinates is $L_v^{-1} := L_{-v}$. It results

$$I_2 = a^2 \begin{pmatrix} 1 - v^2 & 0 \\ 0 & 1 - v^2 \end{pmatrix},$$

that is $a^2 = \dfrac{1}{1 - v^2}$.

To determine the right sign of a, we use the same Cayley theorem: For a 2×2 real matrix B, it is

$$B^2 - 2\,TrB \cdot B + \det B \cdot I_2 = O_2.$$

In our case $TrL_v = 2a = \lambda_1 + \lambda_2 > 0$.

For those who do not understand this result, we invite to look at the characteristic equation

$$\det(B - \lambda I_2) = 0.$$

The final form of Lorentz transformation (corresponding to the velocity v), in geometric coordinates, is

$$L_v = \frac{1}{\sqrt{1 - v^2}} \begin{pmatrix} 1 & v \\ v & 1 \end{pmatrix}.$$

We can write how the transformation looks like in geometric coordinates in four dimensions:

$$\begin{cases} t = \dfrac{\tau + \bar{x}^3\, v}{\sqrt{1 - v^2}} \\ x^1 = \bar{x}^1 \\ x^2 = \bar{x}^2 \\ x^3 = \dfrac{\tau v + \bar{x}^3}{\sqrt{1 - v^2}}. \end{cases}$$

In the same case as in physical coordinates, let us consider three inertial reference frames, S', S and R, such that S' is moving at constant speed w with respect to S and S is moving at constant speed v with respect to R. Here v, w are in $(-1, 1)$. The two corresponding Lorentz transformations are $L_w = \dfrac{1}{\sqrt{1 - w^2}} \begin{pmatrix} 1 & w \\ w & 1 \end{pmatrix}$ and $L_v = \dfrac{1}{\sqrt{1 - v^2}} \begin{pmatrix} 1 & v \\ v & 1 \end{pmatrix}$.

Exercise 8.2.2 *What is the speed of S' with respect to R?*

Hint. We must find the linear map between S' and R, that is $L_v \cdot L_w$.

A similar computation as the one made in physical coordinates leads to

$$
L_v \cdot L_w = \frac{1}{\sqrt{1-v^2}} \frac{1}{\sqrt{1-w^2}} \begin{pmatrix} 1 & v \\ v & 1 \end{pmatrix} \cdot \begin{pmatrix} 1 & w \\ w & 1 \end{pmatrix} =
$$

$$
= \frac{1+vw}{\sqrt{(1-v^2)(1-w^2)}} \begin{pmatrix} 1 & \dfrac{v+w}{1+vw} \\ \dfrac{v+w}{1+vw} & 1 \end{pmatrix} =
$$

$$
= \frac{1}{\sqrt{1-\left(\dfrac{v+w}{1+vw}\right)^2}} \begin{pmatrix} 1 & \dfrac{v+w}{1+vw} \\ \dfrac{v+w}{1+vw} & 1 \end{pmatrix} = L_{v\oplus w},
$$

where

$$
v \oplus w = \frac{v+w}{1+vw}.
$$

The last formula can be called the *addition of relativistic velocities in geometric coordinates.*

Exercise 8.2.3 *Show that the set $K = (-1, 1)$ endowed with the operation*

$$
v \oplus w = \frac{v+w}{1+vw}
$$

is an Abelian group.

Exercise 8.2.4 *Show that the set of Lorentz transformations $\{L_v \in M_{2\times2}(\mathbb{R})|v \in (-1, 1)\}$ endowed with the standard product of matrices is an Abelian group.*

8.2.3 The Minkowski Geometry of Inertial Frames in Geometric Coordinates and Consequences: Time Dilation and Length Contraction

Let us observe that the addition of velocities was deduced using Einstein's postulates and more, it is related to the Minkowski Geometry attached to S and R frames. Why? Because if we choose

$$
v = \tanh \alpha \; ; \; w = \tanh \beta,
$$

we obtain the known geometric formula

$$\tanh(\alpha + \beta) = \frac{\tanh \alpha + \tanh \beta}{1 + \tanh \alpha \tanh \beta}$$

for the addition of velocities in geometric coordinates.

The Lorentz transformation corresponding to the constant speed v is now

$$L_{\tanh \alpha} = \begin{pmatrix} \cosh \alpha & \sinh \alpha \\ \sinh \alpha & \cosh \alpha \end{pmatrix}.$$

It is well known that the matrices $L_{\tanh \alpha}$ are hyperbolic rotations in the two-dimensional Minkowski space denoted by $\mathbb{M}^2$, where the Minkowski product of the vectors $x = (t_1, x_1^3)$ and $y = (t_2, x_2^3)$ is defined by

$$\langle x, y \rangle_M := t_1 t_2 - x_1^3 x_2^3.$$

It is also known that each matrix $L_{\tanh \alpha}$ preserves the Minkowski product.

The last property suggests another way to think at the Lorentz transformations in the case of geometric coordinates: they preserve the quantity $t^2 - (x^3)^2$.

Exercise 8.2.5 *Show that Lorentz transformation implies the equality*

$$\tau^2 - (\bar{x}^3)^2 = t^2 - (x^3)^2.$$

Hint.

$$t^2 - (x^3)^2 = \left(\frac{\tau + \bar{x}^3 \, v}{\sqrt{1 - v^2}} \right)^2 - \left(\frac{\tau \, v + \bar{x}^3}{\sqrt{1 - v^2}} \right)^2 = \tau^2 - (\bar{x}^3)^2.$$

It results

Corollary 8.2.6 *The Lorentz transformations preserves the square of the Minkowski norm of vectors.*

Theorem 8.2.7 (Time dilation) *A clock slows down when it is moving at constant speed.*

Proof Denote by $\Delta \tau$ the unit interval of a clock moving at constant speed v. It means to consider the unit of τ axis in S to be $\Delta \tau$. Denote by Δt the corresponding element of $\Delta \tau$ after a Lorentz transformation L_v in geometric coordinates. We have

$$L_v \cdot \begin{pmatrix} \Delta \tau \\ 0 \end{pmatrix} = \frac{1}{\sqrt{1 - v^2}} \begin{pmatrix} 1 & v \\ v & 1 \end{pmatrix} \cdot \begin{pmatrix} \Delta \tau \\ 0 \end{pmatrix} = \begin{pmatrix} \Delta t \\ * \end{pmatrix},$$

where $*$ meaning is related to the fact we are not interested in. Therefore

$$\Delta t = \frac{\Delta \tau}{\sqrt{1 - v^2}},$$

that is

$$\Delta\tau < \Delta t.$$

$\square$

Example 8.2.8 Let us consider two twins separated. The first one is sent in space with a cosmic vehicle having the constant speed $v = 4/5$. The other one remains on Earth. When they separated they are 20 years old. After 15 years in space, according with his time, the brother from space returned. He is now, according to his time, 35 years old. How old is the brother remained on Earth, according to his perspective? The factor $\sqrt{1 - v^2}$ is $3/5$. From the formula $\Delta t = \dfrac{\Delta\tau}{\sqrt{1 - v^2}}$, after we replace, we obtain $3\Delta t = 5\Delta\tau$. Now, for the observer in S fifteen years have passed, that is $\Delta\tau = 15$. It results $\Delta t = 25$. Therefore his brother is 45 years old.

Theorem 8.2.9 (Length contraction) *The lengths are contracting when the frame is moving at constant speed.*

Proof Denote by $\Delta\bar{l}$ the unit length of S. Let Δl be the corresponding element of $\Delta\bar{l}$ after a Lorentz transformation L_v. In order to compare the two lengths, we compute

$$L_v \cdot \begin{pmatrix} 0 \\ \Delta\bar{l} \end{pmatrix} = \frac{1}{\sqrt{1 - v^2}} \begin{pmatrix} 1 & v \\ v & 1 \end{pmatrix} \cdot \begin{pmatrix} 0 \\ \Delta\bar{l} \end{pmatrix} = \begin{pmatrix} * \\ \Delta l \end{pmatrix},$$

where $*$ meaning is related to the fact we are not interested in. It results

$$\Delta l = \frac{\Delta\bar{l}}{\sqrt{1 - v^2}},$$

that is

$$\Delta\bar{l} < \Delta l.$$

$\square$

Example 8.2.10 A cosmic vehicle is 125 m long at rest. Suppose it is sent in space and it is moving at constant speed $v = 3/5$. How long is this moving cosmic vehicle for an observer at rest? We apply $\Delta l = \dfrac{\Delta\bar{l}}{\sqrt{1 - v^2}}$ formula for $v = 3/5$ and $\Delta l = 125$. It results $\Delta\bar{l} = 100$ m.

8.2.4   Relativistic Mass, Rest Mass and Energy

Newton's second law involves the concept of inertial mass. As we have seen at that time, the mass was considered as a constant. We have discussed about the inertial mass and the gravitational mass and how the mass is part of the so called quantity of motion, also known as momentum. In Classical Mechanics momentum means inertial mass in motion and redefined in a relativistic way, will lead to important consequences.

Let us think at an object at rest, having a rest mass denoted by $m_0 \neq 0$. Is the mass of the object "moving at constant speed" the same as its rest mass? The answer is related to how the relativistic momentum is changing with respect to the Lorentz transformations.

Let us denote by $\mathbb{P} = \begin{pmatrix} m \\ mv \end{pmatrix}$ the *relativistic momentum* of a classical body moving at constant speed v. The second component of the relativistic momentum is the classical momentum.

The relativistic momentum of a classical body at rest in S has to be $\mathbb{P}_0 = \begin{pmatrix} m_0 \\ 0 \end{pmatrix}$.

According to the theory we are developing, the formula of the relativistic momentum at constant speed v is obtained from the relativistic momentum at rest, changed with respect to the Lorentz transformation L_v. This was the key point where Einstein applied, in a brilliant way, the idea that all physical formulas have to be invariant under Lorentz transformations. The consequences can be seen in the following two theorems.

Theorem 8.2.11 *If $m_0 \neq 0$ is the rest mass of a body moving at constant speed v, then*

$$m = m(v) = \frac{m_0}{\sqrt{1 - v^2}}.$$

Proof Using the Lorentz transformation L_v we have $\mathbb{P} = L_v \cdot \mathbb{P}_0$. It results

$$\begin{pmatrix} m \\ mv \end{pmatrix} = \frac{1}{\sqrt{1 - v^2}} \begin{pmatrix} 1 & v \\ v & 1 \end{pmatrix} \cdot \begin{pmatrix} m_0 \\ 0 \end{pmatrix},$$

which leads to the so called *relativistic mass* formula

$$m = m(v) = \frac{m_0}{\sqrt{1 - v^2}}.$$

$\square$

We may observe that the mass of an object is increasing when the object travel at constant speed v. Another consequence is related to the fact that an object having its rest mass $m_0 \neq 0$ can not reach the speed of light.

Definition 8.2.12 $m(v)$ is called relativistic mass corresponding to the constant speed v of an object having the rest mass m_0.

The previous obtained formula has sense when $m_0 \neq 0$. The physicists know that there is no rest mass for the photon. Therefore this formula does not work for photon or for any other physical particle with no rest mass.

The following theorem explains why it is a good choice to consider the relativistic momentum if we intend to show how the mass is changing when it is moving at constant speed. Even if the proof is done using the geometric coordinates, the reader can change it to adapt the result to physical coordinates.

Theorem 8.2.13 *The relativistic mass formula is preserved by the Lorentz transformations.*

Proof If we consider the inertial frame S, moving at constant speed v with respect to R, we have

$$L_v \cdot \begin{pmatrix} m_0 \\ 0 \end{pmatrix} = \begin{pmatrix} \dfrac{m_0}{\sqrt{1 - v^2}} \\ \dfrac{m_0 v}{\sqrt{1 - v^2}} \end{pmatrix}.$$

In the same way, for the inertial frame S, moving at constant speed V with respect to R_1, we have

$$L_V \cdot \begin{pmatrix} m_0 \\ 0 \end{pmatrix} = \begin{pmatrix} \dfrac{m_0}{\sqrt{1 - V^2}} \\ \dfrac{m_0 V}{\sqrt{1 - V^2}} \end{pmatrix}.$$

If the frame R is moving at constant speed w with respect to R_1, we have to compute $L_w \cdot L_v \begin{pmatrix} m_0 \\ 0 \end{pmatrix}$ and we wish the result to be coincident with $L_V \cdot \begin{pmatrix} m_0 \\ 0 \end{pmatrix}$. We have

$$L_w \cdot L_v \cdot \begin{pmatrix} m_0 \\ 0 \end{pmatrix} = L_w \cdot \begin{pmatrix} \dfrac{m_0}{\sqrt{1 - v^2}} \\ \dfrac{m_0 v}{\sqrt{1 - v^2}} \end{pmatrix} = \frac{1}{\sqrt{1 - w^2}} \frac{m_0}{\sqrt{1 - v^2}} \begin{pmatrix} 1 & w \\ w & 1 \end{pmatrix} \begin{pmatrix} 1 \\ v \end{pmatrix} =$$

$$= \frac{1}{\sqrt{1 - w^2}} \frac{m_0}{\sqrt{1 - v^2}} \begin{pmatrix} 1 + wv \\ w + v \end{pmatrix} = \frac{1 + wv}{\sqrt{1 - w^2}} \frac{m_0}{\sqrt{1 - v^2}} \begin{pmatrix} 1 \\ \dfrac{w + v}{1 + wv} \end{pmatrix} =$$

$$= \frac{m_0}{\sqrt{1 - \left(\dfrac{w + v}{1 + wv} \right)^2}} \begin{pmatrix} 1 \\ \dfrac{w + v}{1 + wv} \end{pmatrix} = L_{w \oplus v} \begin{pmatrix} m_0 \\ 0 \end{pmatrix} = L_V \begin{pmatrix} m_0 \\ 0 \end{pmatrix},$$

that is $V = w \oplus v = \dfrac{w + v}{1 + wv}$.   □

We are close to prove a very important consequence of the previous relativistic mass formula:

Theorem 8.2.14 *In geometric coordinates, mass means energy.*

Proof Denote by f', f'' the first and the second derivative of a real function f. It is easy to prove that

$$f(x) = f(0) + \frac{x}{1!}f'(0) + \frac{x^2}{2!}f''(0) + B[x^3],$$

where $B[x^3]$ contains only terms in x with powers greater than 3.

If we neglect the B terms, when we consider the real function

$$f(v) = \frac{1}{\sqrt{1 - v^2}}$$

and the formula of the relativistic mass, we can write

$$m(v) = \frac{m_0}{\sqrt{1 - v^2}} = m_0 + \frac{1}{2}m_0 v^2.$$

Looking at both members we can observe how, in geometric coordinates, the relativistic mass is related to the rest mass and the kinetic energy, that is the statement: "mass is energy" is confirmed.   □

8.3   Consequences of Lorentz Physical Transformations: Time Dilation, Length Contraction, Relativistic Mass and Rest Energy

In the previous section, we used Lorentz transformations in geometric coordinates which can be called *Lorentz geometric transformations*. When we obtained, for the first time, the Lorentz transformations, we worked in physical coordinates. therefore the Lorentz transformations found there can be called *Lorentz physical transformations*. How can we adapt the previous results in the case of physical coordinates?

8.3.1   *The Minkowski Geometry of Inertial Frames in Physical Coordinates and Consequences: Time Dilation and Length Contraction*

If we choose
$$v = c\tanh\alpha \; ; \; w = c\tanh\beta,$$

we obtain the known geometric formula

$$c \cdot \tanh(\alpha + \beta) = c \cdot \frac{\tanh\alpha + \tanh\beta}{1 + \tanh\alpha \tanh\beta}$$

for the velocities addition, the Lorentz transformation corresponding to the constant speed v being

$$L_{c \tanh \alpha} = \begin{pmatrix} \cosh \alpha & \dfrac{1}{c} \sinh \alpha \\ c \sinh \alpha & \cosh \alpha \end{pmatrix}.$$

In the two-dimensional Minkowski space, denoted by $\mathbb{M}^2$, where the Minkowski product of the vectors $x = (t_1, x_1^3)$ and $y = (t_2, x_2^3)$ is defined by

$$\langle x, y \rangle_M := c^2 t_1 t_2 - x_1^3 x_2^3,$$

each matrix $L_{c \tanh \alpha}$ preserves the Minkowski product.

Indeed, for $j \in \{1, 2\}$ we have

$$L_{c \tanh \alpha} \cdot \begin{pmatrix} \tau_j \\ \bar{x}_j^3 \end{pmatrix} = \begin{pmatrix} \cosh \alpha & \dfrac{1}{c} \sinh \alpha \\ c \sinh \alpha & \cosh \alpha \end{pmatrix} \cdot \begin{pmatrix} \tau_j \\ \bar{x}_j^3 \end{pmatrix} = \begin{pmatrix} \tau_j \cosh \alpha + \dfrac{1}{c} \bar{x}_j^3 \sinh \alpha \\ c \tau_j \sinh \alpha + \bar{x}_j^3 \cosh \alpha \end{pmatrix},$$

and

$$c^2 \left(\tau_1 \cosh \alpha + \frac{1}{c} \bar{x}_1^3 \sinh \alpha \right) \left(\tau_2 \cosh \alpha + \frac{1}{c} \bar{x}_2^3 \sinh \alpha \right) -$$

$$- \left(c \tau_1 \sinh \alpha + \bar{x}_1^3 \cosh \alpha \right) \left(c \tau_1 \sinh \alpha + \bar{x}_1^3 \cosh \alpha \right) = c^2 \tau_1 \tau_2 - \bar{x}_1^3 \bar{x}_2^3.$$

The last property suggests another way to think at the Lorentz transformations in the case of physical coordinates: they preserve the quantity $c^2 t^2 - (x^3)^2$.

Exercise 8.3.1 *Show that*

$$c^2 \tau^2 - (\bar{x}^3)^2 = c^2 t^2 - (x^3)^2.$$

Hint.

$$c^2 t^2 - (x^3)^2 = c^2 \left(\frac{\tau + \bar{x}^3 \, v/c^2}{\sqrt{1 - v^2/c^2}} \right)^2 - \left(\frac{\tau v + \bar{x}^3}{\sqrt{1 - v^2/c^2}} \right)^2 = c^2 \tau^2 - (\bar{x}^3)^2.$$

Now it becomes clear how the physical coordinates can be transformed into "geometric physical coordinates": The Ox^0 axis has the units done with respect to ct in R. In S, the corresponding axis becomes $c\tau$.

In this way, the unit of measure for the first axis is a length, the same as the unit for the spatial axes.

Theorem 8.3.2 *Lorentz physical transformations preserves the square of the Minkowski norm of vectors.*

However, in the case in which we are not interested in highlighting the Minkowski Geometry, we prefer to work in our initial R and S systems of coordinates.

Consider an infinitesimal time-like interval between the points (t, x) and $(t + dt, x + dx)$ and its arclength expressed in the form suggested by the previous invariant, that is

$$ds^2 = c^2(dt)^2 - (dx)^2.$$

We denoted by x the x^3 coordinate to make the notations easier. The same interval can be seen in a frame such that, at each time τ, the moving point which describes the interval is at rest. Denote by (τ, x_τ) the world line whose coordinates express the moving point at rest. Taking into account the conservation law seen before, we have

$$ds^2 = c^2(dt)^2 - (dx)^2 = c^2(d\tau)^2 - (dx_\tau)^2 = c^2(d\tau)^2.$$

Therefore

$$ds = cd\tau,$$

that is we can define

$$\Delta\tau = \int_l d\tau = \int_l \frac{ds}{c},$$

where l is the notation for the chosen time-like infinitesimal interval. We observe

$$\Delta\tau = \int_l \frac{\sqrt{c^2 dt^2 - dx^2}}{c} = \int_l \sqrt{1 - \frac{1}{c^2}\frac{dx^2}{dt^2}}\, dt = \int_l \sqrt{1 - \frac{v^2(t)}{c^2}}\, dt,$$

where $v(t)$ is the usual speed.

Definition 8.3.3 $\Delta\tau$ is called a proper time interval.

Therefore, we can say that proper time measured along the time-like world line above is the time measured by a clock following point by point the considered world line. Let us give now an important property of the proper time $\Delta\tau$ in Special Relativity:

Theorem 8.3.4 (Time dilation in physical coordinates) *A clock slows down when it is moving at constant speed.*

Proof Denote by $\Delta\tau$ the unit interval of a clock moving at constant speed v. This clock measures the proper time defined above. It is like you consider the unit of τ axis in S to be $\Delta\tau$. We are interested in knowing the connection between the proper time and the time coordinate t of the frame at rest, R. Denote by Δt the corresponding element of $\Delta\tau$ after a Lorentz transformation L_v in geometric coordinates. We have

$$L_v \cdot \begin{pmatrix} \Delta\tau \\ 0 \end{pmatrix} = \frac{1}{\sqrt{1 - v^2/c^2}} \begin{pmatrix} 1 & v/c^2 \\ v & 1 \end{pmatrix} \cdot \begin{pmatrix} \Delta\tau \\ 0 \end{pmatrix} = \begin{pmatrix} \Delta t \\ * \end{pmatrix},$$

where $*$ meaning is related to the fact we are not interested in. Therefore

$$\Delta t = \frac{\Delta \tau}{\sqrt{1 - v^2/c^2}},$$

that is

$$\Delta \tau < \Delta t.$$

$\square$

Theorem 8.3.5 (Length contraction in physical coordinates) *The length are contracting when the frame is moving at constant speed.*

Proof Denote by $\Delta \bar{l}$ the unit length of S. Let Δl be the corresponding element of $\Delta \bar{l}$ after a Lorentz transformation L_v. In order to compare the two lengths, we compute

$$L_v \cdot \begin{pmatrix} 0 \\ \Delta \bar{l} \end{pmatrix} = \frac{1}{\sqrt{1 - v^2/c^2}} \begin{pmatrix} 1 & v/c^2 \\ v & 1 \end{pmatrix} \cdot \begin{pmatrix} 0 \\ \Delta \bar{l} \end{pmatrix} = \begin{pmatrix} * \\ \Delta l \end{pmatrix},$$

where $*$ meaning is related to the fact we are not interested in. It results

$$\Delta l = \frac{\Delta \bar{l}}{\sqrt{1 - v^2/c^2}},$$

that is

$$\Delta \bar{l} < \Delta l.$$

$\square$

8.3.2   Relativistic Mass, Rest Mass and Rest Energy in Physical Coordinates

Let us see how it looks like the relativistic mass in the case of physical coordinates. We start from an object at rest, having a rest mass denoted by $m_0 \neq 0$ with its relativistic momentum as in the case of geometrical coordinates in S, $\mathbb{P}_0 = \begin{pmatrix} m_0 \\ 0 \end{pmatrix}$.

Let us denote by $\mathbb{P} = \begin{pmatrix} m \\ mv \end{pmatrix}$ the relativistic momentum of a classical body moving at constant speed v.

Theorem 8.3.6 *If $m_0 \neq 0$ is the rest mass of a body moving at constant speed v, then*

$$m = m(v) = \frac{m_0}{\sqrt{1 - v^2/c^2}}.$$

Proof Using the Lorentz transformation L_v, we have $\mathbb{P} = L_v \cdot \mathbb{P}_0$, i.e.

$$\begin{pmatrix} m \\ mv \end{pmatrix} = \frac{1}{\sqrt{1 - v^2/c^2}} \begin{pmatrix} 1 & v/c^2 \\ v & 1 \end{pmatrix} \cdot \begin{pmatrix} m_0 \\ 0 \end{pmatrix},$$

which leads to the so called relativistic mass, now in physical coordinates,

$$m = m(v) = \frac{m_0}{\sqrt{1 - v^2/c^2}}.$$

$\square$

As in the case of geometrical coordinates, the previous formula holds when $m_0 \neq 0$.

We are talking about the rest energy, of course, in the same case $m_0 \neq 0$. The discussion is almost the same as when we proved that, in geometric coordinates, mass means energy.

If we consider the real function

$$f(v) = \frac{1}{\sqrt{1 - v^2/c^2}}$$

and the formula of the relativistic mass, we can neglect the B terms because $1/c^4$ modify a given quantity in an irrelevant mode. We may write

$$\frac{m_0}{\sqrt{1 - v^2/c^2}} = m_0 + \frac{1}{2}m_0 v^2/c^2.$$

Let us define the *kinetic relativistic energy* by

$$E(v) := \frac{m_0 c^2}{\sqrt{1 - v^2/c^2}}.$$

The previous formula becomes

$$E(v) = m_0 c^2 + \frac{1}{2}m_0 v^2.$$

We may call *rest energy* the formula $E := m_0 c^2$; it makes sense when $m_0 \neq 0$.

A comment. It is useful, at this point, after the discussion about the relativistic mass, saying some words about the light energy which is not 0, even if the rest mass of photons is 0. To understand why, we have to accept the alternative way to consider the light as explained by Maxwell equations, that is light is an electromagnetic wave. We have also to accept the dual behavior of light and to define the photon as the particle attached to the wave.[2] The equation of photon energy is $E = hf = hc/\lambda$, where h is the Planck constant, f is the photon frequency, λ is the photon wavelength and, of course, c is the speed of light in vacuum. Therefore, in the case of a photon, we have a relativistic equivalent of mass given by the formula E/c^2.

8.4 Maxwell's Equations

Maxwell's equations are the "core" of Special Relativity. Essentially, this theory has been developed in view of explaining their invariance under Lorentz transformations. In order to discuss Maxwell's equations, which describes the electromagnetic field, we need some preliminary algebraic result.

Theorem 8.4.1 *If*

$$A = (A_1, A_2, A_3), \ B = (B_1, B_2, B_3), \ C = (C_1, C_2, C_3),$$

$$B \times C := \begin{vmatrix} \vec{i} & \vec{j} & \vec{k} \\ B_1 & B_2 & B_3 \\ C_1 & C_2 & C_3 \end{vmatrix},$$

$$A \cdot B := A_1 B_1 + A_2 B_2 + A_3 B_3, \ A \cdot C := A_1 C_1 + A_2 C_2 + A_3 C_3,$$

then

$$A \times (B \times C) = (A \cdot C)B - (A \cdot B)C.$$

Proof We have

$$(A \cdot C)B - (A \cdot B)C =$$

$$= (A_1 C_1 + A_2 C_2 + A_3 C_3)(B_1, B_2, B_3) - (A_1 B_1 + A_2 B_2 + A_3 B_3)(C_1, C_2, C_3) =$$

$$= \begin{vmatrix} \vec{i} & \vec{j} & \vec{k} \\ A_1 & A_2 & A_3 \\ B_2 C_3 - B_3 C_2 & -B_1 C_3 + B_3 C_1 & B_1 C_2 - B_2 C_1 \end{vmatrix} = A \times (B \times C).$$

$\square$

Now, consider both the gradient operator and the Laplace operator in spatial coordinates denoted by (x^1, x^2, x^3), that is

---

[2]The dual nature of light, and of any particle, is better framed in the context of Quantum Mechanics in relation to the concept of wave-particle. For a discussion, see [38].

$$\nabla := \left(\frac{\partial}{\partial x^1}, \frac{\partial}{\partial x^2}, \frac{\partial}{\partial x^3} \right),$$

$$\nabla^2 := \frac{\partial^2}{(\partial x^1)^2} + \frac{\partial^2}{(\partial x^3)^2} + \frac{\partial^2}{(\partial x^3)^2}.$$

The last formula can be also seen written in the formal way

$$\nabla^2 := \nabla \cdot \nabla$$

We formally define

$$\nabla \cdot A := \frac{\partial A_1}{\partial x^1} + \frac{\partial A_2}{\partial x^2} + \frac{\partial A_3}{\partial x^3}$$

and

$$\nabla \times A := \begin{vmatrix} \vec{i} & \vec{j} & \vec{k} \\ \frac{\partial}{\partial x^1} & \frac{\partial}{\partial x^2} & \frac{\partial}{\partial x^3} \\ A_1 & A_2 & A_3 \end{vmatrix} = \left(\frac{\partial A_3}{\partial x^2} - \frac{\partial A_2}{\partial x^3}, \frac{\partial A_1}{\partial x^3} - \frac{\partial A_3}{\partial x^1}, \frac{\partial A_2}{\partial x^1} - \frac{\partial A_1}{\partial x^2} \right).$$

Using these operators, a consequence of the above theorem is

Corollary 8.4.2

$$\nabla \times (\nabla \times A) = (\nabla \cdot A)\nabla - (\nabla \cdot \nabla)A.$$

Another comment is in order. We know the meaning of $\nabla^2 \phi$, where ϕ is a scalar function. The meaning of $\nabla^2 A$ is related to the fact that ∇^2 acts on each component of A, i.e.

$$\nabla^2 A := (\nabla^2 A_1, \nabla^2 A_2, \nabla^2 A_3).$$

Therefore we can write

$$\nabla \times (\nabla \times A) = (\nabla \cdot A)\nabla - \nabla^2 A.$$

If $\nabla \cdot A = 0$, the previous formula becomes

Corollary 8.4.3

$$\nabla \times (\nabla \times A) = - \nabla^2 A.$$

We will use this result later.
Denote by

$$E = E(t, x^1, x^2, x^3) := (E_1(t, x^1, x^2, x^3), E_2(t, x^1, x^2, x^3), E_3(t, x^1, x^2, x^3))$$

the electric force vector and by

$$H = H(t, x^1, x^2, x^3) := (H_1(t, x^1, x^2, x^3), H_2(t, x^1, x^2, x^3), H_3(t, x^1, x^2, x^3))$$

the magnetic force vector;

In geometric units, the *Maxwell equations*, in the frame R considered as an empty space, are

$$\begin{cases} \nabla \cdot E = 0 \\ \nabla \times E = -\dfrac{\partial H}{\partial t} \\ \nabla \cdot H = 0 \\ \nabla \times H = \dfrac{\partial E}{\partial t} \end{cases}$$

The first equation reveals the existence of an electric field in the absence of electric charge. If we are not in vacuum, the first equation is $\nabla \cdot E = \rho$, where ρ is the electric charge, therefore the first equation describes how an electric charge acts as source for the electric force, here seen as an electric field.

The second equation $\nabla \times E = -\dfrac{\partial H}{\partial t}$ shows how a time varying magnetic field gives rise to an electric field.

The third equation $\nabla \cdot H = 0$ shows that there are no magnetic charges.

The forth equation $\nabla \times H = \dfrac{\partial E}{\partial t}$ shows how the time variation of electric field creates the magnetic field.

Let us consider the derivative with respect t of the second equation.

$$-\frac{\partial^2 H}{\partial t^2} = \frac{\partial}{\partial t}(\nabla \times E) = \frac{\partial}{\partial t} \begin{vmatrix} \vec{i} & \vec{j} & \vec{k} \\ \dfrac{\partial}{\partial x^1} & \dfrac{\partial}{\partial x^2} & \dfrac{\partial}{\partial x^3} \\ E_1 & E_2 & E_3 \end{vmatrix} = \begin{vmatrix} \vec{i} & \vec{j} & \vec{k} \\ \dfrac{\partial}{\partial x^1} & \dfrac{\partial}{\partial x^2} & \dfrac{\partial}{\partial x^3} \\ \dfrac{\partial E_1}{\partial t} & \dfrac{\partial E_2}{\partial t} & \dfrac{\partial E_3}{\partial t} \end{vmatrix} = \nabla \times \frac{\partial E}{\partial t}.$$

Using the last Maxwell equation and the above results, we find

$$-\frac{\partial^2 H}{\partial t^2} = \nabla \times \frac{\partial E}{\partial t} = \nabla \times (\nabla \times H) = -\nabla^2 H,$$

that is

$$\frac{\partial^2 H}{\partial t^2} = \nabla^2 H.$$

If we denote by

$$\Box := \frac{\partial^2}{\partial t^2} - \nabla^2$$

the d'Alembert operator, the previous equation is

$$\Box H = 0.$$

This is the wave equation corresponding to the magnetic field. Therefore, for each component H_i, $i \in \{1, 2, 3\}$ we have

$$\frac{\partial^2 H_i}{\partial t^2} = \nabla^2 H_i = \frac{\partial^2 H_i}{(\partial x^1)^2} + \frac{\partial^2 H_i}{(\partial x^2)^2} + \frac{\partial^2 H_i}{(\partial x^3)^2}.$$

Now, let us consider the derivative with respect t of the last equation.

$$\frac{\partial^2 E}{\partial t^2} = \frac{\partial}{\partial t}(\nabla \times H) = \frac{\partial}{\partial t}\begin{vmatrix} \vec{i} & \vec{j} & \vec{k} \\ \frac{\partial}{\partial x^1} & \frac{\partial}{\partial x^2} & \frac{\partial}{\partial x^3} \\ H_1 & H_2 & H_3 \end{vmatrix} = \begin{vmatrix} \vec{i} & \vec{j} & \vec{k} \\ \frac{\partial}{\partial x^1} & \frac{\partial}{\partial x^2} & \frac{\partial}{\partial x^3} \\ \frac{\partial H_1}{\partial t} & \frac{\partial H_2}{\partial t} & \frac{\partial H_3}{\partial t} \end{vmatrix} = \nabla \times \frac{\partial H}{\partial t}.$$

Using the second Maxwell's equation and the consequence, we find that

$$\frac{\partial^2 E}{\partial t^2} = \nabla \times \frac{\partial H}{\partial t} = -\nabla \times (\nabla \times E) = \nabla^2 E,$$

i.e.

$$\Box E = 0.$$

This one is the wave equation corresponding to the electric field. We have now a picture of the electromagnetic field described by the Maxwell equations: The two waves equations of electric and magnetic field are interconnected by the four Maxwell equations. We understand that one field can not exist without the other. Each one generates the other.

Are these wave equations invariant under Lorentz transformations? The answer is yes, but we need to perform more steps in order to achieve these results.

In the same way as before, for each component E_i, $i \in \{1, 2, 3\}$, we have

$$\frac{\partial^2 E_i}{\partial t^2} = \nabla^2 E_i = \frac{\partial^2 E_i}{(\partial x^1)^2} + \frac{\partial^2 E_i}{(\partial x^2)^2} + \frac{\partial^2 E_i}{(\partial x^3)^2}.$$

To simplify, let us suppose that the electric field E depends only on the variables t and x^3, as in the case of a plane wave. The previous equations become

$$\frac{\partial^2 E_i}{\partial t^2} - \frac{\partial^2 E_i}{(\partial x^3)^2} = 0.$$

To continue, let us choose a component only, say $i = 1$. Since for the other two components, the following computations are the same, we prefer instead to use E_1, to denote this chosen component by the letter $\mathbb{E}$. The previous equation becomes

$$\frac{\partial^2 \mathbb{E}}{\partial t^2} - \frac{\partial^2 \mathbb{E}}{(\partial x^3)^2} = 0.$$

How this simple equation looks like in S, frame considered with coordinates $\tau, \bar{x}_3$, if S is supposed to move at constant speed v along the x_3 axis in R? We have to use the Lorentz inverse transformation L_{-v}, that is

$$\begin{cases} \tau = \dfrac{t - x^3\, v}{\sqrt{1 - v^2}} \\ \bar{x}^3 = \dfrac{-t\, v + x^3}{\sqrt{1 - v^2}}. \end{cases}$$

Denote by $\bar{\mathbb{E}}(\tau, \bar{x}^3) = \bar{\mathbb{E}}\left(\dfrac{t - x^3\, v}{\sqrt{1 - v^2}}, \dfrac{-t\, v + x^3}{\sqrt{1 - v^2}} \right) := \mathbb{E}(t, x^3)$ the corresponding component of the electric field in S, which, obviously have to be the same as in R. We would like to prove that

$$\frac{\partial^2 \mathbb{E}}{\partial t^2} - \frac{\partial^2 \mathbb{E}}{(\partial x^3)^2} = \frac{\partial^2 \bar{\mathbb{E}}}{\partial \tau^2} - \frac{\partial^2 \bar{\mathbb{E}}}{(\partial \bar{x}^3)^2}.$$

We have

$$\frac{\partial \mathbb{E}}{\partial t} = \frac{\partial \bar{\mathbb{E}}}{\partial \tau} \frac{\partial \tau}{\partial t} + \frac{\partial \bar{\mathbb{E}}}{\partial \bar{x}^3} \frac{\partial \bar{x}^3}{\partial t} = \frac{\partial \bar{\mathbb{E}}}{\partial \tau} \frac{1}{\sqrt{1 - v^2}} + \frac{\partial \bar{\mathbb{E}}}{\partial \bar{x}^3} \frac{-v}{\sqrt{1 - v^2}}$$

and

$$\frac{\partial^2 \mathbb{E}}{\partial t^2} = \frac{1}{\sqrt{1 - v^2}} \left(\frac{\partial^2 \bar{\mathbb{E}}}{\partial \tau^2} \frac{\partial \tau}{\partial t} + \frac{\partial^2 \bar{\mathbb{E}}}{\partial \tau \partial \bar{x}^3} \frac{\partial \bar{x}^3}{\partial t} \right) - \frac{v}{\sqrt{1 - v^2}} \left(\frac{\partial^2 \bar{\mathbb{E}}}{\partial \bar{x}^3 \partial \tau} \frac{\partial \tau}{\partial t} + \frac{\partial^2 \bar{\mathbb{E}}}{(\partial \bar{x}^3)^2} \frac{\partial \bar{x}^3}{\partial t} \right),$$

that is

$$\frac{\partial^2 \mathbb{E}}{\partial t^2} = \frac{1}{1 - v^2} \frac{\partial^2 \bar{\mathbb{E}}}{\partial \tau^2} - \frac{2v}{1 - v^2} \frac{\partial^2 \bar{\mathbb{E}}}{\partial \bar{x}^3 \partial \tau} + \frac{v^2}{1 - v^2} \frac{\partial^2 \bar{\mathbb{E}}}{(\partial \bar{x}^3)^2}.$$

In the same way

$$\frac{\partial^2 \mathbb{E}}{(\partial x^3)^2} = \frac{v^2}{1 - v^2} \frac{\partial^2 \bar{\mathbb{E}}}{\partial \tau^2} - \frac{2v}{1 - v^2} \frac{\partial^2 \bar{\mathbb{E}}}{\partial \bar{x}^3 \partial \tau} + \frac{1}{1 - v^2} \frac{\partial^2 \bar{\mathbb{E}}}{(\partial \bar{x}^3)^2},$$

therefore the desired relation is obtained by subtracting the two expressions. Now, from

$$\frac{\partial^2 \mathbb{E}}{\partial t^2} - \frac{\partial^2 \mathbb{E}}{(\partial x^3)^2} = 0$$

in R, we obtain

$$\frac{\partial^2 \bar{\mathbb{E}}}{\partial \tau^2} - \frac{\partial^2 \bar{\mathbb{E}}}{(\partial \bar{x}^3)^2} = 0$$

in S, that is the corresponding equation is the same as it has to be. Therefore, in a moving inertial frame, the Maxwell equations are the same as in a frame at rest. We have proved

Theorem 8.4.4 *Lorentz transformations preserve Maxwell's equations.*

If the reader try to prove if the equality

$$\frac{\partial^2 \mathbb{E}}{\partial t^2} - \frac{\partial^2 \mathbb{E}}{(\partial x^3)^2} = \frac{\partial^2 \bar{\mathbb{E}}}{\partial \tau^2} - \frac{\partial^2 \bar{\mathbb{E}}}{(\partial \bar{x}^3)^2}$$

holds for the inverse of Galilean transformations $\bar{\mathbb{E}}(\tau, \bar{x}^3) = \bar{\mathbb{E}}\left(t, -vt + x^3\right) := \mathbb{E}(t, x^3)$, the answer is no, that is the Galilean transformations fail for the Maxwell equations. This can be easily shown. If the reader computes

$$\frac{\partial \mathbb{E}}{\partial t} = \frac{\partial \bar{\mathbb{E}}}{\partial \tau}\frac{\partial \tau}{\partial t} + \frac{\partial \bar{\mathbb{E}}}{\partial \bar{x}^3}\frac{\partial \bar{x}^3}{\partial t} = \frac{\partial \bar{\mathbb{E}}}{\partial \tau} - v\frac{\partial \bar{\mathbb{E}}}{\partial \bar{x}^3}$$

and

$$\frac{\partial^2 \mathbb{E}}{\partial t^2} = \left(\frac{\partial^2 \bar{\mathbb{E}}}{\partial \tau^2}\frac{\partial \tau}{\partial t} + \frac{\partial^2 \bar{\mathbb{E}}}{\partial \tau \partial \bar{x}^3}\frac{\partial \bar{x}^3}{\partial t}\right) - v\left(\frac{\partial^2 \bar{\mathbb{E}}}{\partial \bar{x}^3 \partial \tau}\frac{\partial \tau}{\partial t} + \frac{\partial^2 \bar{\mathbb{E}}}{(\partial \bar{x}^3)^2}\frac{\partial \bar{x}^3}{\partial t}\right) =$$

$$= \frac{\partial^2 \bar{\mathbb{E}}}{\partial \tau^2} - 2v\frac{\partial^2 \bar{\mathbb{E}}}{\partial \bar{x}^3 \partial \tau} + v^2\frac{\partial^2 \bar{\mathbb{E}}}{(\partial \bar{x}^3)^2}.$$

Then,

$$\frac{\partial \mathbb{E}}{\partial x^3} = \frac{\partial \bar{\mathbb{E}}}{\partial \bar{x}^3}$$

and

$$\frac{\partial^2 \mathbb{E}}{(\partial x^3)^2} = \frac{\partial^2 \bar{\mathbb{E}}}{(\partial \bar{x}^3)^2},$$

that is

$$\frac{\partial^2 \mathbb{E}}{\partial t^2} - \frac{\partial^2 \mathbb{E}}{(\partial x^3)^2} = \frac{\partial^2 \bar{\mathbb{E}}}{\partial \tau^2} - \frac{\partial^2 \bar{\mathbb{E}}}{(\partial \bar{x}^3)^2} - 2v\frac{\partial^2 \bar{\mathbb{E}}}{\partial \bar{x}^3 \partial \tau} + v^2\frac{\partial^2 \bar{\mathbb{E}}}{(\partial \bar{x}^3)^2} \neq \frac{\partial^2 \bar{\mathbb{E}}}{\partial \tau^2} - \frac{\partial^2 \bar{\mathbb{E}}}{(\partial \bar{x}^3)^2}.$$

Theorem 8.4.5 *Galilei's transformations do not preserve Maxwell's equations.*

The final conclusion is: Classical Mechanics through Galilei's transformations does not preserve Maxwell's equations while the Special Relativity, through Lorentz transformations, does it.

8.5 Doppler's Effect in Special Relativity

We have proved that the speed of light does not depend on the speed of the source of light. Let us now focus on the frequency of light signals. We prove that the frequency of light signals depends on the speed of the source, that is, we show that light frequency is increasing when the source is approaching to the observer O at rest in R, then, when the source is moving away, the light frequency is decreasing. This is the so called *Doppler's effect* or *relativistic Doppler's effect*.

Definition 8.5.1 Doppler's effect is a change in frequency of light-wave when a source is moving at constant speed with respect to the frequency perceived by an observer at rest.

Therefore we have two different formulas, one to estimate the frequency of the source which is approaching, and another one for the frequency in the case when the source is moving away. Let us translate this in a mathematical way.

Consider, as usual, two local frames of geometric coordinates, one moving at constant speed v, denoted by S, and another one considered at rest, denoted by R. The first local frame S is described by the coordinates $(\tau = \bar{x}^0, \bar{x}^3)$, while the frame R is described by the coordinates $(t = x^0, x^3)$.

Consider a source of light in S which, for each $\Delta\tau$ seconds, releases a light signal. If the *frequency* is denoted by ν, the connection between the two physical quantities is

$$\nu = \frac{1}{\Delta\tau}.$$

This formula is related to the behavior of a light wave.

The quantity $\Delta\tau$ is the *period* of a light wave in the frame S. The light wave, with frequency ν, is imagined as emitted light signals of duration $\Delta\tau$ seconds. They produces light cones with the vertexes on the τ-axis.

So, the source is moving in S along the τ-axis and, in R, this τ axis becomes the line $x^3 = vt$. The observer, at the origin O of R, perceives the source first as approaching, then as moving away.

To simplify, let us consider the moment when the two origins are coincident and, on the τ-axis, we draw $\Delta\tau$ intervals to the left and to the right. The light cones, considered in S, determine two kinds of equal intervals Δt on the t-axis in R. Until the origin, we denote them Δt_{app}, after we denote them by Δt_{ma}, each one determining its corresponding frequency in R. The subscript *app* and *ma* are obviously from the words "approaching" and "moving away."

Therefore, two kinds of frequencies appear in R, that is

$$f_{app} := \frac{1}{\Delta t_{app}}$$

and

$$f_{ma} := \frac{1}{\Delta t_{ma}}.$$

Theorem 8.5.2 *According to the above conditions, we have:*
(i) If the source is approaching,

$$f_{app} = \nu \sqrt{\frac{1+v}{1-v}} > \nu.$$

(ii) If the source is moving away,

$$f_{ma} = \nu \sqrt{\frac{1-v}{1+v}} < \nu.$$

Proof Denote by $(0,0)$ and (b, vb), the coordinates at the ends of the first interval $\Delta \tau$ on the τ-axis, as seen in R. The light-ray emitted at the point (b, vb) reaches the t-axis at $(b + vb, 0)$. Of course, in this case, we used the photon corresponding to speed -1. Therefore

$$\Delta t_{ma} = b(1 + v).$$

Now, consider the points $(0,0)$ and $(-b, -vb)$ as the coordinates of a $\Delta \tau$ interval when the source is approaching. The light-ray emitted at the point $(-b, -vb)$ reaches the t-axis at $(-b + vb, 0)$ because we used the photon corresponding to speed 1. In this case

$$\Delta t_{app} = b(1 - v).$$

If we consider the Minkowski arc length corresponding to a $\Delta \tau$ interval, we have

$$\Delta \tau^2 = b^2 - b^2 v^2,$$

that is

$$b = \frac{\Delta \tau}{\sqrt{1 - v^2}}.$$

Now using this last formula and the two formulas for $f_{app} = \dfrac{1}{\Delta t_{app}}$ and $f_{ma} = \dfrac{1}{\Delta t_{ma}}$, the statement is proved. $\qquad \square$

Let us observe that we can write the two above formulas in the form

$$f = \nu \sqrt{\frac{1 - v}{1 + v}},$$

if we perceive the approaching wave as moving away with speed $-v$.

8.6 Gravity in Special Relativity: The Case of the Constant Gravitational Field

The fact that Special Relativity had to be improved towards General Relativity is essentially due to two main reasons: From one side, Einstein, according to the Mach criticisms [39], realized that the laws of Physics must be written in the same way for *any* (inertial or non-inertial) observer (Invariance Principle). Secondly, considering the gravitational phenomena, he realized that one needs to introduce accelerating frames. According to these observations, Special Relativity is inadequate to enclose gravity.

In order to discuss gravity in the framework of Special Relativity and show their basic incompatibility, let us begin considering a very simple result.

In a Minkowski space, for every t, let $v(t)$ be a vector of constant norm.

It results $\langle v(t), v(t) \rangle_M = k$. If we consider the derivative with respect to t, we obtain

$$\langle \dot{v}(t), v(t) \rangle_M = 0.$$

We have proved the following

Proposition 8.6.1 *(i) The derivative of a constant norm vector is a vector orthogonal on the given vector, that is $\dot{v}(t) \perp_M v(t)$,*
(ii) The vectors $\dot{v}(t)$ and $v(t)$ are Minkowski type different, that is, if $v(t)$ is space-like vector, the derivative $\dot{v}(t)$ is time-like vector, and vice versa.

A second very important observation is this one:

In a local frame S of coordinates $(\tau = \bar{x}^0, \bar{x}^3)$, let us consider an event $E(\tau, \bar{x}^3)$, $\bar{x}^3 > 0$. There are only two events on the τ-axis, say $E_1(\tau_1, 0)$ and $E_2(\tau_2, 0)$ with $\tau_1 < \tau_2$, such that the event E is connected to the events E_1 and E_2 by light-rays. Indeed, considering that the slopes of the lines $E_1 E$ and $E_2 E$ have to be 1 and -1 respectively, the connections among the coordinates are

$$\tau = \frac{\tau_1 + \tau_2}{2}; \quad \bar{x}^3 = \frac{\tau_2 - \tau_1}{2},$$

or equivalently

$$\tau_1 = \tau - \bar{x}^3; \quad \tau_2 = \bar{x}^3 + \tau.$$

Therefore we have proved

Proposition 8.6.2 *Suppose the event* $E(\tau, \bar{x}^3)$, $\bar{x}^3 > 0$ *is connecting the events* E_1 *and* E_2 *by light-rays. If the coordinates are* $E_1(\tau_1, 0)$, $E_2(\tau_2, 0)$, $\tau_1 < \tau_2$, *then, between the above coordinates there are the relations*

$$\tau_1 = \tau - \bar{x}^3; \quad \tau_2 = \bar{x}^3 + \tau.$$

The physical image is the following: a light-ray from E_1 reaches E and is reflected to E_2. The coordinates are like in the previous proposition.

Let us now suppose that τ-axis is seen in the frame R as a curve. To move forward, let us suppose that τ-axis is parameterized by

$$\tau - axis : \begin{cases} t(\tau) = \dfrac{1}{\alpha} \sinh \alpha \tau \\ x^3(\tau) = \dfrac{1}{\alpha} \cosh \alpha \tau. \end{cases}$$

Consider the event E, now in coordinates of R, that is $E(t, x^3)$.

The events E_1 and E_2 belong now to the curve which represents the τ-axis in R, that is $E_1(t_1, x_1^3)$ with $t_1 = \dfrac{1}{\alpha} \sinh \alpha \tau_1$; $x_1^3 = \dfrac{1}{\alpha} \cosh \alpha \tau_1$

and

$E_2(t_2, x_2^3)$ with $t_2 = \dfrac{1}{\alpha} \sinh \alpha \tau_2$; $x_2^3 = \dfrac{1}{\alpha} \cosh \alpha \tau_2$,

in such a way that a light-ray from E_1 reaches E and is reflected to E_2.

Since the slopes $E_1 E$ and $E_2 E$ have to be 1 and -1 respectively, we have

$$\frac{x^3 - x_1^3}{t - t_1} = 1; \quad \frac{x^3 - x_2^3}{t - t_2} = -1.$$

It results the system of equations

$$\begin{cases} -t + x^3 = -t_1 + x_1^3 \\ t + x^3 = t_2 + x_2^3 \end{cases}$$

with the solution

$$\begin{cases} t = \dfrac{t_1 + t_2 + x_2^3 - x_1^3}{2} \\ x^3 = \dfrac{-t_1 + t_2 + x_2^3 + x_1^3}{2}. \end{cases}$$

The first formula becomes

$$t = \frac{\sinh \alpha \tau_1 + \sinh \alpha \tau_2 + \cosh \alpha \tau_2 - \cosh \alpha \tau_1}{2\alpha} = \frac{e^{\alpha \tau_2} - e^{-\alpha \tau_1}}{2\alpha},$$

that is

$$t = \frac{e^{\alpha(\tau + \bar{x}^3)} - e^{-\alpha(\tau - \bar{x}^3)}}{2\alpha} = \frac{e^{\alpha \bar{x}^3}}{\alpha} \sinh \alpha \tau.$$

In the same way

$$x^3 = \frac{e^{\alpha \bar{x}^3}}{\alpha} \cosh \alpha \tau,$$

that is we found out a coordinate transformation $G : S \to R$,

$$G : \begin{cases} t(\tau, \bar{x}^3) = \dfrac{e^{\alpha \bar{x}^3}}{\alpha} \sinh \alpha \tau \\ x^3(\tau, \bar{x}^3) = \dfrac{e^{\alpha \bar{x}^3}}{\alpha} \cosh \alpha \tau. \end{cases}$$

This is the proof of the following

Theorem 8.6.3 *Consider a system R of coordinates (t, x^3) in which the τ-axis is the curve parameterized by*

$$\begin{cases} t(\tau) = \dfrac{1}{\alpha} \sinh \alpha \tau \\ x^3(\tau) = \dfrac{1}{\alpha} \cosh \alpha \tau. \end{cases}$$

Suppose it exists three events E_1, E, E_2 such that a light-ray from E_1 reaches E and is reflected to E_2. Then, between the coordinates of $E_1(t_1, x_1^3)$, $E_2(t_2, x_2^3)$ with

$$t_1 = \frac{1}{\alpha} \sinh \alpha \tau_1 \, , x_1^3 = \frac{1}{\alpha} \cosh \alpha \tau_1 \, ; \quad t_2 = \frac{1}{\alpha} \sinh \alpha \tau_2 \, , x_2^3 = \frac{1}{\alpha} \cosh \alpha \tau_2 \, ;$$

and the coordinates of the event $E(t, x^3)$, there are the relations

$$\begin{cases} t(\tau, \bar{x}^3) = \dfrac{e^{\alpha \bar{x}^3}}{\alpha} \sinh \alpha \tau \\ x^3(\tau, \bar{x}^3) = \dfrac{e^{\alpha \bar{x}^3}}{\alpha} \cosh \alpha \tau, \end{cases}$$

where

$$\tau_1 = \tau - \bar{x}^3; \quad \tau_2 = \bar{x}^3 + \tau.$$

Now, we consider a local frame S with coordinates $(\tau = \bar{x}^0, \bar{x}^1, \bar{x}^2, \bar{x}^3)$ in which a constant gravitational field exists. This constant gravitational field can be imagined as a vector $-\vec{\alpha}$ acting along the $\bar{x}^3$ axis in its negative direction, therefore as a vector with the spatial coordinates $(0, 0, -\alpha)$.

Let us consider another frame of coordinates R, whose coordinates are $(t = x^0, x^1, x^2, x^3)$. This frame is in free fall in the previous constant gravitational field.

We may assume that, at $\tau = t = 0$, the two frames can be seen together with axes corresponding in notation of indexes. Let us suppose that the second frame R is moving along the $\bar{x}^3$ axis in its negative direction. So, we can think of a transformation which describes the constant gravitational field in S, involving only the pairs of axis $(\tau, \bar{x}^3)$ of S and (t, x^3) of R.

To obtain it, we change the perspective: We consider R at rest and the frame S accelerating along the x^3 axis with the constant acceleration $(0, \alpha)$.

When we determine the Lorentz transformation, our first concern is describing the τ axis in R when S is moving at constant speed v. The question is: If S is $\vec{\alpha}$-accelerating with respect to R, what becomes the τ axis of S in R?

Let us think of a line as a trajectory of a moving point. The speed is constant along the line, that is the vector speed of a given line has constant norm; in the same time, the acceleration vector is null. If we consider the current point $(\tau, 0)$ on τ-axis, the speed vector is $\vec{V} = (1, 0)$ and the acceleration vector is $\vec{A} = (0, 0)$. Therefore the τ-axis at rest is characterized by

$$\| \vec{V} \|_M = 1; \ \| \vec{A} \|_M = 0.$$

Looking at the accelerated frame S, the R observer sees a modified τ-axis, denoted now

$$c(\tau) := (t(\tau), x^3(\tau))$$

and characterized by

$$\| \vec{\dot{c}} (\tau) \|_M = 1; \ \| \vec{\ddot{c}} (\tau) \|_M = \alpha.$$

Now, we observe that, according to the first proposition, the speed vector $\vec{\dot{c}}$ is a time-like one, while the acceleration vector $\vec{\ddot{c}}$ is a space-like one. The two conditions become the system of differential equations

$$\begin{cases} (\dot{t}(\tau))^2 - (\dot{x}^3(\tau))^2 = 1 \\ -(\ddot{t}(\tau))^2 + (\ddot{x}^3(\tau))^2 = \alpha^2 \end{cases}$$

with the general solution

$$\begin{cases} t(\tau) = \dfrac{1}{\alpha} \sinh \alpha(\tau + \tau_0) + t_0 \\ x^3(\tau) = \overset{-}{+} \dfrac{1}{\alpha} \cosh \alpha(\tau + \tau_0) + x_0^3. \end{cases}$$

From the Euclidean point of view, we deal with the hyperbola

$$(t - t_0)^2 - (x^3 - x_0^3)^2 = \frac{1}{\alpha^2}$$

having the center at (t_0, x_0^3) and the parallel asymptotes along the light cone. This is a good exercise for the reader.

Of course, in R, where the Minkowski Geometry is acting, this curve is a Minkowski space-like circle. From symmetry reason, we may choose the center of this hyperbola at $(0, 0)$, $\tau_0 = 0$ and the sign $+$. That is, we have an image of the τ-axis of S in R,

$$\begin{cases} t(\tau) = \dfrac{1}{\alpha} \sinh \alpha\tau \\ x^3(\tau) = \dfrac{1}{\alpha} \cosh \alpha\tau. \end{cases}$$

We have proved the following

Theorem 8.6.4 *If a coordinates frame S is $\vec{\alpha}$-accelerated with respect to a frame at rest R, then the image of the τ-axis of S in R is a curve $c(\tau) := (t(\tau), x^3(\tau))$ characterized by the equations*

$$\begin{cases} t(\tau) = \dfrac{1}{\alpha} \sinh \alpha\tau \\ x^3(\tau) = \dfrac{1}{\alpha} \cosh \alpha\tau. \end{cases}$$

Now, we have the complete image: It exists a local change of coordinates between S and R described by the transformation G, $G : S \to R$

$$G : \begin{cases} t(\tau, \bar{x}^3) = \dfrac{e^{\alpha \bar{x}^3}}{\alpha} \sinh \alpha\tau \\ x^3(\tau, \bar{x}^3) = \dfrac{e^{\alpha \bar{x}^3}}{\alpha} \cosh \alpha\tau. \end{cases}$$

The transformation G, which was defined by using the idea of accelerating frame, allows us to understand how the constant gravitational field $- \vec{\alpha}$ in the frame of coordinates S can be seen via the system of coordinates R. Consequently, in the future, we will be able to compute the metric of S.

Exercise 8.6.5 *Show that the inverse transformation* $G^{-1} : R \rightarrow S$ *is*

$$G^{-1} : \begin{cases} \tau(t, x^3) = \dfrac{1}{\alpha} \tanh^{-1}\left(\dfrac{t}{x^3}\right) \\ \bar{x}^3(t, x^3) = \dfrac{1}{2\alpha} \ln\left[\alpha^2 \left((x^3)^2 - t^2\right)\right]. \end{cases}$$

8.6.1 Doppler's Effect in Constant Gravitational Field and Consequences

We know, up to this point, that frames at rest and frames moving at constant speed
are inertial frames. The laws of mechanics and the new laws of Special Relativity
have the same form and hold in such frames. There are no evidences that frames
in which acts a constant gravitational field are non-inertial frames. Are they really
inertial frames? The answer is related to the Doppler effect in a constant gravitational
field.

We are interested in finding out how the frequency of light in S is affected by the
constant gravitational field $- \vec{\alpha}$ which acts in S.

To obtain a formula which connects the frequency of the light and $- \vec{\alpha}$, we need
to change the perspective as we have done before. We use two frames of coordinates
S and R. Instead of looking at the frame of coordinates R in free fall in the previous
constant gravitational field, we look at R at rest and at the frame S accelerated along
the x^3 axis with the constant acceleration $\vec{\alpha}$.

Our study is done again in the two corresponding slices of S and R, taking into
account the coordinates $(\tau = \bar{x}^0, \bar{x}^3)$, respectively $(t = x^0, x^3)$.

Let us pose the problem.

From the origin $O(0, 0)$ of S is emitted a light signal with frequency ν. Consider
$C(0, h)$, a point on $\bar{x}^3$ axis at height h. The level h is reached by the light-ray at the
point $H(h, h)$. In order to obtain the frequency at the level h, we need to consider
the frame R. Denote by f_h the frequency of the light-ray in R corresponding to the
level h in S. We have

Theorem 8.6.6

$$f_h = \nu e^{-\alpha h}.$$

Proof Let us remember the transformation G,

$$G : \begin{cases} t(\tau, \bar{x}^3) = \dfrac{e^{\alpha \bar{x}^3}}{\alpha} \sinh \alpha\tau \\ x^3(\tau, \bar{x}^3) = \dfrac{e^{\alpha \bar{x}^3}}{\alpha} \cosh \alpha\tau. \end{cases}$$

The points O and C, from S, are seen through G in R with the coordinates $(t_0, x_0^3) = \left(0, \dfrac{1}{\alpha}\right)$ and $(t_C, x_C^3) = \left(0, \dfrac{e^{\alpha h}}{\alpha}\right)$ respectively.

The equivalent of the point H in R has the coordinates $\left(\dfrac{e^{\alpha h}}{\alpha} \sinh \alpha h, \dfrac{e^{\alpha h}}{\alpha} \cosh \alpha h\right)$. Since, through $\left(0, \dfrac{1}{\alpha}\right)$, the new τ-axis passes in R, that is the curve $c(\tau) = \left(\dfrac{1}{\alpha} \sinh \alpha\tau, \dfrac{1}{\alpha} \cosh \alpha\tau\right)$, equivalent to the line $\bar{x}^3 = h$, is the curve $c_h(\tau) = \left(\dfrac{e^{\alpha h}}{\alpha} \sinh \alpha\tau, \dfrac{e^{\alpha h}}{\alpha} \cosh \alpha\tau\right)$.

The speed vector at h has the components $\left(e^{\alpha h} \cosh \alpha h, e^{\alpha h} \sinh \alpha h\right)$, that is

$$v_h = \tanh \alpha h.$$

We replace this formula in the general formula found before for the relativistic Doppler's effect and it results

$$f_h = \nu \sqrt{\frac{1 - v_h}{1 + v_h}},$$

that is

$$f_h = \nu \sqrt{\frac{1 - \tanh h}{1 + \tanh h}} = \nu e^{-\alpha h}.$$

$\square$

Let us observe that if we denote by

$$\Delta\tau = \frac{1}{\nu}$$

the corresponding period in S, and by

$$\Delta t = \frac{1}{f_h}$$

the corresponding period in R, we obtain a formula connecting the two periods, that is

$$\Delta\tau = e^{-\alpha h} \Delta t.$$

If $h > 0$, that is, if the point C belongs to the upper half-plane of S, comparing the periods in S with the one in R, we have

$$\Delta\tau < \Delta t.$$

If $h < 0$, that is if the point C is in the complementary half-plane of S, we have

$$\Delta\tau > \Delta t.$$

If h is very close to 0, we may consider the approximation

$$e^{-\alpha h} = 1 - \alpha h.$$

From a physical point of view, $-\alpha h$ corresponds to a potential energy for an object whose mass is 1. Therefore we can write the formula

$$\Delta t = \frac{1}{1 - \alpha h}\Delta\tau$$

written with respect to the potential energy.

Now, let us take into account two clocks, one in O and one in C. Suppose the first one ticking at each $\Delta\tau$ seconds. The second clock at C is ticking in Δt seconds.

The results $\Delta\tau < \Delta t$ if $h > 0$ and $\Delta\tau > \Delta t$, if $h < 0$ hold.

This situation shows that S cannot be an inertial reference frame. In an inertial reference frame, the position cannot affect the way in which time is running. In the entire frame S, we should have a same result.

Therefore we have

Corollary 8.6.7 *The frames in which a constant gravitational field is acting are not inertial frames.*

A further remark is the following. Let us suppose we are on the surface of a planet. Consider $0 < h_1 < h_2$. It results $-\alpha h_1 > -\alpha h_2$, that is $1 - \alpha h_1 > 1 - \alpha h_2$. Suppose that h_1, h_2 are so small than the quantities $1 - \alpha h_1$ and $1 - \alpha h_2$ are positive. We obtain

$$\Delta t_1 = \frac{1}{1 - \alpha h_1}\Delta\tau < \frac{1}{1 - \alpha h_2}\Delta\tau = \Delta t_2.$$

Therefore, while h is decreasing, the clock, from C is approaching O and it is ticking slower and slower, *that is the gravity slows down the clocks.* This effect is taken into consideration in the case of GPS systems where we need to have same times at ground level and at the GPS satellite level.[3]

---

[3]The acronym GPS stays for Global Positioning System. It is a satellite-based radio-navigation system that provides geolocation and time information to a receiver anywhere on or near the Earth where there is an unobstructed line of sight to the fleet of GPS satellites. Obstacles, such as mountains, block or weaken the GPS signals.

8.6.2   Bending of Light-Rays in a Constant Gravitational Field

Theorem 8.6.8 *The light-rays are bending in a constant gravitational field* $-\vec{\alpha}$.

Proof The main idea of the proof that the light is bending in a constant gravitational field is related to the fact that the projection of a line to a plane is a line or a point. For the proof, the trajectory of a photon included in a given plane, in our case $x^3 = \dfrac{1}{\alpha}$, is transferred into the frame of coordinates $(\tau = \bar{x}^0, \bar{x}^2, \bar{x}^3)$ and then it is projected to the plane $(\bar{x}^2, \bar{x}^3)$. The result is neither a line nor a point. Therefore the light-ray is bent by the constant gravitational field.

Let us focus on G^{-1}, now defined for a three dimensional slice in R. The result is

$$
G^{-1} : \begin{cases} \tau(t, x^2, x^3) = \dfrac{1}{\alpha} \tanh^{-1}\left(\dfrac{t}{x^3}\right) \\[2mm] \bar{x}^2(t, x^2, x^3) = x^2 \\[2mm] \bar{x}^3(t, x^2, x^3) = \dfrac{1}{2\alpha} \ln\left[\alpha^2\left((x^3)^2 - t^2\right)\right] \end{cases}
$$

and we look at the image of the plane $x^3 = \dfrac{1}{\alpha}$. In the next formulas, we suppress the (t, x^2) coordinates, therefore

$$
G^{-1}\left(x^3 = \dfrac{1}{\alpha}\right) : \begin{cases} \tau = \dfrac{1}{\alpha} \tanh^{-1}(\alpha t) \\[2mm] \bar{x}^2 = x^2 \\[2mm] \bar{x}^3 = \dfrac{1}{2\alpha} \ln\left(1 - \alpha^2 t^2\right). \end{cases}
$$

We observe: $G^{-1}\left(x^3 = \dfrac{1}{\alpha}\right)$ is a cylinder containing the $\bar{x}^2$ axis.

If we consider the trajectory of a photon in the $x^3 = \dfrac{1}{\alpha}$ plane, this has to be the line $c(s) = \left(s, s, \dfrac{1}{\alpha}\right)$. The system $G^{-1}(c(s))$ is described by the equations

$$
G^{-1}(c(s)) : \begin{cases} \tau = \dfrac{1}{\alpha} \tanh^{-1}(\alpha s) \\[2mm] \bar{x}^2 = s \\[2mm] \bar{x}^3 = \dfrac{1}{2\alpha} \ln\left(1 - \alpha^2 s^2\right). \end{cases}
$$

f

If we project the previous trajectory of a photon, that is trajectory seen in S to the plane $(\bar{x}^2, \bar{x}^3)$, the result, denoted by $\bar{c}(s)$, is parameterized as

$$\bar{c}(s) = \left(s, \ \frac{1}{2\alpha} \ln \left(1 - \alpha^2 s^2 \right) \right).$$

It is obvious that $\bar{c}(s)$ is neither a point nor a line.   □

8.6.3 The Basic Incompatibility Between Gravity and Special Relativity

We can conclude this chapter pointing out the basic incompatibility between gravity and Special Relativity.

Let us suppose we are in a local frame S, where a constant gravitational field $-\overrightarrow{\alpha}$ is acting and let us consider a photon emitted at the origin O from a source moving along the τ-axis. Taking into account the frequency ν and the formula $\Delta\tau = \frac{1}{\nu}$, the next photon is emitted by the source at the point $A(\Delta\tau, 0)$. The frame S is not an inertial one and we have proved that the trajectories of photons are bending, that is they are not straight lines but curves. This means that there is a specific curve starting at the emitting point of the photon, in our case O, which reaches the line $\bar{x}^3 = h$ at a point denoted by M. The second photon, emitted in A has an identical trajectory to the one emitted at O. This second trajectory reaches the line $\bar{x}^3 = h$ in a point denoted by N.

The quadrilateral $OANM$ has the property $\Delta\tau = OA = MN$. The length MN is the period Δt corresponding to the frequency f_h in R.

We have

$$\Delta\tau = \Delta t,$$

instead of

$$\Delta\tau = e^{-\alpha h} \Delta t.$$

This contradiction shows that the gravity cannot be integrated into the framework of Special Relativity. Another theory has to be developed in order to fix this shortcoming. This is General Relativity.

Chapter 9
General Relativity and Relativistic Cosmology

Quod erat demostrandum.

An imaginary discussion between Newton and Einstein could be the following.

....*...

Isaac Newton: *Dear Prof. Einstein, my Universe is very simple. I can describe it using vectors and calculus. Between any two objects, a gravitational force is acting and, according to the masses of objects and the distance between them, the gravitational force law is $F = G\dfrac{mM}{r^2}$. The gravitational field, in this case, is $A = \dfrac{GM}{r^2}$. However, there exists an artifact, the gravitational potential $\Phi = \dfrac{GM}{r}$. After me, the brilliant experimental physicist, Henry Cavendish, measured the gravitational constant $G = 6.67 \times 10^{-11} \mathrm{N\,m^2/kg^2}$, considered "universal". The potential is related to the gravitational field through the formula $\nabla \Phi = -\overrightarrow{A}$, the vacuum field equation is $\nabla^2 \Phi = 0$, as established by Pierre Simon Laplace, and the general gravitational field equation is $\nabla^2 \Phi = 4\pi G \rho$ as pointed out by Siméon Denis Poisson, once the density of matter is known. The objects are moving in this gravitational field according to $\overrightarrow{F} = m\,\overrightarrow{A}$ and the trajectories are conics because my gravitational universal law gives a mathematical proof for the Kepler laws. What do you think?*

    Albert Einstein: *Very simple indeed, Sir Isaac! Conversely, my Universe is geometric and has four dimensions, it is called spacetime! I need more mathematics to describe it. Differential Geometry is essential, but, my dear Sir, this was invented after you passed away! My Universe is expressed by a metric $ds^2 = g_{ij}dx^i dx^j$, where the coefficients g_{ij} play the role of your gravitational potential Φ. The Christoffel symbols Γ^i_{jk} are related to your gravitational field A. This means that "my gravitational field" has more variables and structures than yours. The vacuum field equations are*

$$R_{ij} = 0$$

W. Boskoff and S. Capozziello, *A Mathematical Journey to Relativity*, UNITEXT for Physics, https://doi.org/10.1007/978-3-030-47894-0_9

and my general field equations are

$$R_{ij} - \frac{1}{2} R \, g_{ij} = \frac{8\pi G}{c^4} T_{ij}.$$

Starting from them, I can recover the Laplace and Poisson equations in the weak field limit so, my dear Sir Isaac,....I am coherent with your picture! The metric I mentioned before is the one that satisfies the field equations. Objects are always moving on geodesics of the metric, therefore their equations are

$$\frac{d^2 x^r}{dt^2} + \Gamma^r_{pq} \frac{dx^p}{dt} \frac{dx^q}{dt} = 0$$

These geodesic equations are my way of saying $\vec{F} = m \, \vec{A}$ that I recover, indeed, in the weak field limit. To conclude, one of my collaborators, John Archibald Wheeler, said that the better description of my theory can be reduced to the sentence **"Spacetime tells matter how to move; matter tells spacetime how to curve"** *[34].*

.... * ....

Let us insist on the the last sentence. How the space is curved appears from the Einstein field equations

$$R_{ij} - \frac{1}{2} R \, g_{ij} = \frac{8\pi G}{c^4} T_{ij}.$$

In the left-hand side, we have the "Geometry": Metric g_{ij} and its derivatives are involved; in the right-hand side, we have a tensor depending on matter, the so called energy-momentum tensor. Once we have a metric g_{ij}, according to the Equivalence Principle, we have also the geodesics of the metric as we will discuss below. Which is the meaning of the geodesics described by the equations

$$\frac{d^2 x^r}{dt^2} + \Gamma^r_{pq} \frac{dx^p}{dt} \frac{dx^q}{dt} = 0 \, ?$$

The simplest answers is: They are trajectories of objects moving accordingly to the Geometry of spacetime.

We start this chapter with some general considerations on what a good theory of gravity should do and enunciating the basic principles on which General Relativity lies. After, we take into account the differences between the Classical Newtonian Mechanics and the Einstein picture of gravity based on Geometry. We discuss how it works looking at the differences between the constant gravitational field, as conceived in Classical Mechanics, and the General Relativity counterpart. Finally, we provide Einstein's field equations from the Einstein–Hilbert variational principle and briefly discuss possible generalizations like the so called $f(R)$ gravity.

The Schwarzschild solution of the Einstein vacuum field equations is presented. The orbits of planets and the bending of light rays are computed in the framework of Schwarzschild metric.

Even if it does not verify the field equations, the Einstein metric is presented because Einstein used it to compute the orbits of planets and the bending of light rays. The full computation for the perihelion drift is presented. The same, in both metrics, is presented the bending of light rays passing near our Sun.

Fermi's viewpoint on Einstein's vacuum field equations is presented with implications related to the study of the week gravitational field; the classical counterparts of the relativistic equations are obtained in this way. We analyze Einstein static universe and the basic considerations on the cosmological constant, as a part related to the standard approach to the General Relativity.

A "cosmological metric" is discussed when we study the Friedmann–Lemaître–Robertson–Walker metrics of a Universe in expansion. The way we obtain it is related to the way we considered the energy-momentum tensor. An interesting introductory section devoted to black holes mathematics is also presented. To have a more complete view on Relativity, we offer a short introduction on cosmic strings and gravitational waves.

Particular hypothetical universes without global time coordinate, as Gödel's one and without masses, as de Sitter one, are presented to enlarge the possibilities of solutions of Einstein's field equations.

This is the most important chapter of the book. The main references for the topics we are developing can be found in [4, 17–19, 21, 22, 25, 33–35, 40–45].

9.1 What Is a Good Theory of Gravity?

Before entering the details of General Relativity, some considerations are in order. We need them to discuss the change of perspective introduced by the Einstein theory.

As it is well known, General Relativity is based on the fundamental assumption that space and time are entangled into a single spacetime structure assigned on a pseudo-Riemann manifold. Being a dynamical structure, it has to reproduce, in the absence of gravitational field, the Minkowski spacetime.

General Relativity has to match some minimal requirements to be considered a self-consistent physical theory. First of all, it has to reproduce the Newtonian dynamics in the weak-energy limit, hence it must be able to explain the astronomical dynamics related to the orbits of planets and the self-gravitating structures. Moreover, it passed some observational tests in the Solar System that constitute its experimental foundation [46].

However, General Relativity should be able to explain the Galactic dynamics, taking into account the observed baryonic constituents (e.g. luminous components as stars, sub-luminous components as planets, dust and gas), radiation and Newtonian potential which is, by assumption, extrapolated to Galactic scales. Besides, it should address the problem of large scale structure as the clustering of galaxies. On cosmological scales, it should address the dynamics of the Universe, which means to reproduce the cosmological parameters as the expansion rate, the density parameter, and so on, in a self-consistent way. Observations and experiments, essentially, probe the standard baryonic matter, the radiation and an attractive overall interaction, acting at all scales and depending on distance: this interaction is gravity.

In particular, Einstein's General Relativity is based on four main assumptions. They are

The "*Relativity Principle*" - there is no preferred inertial frames, i.e. all frames are good frames for Physics.

The "*Equivalence Principle*" - inertial effects are locally indistinguishable from gravitational effects (which means the equivalence between the inertial and the gravitational masses). In other words, any gravitational field can be locally cancelled.

The "*General Covariance Principle*" - field equations must be "covariant" in form, i.e. they must be invariant in form under the action of spacetime diffeomorphisms.

The "*Causality Principle*" - each point of space-time has to admit a universally valid notion of past, present and future.

On these bases, Einstein postulated that, in a four-dimensional spacetime manifold, the gravitational field is described in terms of the metric tensor field $ds^2 = g_{ij}dx^i dx^j$, with the same signature of Minkowski metric. The metric coefficients have the physical meaning of gravitational potentials. Moreover, he postulated that spacetime is curved by the distribution of the energy-matter sources.

The above principles require that the spacetime structure has to be determined by either one or both of the two following fields: a Lorentzian metric g and a linear connection Γ, assumed by Einstein to be torsionless. The metric g fixes the causal structure of spacetime (the light cones) as well as its metric relations (clocks and rods); the connection Γ fixes the free-fall, i.e. the locally inertial observers. They have, of course, to satisfy a number of compatibility relations which amount to require that photons follow null geodesics of Γ, so that Γ and g can be independent, a priori, but constrained, a posteriori, by some physical restrictions. These, however, do not impose that Γ has necessarily to be the Levi–Civita connection of g [47].

It should be mentioned, however, that there are many shortcomings in General Relativity, both from a theoretical point of view (non-renormalizability, the presence of singularities, and so on), and from an observational point of view. The latter indeed clearly shows that General Relativity is no longer capable of addressing Galactic, extra-galactic and cosmic dynamics, unless the source side of field equations contains some exotic form of matter-energy. These new elusive ingredients, as mentioned above, are usually addressed as *dark matter"* and *dark energy* and constitute up to the 95% of the total cosmological amount of matter-energy [48].

On the other hand, instead of changing the source side of the Einstein field equations, one can ask for a "geometrical view" to fit the missing matter-energy of the observed Universe. In such a case, the dark side could be addressed by extending General Relativity including more geometric invariants into the standard Einstein–Hilbert Action. Such effective Lagrangians can be easily justified at fundamental level by any quantization scheme on curved spacetimes [49]. However, at present stage of the research, this is nothing else but a matter of taste, since no final probe discriminating between dark matter and extended gravity has been found up to now. Finally, the bulk of observations that should be considered is so high that an effective Lagrangian or a single particle will be difficult to account for the whole phenomenology at all astrophysical and cosmic scales.

9.1.1 Metric or Connections?

As we will see below, in the General Relativity formulation, Einstein assumed that the metric g of the space-time is the fundamental object to describe gravity. The connection Γ is constituted by coefficients with no dynamics. Only g has dynamics. This means that the single object g determines, at the same time, the causal structure (light cones), the measurements (rods and clocks) and the free fall of test particles (geodesics). Spacetime is therefore a couple $\{M, g\}$ constituted by a pseudo-Riemannian manifold and a metric. Even if it was clear to Einstein that gravity induces freely falling observers and that the Equivalence Principle selects an object that cannot be a tensor (the connection Γ)—since it can be switched off and set to zero at least in a point)—he was obliged to choose it (the Levi–Civita connection) as being determined by the metric structure itself.

In the Palatini formalism, a (symmetric) connection Γ and a metric g are given and varied independently. Spacetime is a triple $\{M, g, \Gamma\}$ where the metric determines rods and clocks (i.e., it sets the fundamental measurements of spacetime) while Γ determines the free fall. In the Palatini formalism, Γ are differential equations. The fact that Γ is the Levi–Civita connection of g is no longer an assumption but becomes an outcome of the field equations.

The connection is the gravitational field and, as such, it is the fundamental field in the Lagrangian. The metric g enters the Lagrangian with an "ancillary" role. It reflects the fundamental need to define lengths and distances, as well as areas and volumes. It defines rods and clocks that we use to make experiments. It defines also the causal structure of spacetime. However, it has no dynamical role. There is no whatsoever reason to assume g to be the potential for Γ, nor that it has to be a true field just because it appears in the action. We will not develop any more the Palatini formalism in this book. For a detailed discussion see [18].

9.1.2 The Role of Equivalence Principle

The Equivalence Principle is strictly related to the above considerations and could play a very relevant role in order to discriminate among theories. In particular, it could specify the role of g and Γ selecting between the metric and Palatini formulation of gravity. In particular, precise measurements of Equivalence Principle could say us if

Γ is only Levi–Civita or a more general connection disentangled, in principle, from g. Before, we discussed the Equivalence Principle starting from the early Galileo consideration stating that $m_i \equiv m_g$. Besides this result, in General Relativity, Equivalence Principle states that accelerations can be set to zero in given reference frame. According to this result, the free fall along geodesics, given by the connection, is ruled by the metric, as we will discuss below.

Before entering into details, let us discuss some topics related to the Equivalence Principle. Summarizing, the relevance of this principle comes from the following points:

- Competing theories of gravity can be discriminated according to the validity of Equivalence Principle;
- Equivalence Principle holds at classical level but it could be violated at quantum level;
- Equivalence Principle allows to investigate independently geodesic and causal structure of spacetime.

From a theoretical point of view, Equivalence Principle lies at the physical foundation of metric theories of gravity. The first formulation of Equivalence Principle comes out from the theory of gravitation formulates by Galileo and Newton, i.e. the Weak Equivalence Principle (the above Galilean Equivalence Principle) which asserts the inertial mass m_i and the gravitational mass m_g of any physical object are equivalent. The Weak Equivalence Principle statement implies that it is impossible to distinguish, locally, between the effects of a gravitational field from those experienced in uniformly accelerated frames using the simple observation of the free falling particles behavior.

A generalization of Weak Equivalence Principle claims that Special Relativity is locally valid. Einstein realized, after the formulation of Special Relativity, that the mass can be reduced to a manifestation of energy and momentum as discussed in previous chapter. As a consequence, it is impossible to distinguish between a uniform acceleration and an external gravitational field, not only for free-falling particles, but whatever is the experiment. According to this observation, Einstein Equivalence Principle states:

- The Weak Equivalence Principle is valid.
- The outcome of any local non-gravitational test experiment is independent of the velocity of free-falling apparatus.
- The outcome of any local non-gravitational test experiment is independent of where and when it is performed in the Universe.

One defines as "local non-gravitational experiment" an experiment performed in a small-size of a free-falling laboratory. Immediately, it is possible to realize that the gravitational interaction depends on the curvature of spacetime, i.e. the postulates of any metric theory of gravity have to be satisfied. Hence the following statements hold:

- Spacetime is endowed with a metric g_{ij}.
- The world lines of test bodies are geodesics of the metric.
- In local freely falling frames, called local Lorentz frames, the non-gravitational laws of physics are those of Special Relativity.

One of the predictions of this principle is the gravitational red-shift, experimentally verified by Pound and Rebka in 1960 [46]. Notice that gravitational interactions are excluded from the Weak Equivalence Principle and the Einstein Equivalence Principle.

In order to classify alternative theories of gravity, the gravitational Weak Equivalence Principle and the Strong Equivalence Principle has to be introduced. On the other hands, the Strong Equivalence Principle extends the Einstein Equivalence Principle by including all the laws of physics in its terms. That is:

- Weak Equivalence Principle is valid for self-gravitating bodies as well as for test bodies (Gravitational Weak Equivalence Principle).
- The outcome of any local test experiment is independent of the velocity of the free-falling apparatus.
- The outcome of any local test experiment is independent of where and when in the Universe it is performed.

Alternatively, the Einstein Equivalence Principle is recovered from the Strong Equivalence Principle as soon as the gravitational forces are neglected. Many authors claim that the only theory coherent with Strong Equivalence Principle is General Relativity.

A very important issue is the consistency of Equivalence Principle with respect to the Quantum Mechanics. General Relativity is not the only theory of gravitation and, several alternative theories of gravity have been investigated from the 60's of last century [49]. Considering the spacetime to be special relativistic at a background level, gravitation can be treated as a Lorentz-invariant field on the background. Assuming the possibility of General Relativity extensions, two different classes of experiments can be conceived:

- Tests for the foundations of gravitational theories considering the various formulations of Equivalence Principle.
- Tests of metric theories where spacetime is a priori endowed with a metric tensor and where the Einstein Equivalence Principle is assumed always valid.

The subtle difference between the two classes of experiments lies on the fact that Equivalence Principle can be postulated a priori or, in a certain sense, "recovered" from the self-consistency of the theory. What is today clear is that, for several fundamental reasons, extra fields are necessary to describe gravity with respect to the other interactions. Such fields can be scalar fields or higher-order corrections of curvature invariants [49]. For these reasons, two sets of field equations can be considered: The first set couples the gravitational field to the non-gravitational contents of the Universe, i.e. the matter distribution, the electromagnetic fields, etc. The second set of equations gives the evolution of non-gravitational fields. Within the framework of metric theories, these laws depend only on the metric and this is a consequence of the Einstein Equivalence Principle. In the case where Einstein field equations are modified and matter field are minimally coupled with gravity, we are dealing with the so-called *Jordan frame*. In the case where Einstein field equations are preserved and matter field are non-minimally coupled, we are dealing with the so-called *Einstein frame*. Both frames are conformally related but the very final issue is to understand if passing from one frame to the other (and vice versa) is physically significant. See [18] for details. Clearly, Equivalence Principle plays a fundamental role in this discussion. In particular, the question is if it is always valid or it can be violated at quantum level. See [50–52].

After these preliminary considerations, let us start with the geometric construction of General Relativity. However, we recommend the reader to consider again these introductory sections after he/she finishes to read the book because some current problems in General Relativity are reported.

9.2   Gravity Seen Through Geometry in General Relativity

Let us go back to our previous discussion on the gravitational potential in Newtonian Mechanics. We start from the tidal acceleration equations

$$\frac{d^2}{dt^2} \frac{\partial \bar{x}}{\partial q} = -d^2 \Phi_{\bar{x}} \frac{\partial \bar{x}}{\partial q},$$

where the Hessian matrix of the gravitational potential Φ

$$d^2 \Phi_{\bar{x}} = \left(\frac{\partial^2 \Phi(\bar{x})}{\partial x_i \partial x_k} \right)_{i,k}$$

is encapsulated in its trace by the Laplace equation $\nabla^2 \Phi = 0$ in vacuum. In a space endowed with a metric $ds^2 = g_{ij} dx^i dx^j$, it is possible to find the equivalent

$$\frac{\nabla^2}{d\tau^2} \frac{\partial x^h}{\partial q} = -K^h_j \frac{\partial x^j}{\partial q},$$

where

$$K^h_j = R^h_{ijk} \frac{dx^i}{d\tau} \frac{dx^k}{d\tau}$$

plays the role of the Hessian of the gravitational potential.

It seems to be natural to think of the trace of the matrix K_j^h to obtain an equivalent of classical vacuum fields equations $\nabla^2 \Phi = 0$. Since

$$K_h^h = R_{ijh}^h \frac{dx^i}{d\tau} \frac{dx^j}{d\tau},$$

the Ricci tensor has to be involved in General Relativity field equations. It is why Einstein, and then Hilbert, considered a way to express the field equations through the Ricci tensor.

So, let us repeat their main idea. The gravitational field is not constant. There are small variations of the gravitational field induced by some other bodies or by changing the distance r between bodies. If we are on the surface of the Earth, our legs will experience a higher intensity of the gravitational field of the Earth than our head. To understand this, it is enough to look at the formula $A = \dfrac{GM}{r^2}$, M being the mass of the Earth, G being the gravitational constant and r being the radius R of the Earth at the legs level and $r = R + h$ at the level of our head, h being our height. For the same reason, a person at the first floor of a building experiences a greater intensity of the gravitational field comparing with another person which is at the 33th floor of the same building. The Moon makes ocean tides and we see how these are related to the tidal effects.

If we have tides, mathematically they can be treated under the Newtonian standard, the field equation $\nabla^2 \Phi = 0$ being hidden in the trace of the Hessian matrix $d^2\Phi$ involved in the tidal equations

$$\frac{d^2}{dt^2} \frac{\partial \bar{x}}{\partial q} = -d^2 \Phi_{\bar{x}} \frac{\partial \bar{x}}{\partial q}.$$

Tides can be dealt with a geometric approach considering the Ricci tensor of a given metric from the equations

$$\frac{\nabla^2}{d\tau^2} \frac{\partial x^h}{\partial q} = -K_j^h \frac{\partial x^j}{\partial q},$$

where

$$K_j^h = R_{ijk}^h \frac{dx^i}{d\tau} \frac{dx^k}{d\tau}.$$

Einstein had the power to break the standard Newtonian approach, describing gravity with the language of Differential Geometry.

According to Einstein, the components g_{ij} of a metric $ds^2 = g_{ij} dx^i dx^j$ play the role of gravitational potential Φ, which is just one of the potentials in g_{ij}. The Christoffel symbols Γ^i_{jk} play the role of the gravitational field $\vec{A}$. Let us consider a table of analogies containing the two ways of conceiving at the gravity

$$Newton \qquad Einstein$$

$$\Phi \quad \longleftrightarrow \quad g_{ij}$$

$$\vec{A} \quad \longleftrightarrow \quad \Gamma^i_j k$$

$$\nabla^2 \Phi = 0 \quad \longleftrightarrow \quad ?$$

$$\nabla^2 \Phi = 4\pi G \rho \quad \longleftrightarrow \quad ?$$

The first question mark seems to be replaced by $R_{ij} = 0$, but we still have to work to obtain it. At this moment there is no clue regarding the second question mark.

Einstein was the first who realized that the laws of Nature has to be expressed by equations which hold for any system of coordinates, that is, they must be covariant with respect to any change of coordinates.

Taking into account also the discussion of previous sections, *Einstein's Principle of General Covariance* states:

The laws of Nature have to be expressed as equalities of different tensors.

The changes of coordinates become part of the core of General Relativity. Why are they so important? They allow us to describe the laws of Nature from the point of view of different observers or/and they allow us to describe a new state of a given system.

Let us consider a region of space where the gravitational field can be neglected. Consider a spacecraft there. Suppose that there are no other forces acting there. Therefore all objects are moving on straight lines with constant velocity. The spacecraft does the same. Locally, the spacetime system of coordinates (x^0, x^1, x^2, x^3) can be thought to describe an inertial frame, that is the local metric tensor is the Minkowski one

$$g_{ij}(x^0, x^1, x^2, x^3) = \begin{pmatrix} 1 & 0 & 0 & 0 \\ 0 & -1 & 0 & 0 \\ 0 & 0 & -1 & 0 \\ 0 & 0 & 0 & -1 \end{pmatrix}.$$

Since $\Gamma^i_{jk} = 0$, the geodesics equations are

$$\ddot{x}^j(t) = 0, \quad j \in \{0, 1, 2, 3\},$$

i.e. all objects there experience a free fall. So, the law of motion is described by the previous equations which express in fact the equality $\vec{F} = m \cdot \vec{A}$ for $\vec{F} = \vec{0}$.

Let us now suppose the engines of the spacecraft start and the space craft is accelerated. This is described by a map M which switches from the coordinates $(\bar{x}^0, \bar{x}^1, \bar{x}^2, \bar{x}^3)$ to (x^0, x^1, x^2, x^3), i.e. we have to describe the old coordinates with respect the new ones.

We know, from Differential Geometry, how the new metric looks like.

The new components $\bar{g}_{ij}$ are found after the rule $dM^t \cdot (g_{ij}) \cdot dM$. In this new metric, we can compute the new $\bar{\Gamma}^i_{jk}$ and the geodesic equations are

$$\ddot{\bar{x}}^i(t) + \bar{\Gamma}^i_{jk}\dot{\bar{x}}^j(t)\dot{\bar{x}}^j(t) = 0, \quad j \in \{0, 1, 2, 3\}.$$

Since under a change of coordinates, geodesics are transformed into geodesics, and the meaning is kept, the old law of motion becomes the new law of motion, therefore the equations

$$\ddot{\bar{x}}^i(t) = -\bar{\Gamma}^i_{jk}\dot{\bar{x}}^j(t)\dot{\bar{x}}^j(t), \quad j \in \{0, 1, 2, 3\}$$

describes $\vec{F} = m \cdot \vec{A}$ for $\vec{F} \neq \vec{0}$.

Let us try to understand the constant gravitational field under this more general approach.

9.2.1 Einstein's Landscape for the Constant Gravitational Field

We consider a local frame of coordinates S, $(\tau = \bar{x}^0, \bar{x}^1, \bar{x}^2, \bar{x}^3)$ in which acts a constant gravitational field and another frame of coordinates R, whose coordinates are $(t = x^0, x^1, x^2, x^3)$, frame which is in free fall with respect to the previous constant gravitational field. The metric in the second frame is the Minkowski one,

$$g_{ij}(x^0, x^1, x^2, x^3) = \begin{pmatrix} 1 & 0 & 0 & 0 \\ 0 & -1 & 0 & 0 \\ 0 & 0 & -1 & 0 \\ 0 & 0 & 0 & -1 \end{pmatrix}.$$

Let us assume that, for $\tau = t = 0$, the two frames can be seen as an unique frame with axes corresponding in index notation. We assume also that the second frame is moving along the $\bar{x}^3$ axis in its negative direction. We saw already the transformation which involves the constant gravitational field $-\alpha$ in S. It is, in fact, a change of coordinates between S and R. If we consider only the pairs of axis $(\tau, \bar{x}^3)$ and (t, x^3), this is

$$G : \begin{cases} t(\tau, \bar{x}^3) = \dfrac{e^{\alpha\bar{x}^3}}{\alpha} \sinh \alpha\tau \\[2mm] x^3(\tau, \bar{x}^3) = \dfrac{e^{\alpha\bar{x}^3}}{\alpha} \cosh \alpha\tau. \end{cases}$$

with

$$G^{-1} : \begin{cases} \tau(t, x^3) = \dfrac{1}{\alpha} \tanh^{-1}\left(\dfrac{t}{x^3}\right) \\[2mm] \bar{x}^3(t, x^3) = \dfrac{1}{2\alpha} \ln\left[\alpha^2\left((x^3)^2 - t^2\right)\right]. \end{cases}$$

In the considered slice, in R, the metric is

$$\begin{pmatrix} g_{00} & g_{03} \\ g_{30} & g_{33} \end{pmatrix} = \begin{pmatrix} 1 & 0 \\ 0 & -1 \end{pmatrix}.$$

The metric in S is determined by $dG^t \cdot (g_{ij}) \cdot dG$, where

$$dG = dG^t = \begin{pmatrix} e^{\alpha\bar{x}^3}\cosh\alpha\tau & e^{\alpha\bar{x}^3}\sinh\alpha\tau \\ e^{\alpha\bar{x}^3}\sinh\alpha\tau & e^{\alpha\bar{x}^3}\cosh\alpha\tau \end{pmatrix} = e^{\alpha\bar{x}^3}\begin{pmatrix} \cosh\alpha\tau & \sinh\alpha\tau \\ \sinh\alpha\tau & \cosh\alpha\tau \end{pmatrix}.$$

It results

$$\begin{pmatrix} \bar{g}_{00} & \bar{g}_{03} \\ \bar{g}_{30} & \bar{g}_{33} \end{pmatrix} = \begin{pmatrix} e^{2\alpha\bar{x}^3} & 0 \\ 0 & -e^{2\alpha\bar{x}^3} \end{pmatrix}.$$

The metric which describes the constant gravitational field in the corresponding slice of S is

$$ds^2 = e^{2\alpha\bar{x}^3}\left[d\bar{x}^0 - d\bar{x}^3\right].$$

In S, locally, the metric tensor is

$$\bar{g}_{ij}(\bar{x}^0, \bar{x}^1, \bar{x}^2, \bar{x}^3) = \begin{pmatrix} e^{2\alpha\bar{x}^3} & 0 & 0 & 0 \\ 0 & -1 & 0 & 0 \\ 0 & 0 & -1 & 0 \\ 0 & 0 & 0 & -e^{2\alpha\bar{x}^3} \end{pmatrix}.$$

The Christoffel first kind symbols are

$$\Gamma_{30,0} = \Gamma_{03,0} = \alpha e^{2\alpha\bar{x}^3}, \quad \Gamma_{00,3} = \Gamma_{33,3} = -\alpha e^{2\alpha\bar{x}^3}, \quad \Gamma_{30,3} = \Gamma_{03,3} = \Gamma_{00,0} = \Gamma_{33,0} = 0.$$

The Christoffel second kind symbols are

$$\Gamma^0_{30} = \Gamma^0_{03} = \Gamma^3_{00} = \Gamma^3_{33} = \alpha, \quad \Gamma^3_{30} = \Gamma^3_{03} = \Gamma^0_{00} = \Gamma^0_{33} = 0.$$

The geodesic equations, in the considered slice, with respect to the geodesic parameter λ are

$$\begin{cases} \dfrac{d^2\bar{x}^0}{d\lambda^2} = -2\alpha\dfrac{d\bar{x}^0}{d\lambda}\dfrac{d\bar{x}^3}{d\lambda} \\[4mm] \dfrac{d^2\bar{x}^3}{d\lambda^2} = -\alpha\left(\dfrac{d\bar{x}^0}{d\lambda}\right)^2 - \alpha\left(\dfrac{d\bar{x}^3}{d\lambda}\right)^2. \end{cases}$$

In S, we have

$$
\begin{cases}
\dfrac{d^2\bar{x}^0}{d\lambda^2} = -2\alpha \dfrac{d\bar{x}^0}{d\lambda}\dfrac{d\bar{x}^3}{d\lambda} \\[2ex]
\dfrac{d^2\bar{x}^1}{d\lambda^2} = 0 \\[2ex]
\dfrac{d^2\bar{x}^2}{d\lambda^2} = 0 \\[2ex]
\dfrac{d^2\bar{x}^3}{d\lambda^2} = -\alpha \left(\dfrac{d\bar{x}^0}{d\lambda}\right)^2 - \alpha \left(\dfrac{d\bar{x}^3}{d\lambda}\right)^2
\end{cases}
$$

with the general solutions

$$
\begin{cases}
\bar{x}^0 = k_3 + \dfrac{1}{2\alpha}\ln(k_1 + \lambda) - \dfrac{1}{2\alpha}\ln(k_2 - \lambda) \\[2ex]
\bar{x}^1 = k_4\lambda + k_5 \\[1ex]
\bar{x}^2 = k_6\lambda + k_7 \\[1ex]
\bar{x}^3 = \dfrac{1}{\alpha}\ln\alpha + \dfrac{1}{2\alpha}\ln(k_1 + \lambda) + \dfrac{1}{2\alpha}\ln(k_2 - \lambda).
\end{cases}
$$

A very good exercise for the reader is to prove that the above formulas verify the equations of the geodesics.

Let us analyze the trajectories of photons. They come from the equations $x^3 = t + b$ or $x^3 = -t + b$. The constant b is arbitrary and the speed of light is assumed 1. We consider only the case $x^3 = t + b$, the other case can be analyzed in a similar way.

Let us introduce the previous formula in G^{-1}. It results

$$
G^{-1} : \begin{cases}
\tau(t) = \dfrac{1}{\alpha}\tanh^{-1}\left(\dfrac{t}{t+b}\right) = \dfrac{1}{2\alpha}\ln\dfrac{1 + \dfrac{t}{t+b}}{1 - \dfrac{t}{t+b}} = \dfrac{1}{2\alpha}\ln(2t+b) - \dfrac{1}{2\alpha}\ln b \\[3ex]
\bar{x}^3(t) = \dfrac{1}{2\alpha}\ln\left[\alpha^2\left((t+b)^2 - t^2\right)\right] = \dfrac{1}{2\alpha}\ln(2t+b) + \dfrac{1}{\alpha}\ln\alpha,
\end{cases}
$$

that is

$$
\bar{x}^3(t) = \tau(t) + \beta,
$$

where β is a constant. The trajectories of photons are lines having the slope $+1$ (or -1). Of course these lines are geodesics because they come from the geodesics of R.

In the case $x^3 = k$, let us express $\bar{x}^3$ as a function of τ. From

$$\tau(t) = \frac{1}{\alpha} \tanh^{-1}\left(\frac{t}{k}\right)$$

it results

$$t = k \tanh(\alpha\tau),$$

that is

$$\bar{x}^3(\tau) = \frac{1}{2\alpha} \ln\left[\alpha^2 k^2 (1 - \tanh^2(\alpha\tau))\right].$$

Therefore

$$\bar{x}^3(\tau) = \frac{1}{\alpha} \ln(\alpha k) - \frac{1}{\alpha} \ln(\cosh(\alpha\tau)).$$

Since

$$\bar{x}^3(0) = \frac{1}{\alpha} \ln(\alpha k); \quad \frac{d\bar{x}^3}{d\tau}(0) = 0; \quad \frac{d^2\bar{x}^3}{d\tau^2}(0) = -\alpha;$$

the second order approximation of $\bar{x}^3$ is the parabola

$$\bar{x}^3(\tau) = \frac{1}{\alpha} \ln(\alpha k) - \frac{\alpha}{2}\tau^2,$$

which, in the case $k = \frac{1}{\alpha}$ and $\tau = \frac{\tau_1}{v}$, becomes

$$\bar{x}^3(\tau_1) = \frac{\alpha}{2v^2}\tau_1^2.$$

This is the parabola seen in the case of constant gravitational field in Classical Mechanics, that is the trajectory function of time.

Since the second kind Christoffel symbols are constant, it is easy to compute R^0_{303}. We find $R^0_{303} = \Gamma^0_{h0}\Gamma^h_{33} - \Gamma^0_{h3}\Gamma^h_{30} = \alpha^2 - \alpha^2 = 0$.

In fact, all sectional curvatures are 0, but, in general, the geodesics are not straight lines as we saw, they only come from lines of R.

In simple words, we can say that the constant gravitational field bends geodesics of space.

How the constant gravitational field affects the proper time can be found out by looking at the metrics involved in this description. For the frame R, free falling in the constant gravitational field $-\alpha$ of S, the metric is the Minkowski one, i.e.

$$ds^2 = dt^2 - dx^2.$$

The clock ticks in Δt seconds. The constant gravitational field induces in S, as we saw, the metric

$$ds^2 = e^{2\alpha\bar{x}}(d\tau^2 - d\bar{x}^2).$$

Here, the clock ticks in $\Delta\tau$ seconds. Between the observers of R and S, if $\bar{x} \to 0$, $\bar{x} > 0$, there is the connection

$$\Delta t = e^{\alpha\bar{x}}\Delta\tau \geq (1 + \alpha\bar{x})\Delta\tau \geq \Delta\tau,$$

that is the clock of R ticks slower and slower as $\bar{x} \to 0$, $\bar{x} > 0$.

The clock of a person A at the ground level of a building ticks less than the clock of a person B at the 33th floor. Therefore the ground level person A ages slower than the person B. Or, everyone legs are younger than the brain. Of course even at the level of lifetime of a person, the effects are imperceptible.

Therefore, according to Newton, the constant gravitational field landscape exists in $n = 3$ dimensions. The gravitational field is $\overrightarrow{A}$ and the gravitational potential Φ is related to it by the formula

$$\overrightarrow{A} = -\nabla\Phi = (0, 0, -\alpha).$$

The constant gravitational field satisfies the vacuum field equation

$$\nabla^2\Phi = 0,$$

The equations of motion are

$$\frac{d^2x}{dt^2} = 0; \quad \frac{d^2y}{dt^2} = 0; \quad \frac{d^2z}{dt^2} = -\alpha;$$

The solution, in appropriate initial conditions if we consider a plane (t, z), is

$$z(t) = -\frac{\alpha}{2v^2}t^2.$$

Einstein's constant gravitational field landscape exists in four dimensions.

The gravitational potential appears in the coefficients of the metric tensor

$$\bar{g}_{ij}(\bar{x}^0, \bar{x}^1, \bar{x}^2, \bar{x}^3) = \begin{pmatrix} e^{2\alpha\bar{x}^3} & 0 & 0 & 0 \\ 0 & -1 & 0 & 0 \\ 0 & 0 & -1 & 0 \\ 0 & 0 & 0 & -e^{2\alpha\bar{x}^3} \end{pmatrix}.$$

The gravitational field is described by the Christoffel second kind symbols:

$$\Gamma^0_{30} = \Gamma^0_{03} = \Gamma^3_{00} = \Gamma^3_{33} = \alpha, \ \Gamma^3_{30} = \Gamma^3_{03} = \Gamma^0_{00} = \Gamma^0_{33} = 0$$

and satisfies

$$R_{ij} = 0.$$

The equations of motions are the geodesic equations:

$$\begin{cases} \dfrac{d^2 \bar{x}^0}{d\lambda^2} = -2\alpha \dfrac{d\bar{x}^0}{d\lambda} \dfrac{d\bar{x}^3}{d\lambda} \\[3mm] \dfrac{d^2 \bar{x}^1}{d\lambda^2} = 0 \\[3mm] \dfrac{d^2 \bar{x}^2}{d\lambda^2} = 0 \\[3mm] \dfrac{d^2 \bar{x}^3}{d\lambda^2} = -\alpha \left(\dfrac{d\bar{x}^0}{d\lambda} \right)^2 - \alpha \left(\dfrac{d\bar{x}^3}{d\lambda} \right)^2 \end{cases}$$

with the general solutions

$$\begin{cases} \bar{x}^0 = k_3 + \dfrac{1}{2\alpha} \ln(k_1 + \lambda) - \dfrac{1}{2\alpha} \ln(k_2 - \lambda) \\[2mm] \bar{x}^1 = k_4 \lambda + k_5 \\ \bar{x}^2 = k_6 \lambda + k_7 \\ \bar{x}^3 = \dfrac{1}{\alpha} \ln \alpha + \dfrac{1}{2\alpha} \ln(k_1 + \lambda) + \dfrac{1}{2\alpha} \ln(k_2 - \lambda). \end{cases}$$

Locally, the particular solution presented before,

$$\bar{x}^3(\tau) = \frac{1}{\alpha} \ln(\alpha k) - \frac{\alpha}{2} \tau^2,$$

can be approximated by the classical solution

$$\bar{x}^3(\tau_1) = \frac{\alpha}{2v^2} \tau_1^2.$$

This intuitive description of Einstein's pictures can be fully formalized considering the Hilbert approach by which the gravitational field equations come out from a variational principle.

9.3 The Einstein–Hilbert Action and the Einstein Field Equations

Under a change of coordinates $x^r = x^r(\overline{x}^h)$, $r \in \{0, 1, \ldots, n\}$, $h \in \{0, 1, \ldots, n\}$, the second kind Christoffel symbols change according to the rule

$$\frac{\partial^2 x^k}{\partial \overline{x}^i \partial \overline{x}^j} = -\Gamma^k_{rs} \frac{\partial x^r}{\partial \overline{x}^i} \frac{\partial x^s}{\partial \overline{x}^j} + \overline{\Gamma}^r_{ij} \frac{\partial x^k}{\partial \overline{x}^r}.$$

Suppose we vary the metric. It means that the coefficients g_{ij} are changed in some new coefficients $\overline{g}_{ij} := g_{ij} + \delta g_{ij}$. This second metric produces first type and second type Christoffel symbols. Let us denote them by $\gamma_{ij,k}$ and γ^i_{jk}. The same change of coordinates gives for these new Christoffel symbols a similar formula

$$\frac{\partial^2 x^k}{\partial \overline{x}^i \partial \overline{x}^j} = -\gamma^k_{rs} \frac{\partial x^r}{\partial \overline{x}^i} \frac{\partial x^s}{\partial \overline{x}^j} + \overline{\gamma}^r_{ij} \frac{\partial x^k}{\partial \overline{x}^r}.$$

The difference of the previous formulas leads to

Proposition 9.3.1 *The variation difference* $\delta\Gamma^i_{jk} := \Gamma^i_{jk} - \gamma^i_{jk}$ *satisfies*

$$\delta\Gamma^k_{rs} \frac{\partial x^r}{\partial \overline{x}^i} \frac{\partial x^s}{\partial \overline{x}^j} = \delta\overline{\Gamma}^r_{ij} \frac{\partial x^k}{\partial \overline{x}^r},$$

i.e. $\delta\Gamma^i_{jk}$ *is a* $(1, 2)$ *mixed tensor.*

Let g_{ij} be the matrix of the metric $ds^2 = g_{ij} dx^i dx^j$ and let g be the determinant of g_{ij}. Suppose this determinant is negative as in the case of the Minkowski metric.

Theorem 9.3.2 *The formula which expresses the variation of* $\sqrt{-g}$ *is*

$$\delta\sqrt{-g} = -\frac{1}{2}\sqrt{-g}\, g_{ij}\delta g^{ij}.$$

Proof The inverse of the matrix g_{ij} is g^{ij} such that $g^{is}g_{sj} = \delta^i_j$.

Consider a given element g_{ij} of the matrix and denote by M^{ij} the determinant of the matrix obtained from the initial one after we cancel both the line i and the column j.

The corresponding inverse element is $g^{ij} = \dfrac{(-1)^{i+j} M^{ij}}{g}$ and, in this respect, using the column j, the determinant can be thought as $g = \sum_i (-1)^{i+j} g_{ij} M^{ij}$.

Then, for the variation of $\sqrt{-g}$, we have:

$$\delta\sqrt{-g} = \frac{\partial}{\partial g_{ij}}\left(\sqrt{-g}\right)\delta g_{ij} =$$

$$= -\frac{1}{2\sqrt{-g}}\frac{\partial g}{\partial g_{ij}}\delta g_{ij} = -\frac{1}{2\sqrt{-g}}(-1)^{i+j}M^{ij}\delta g_{ij} = -\frac{1}{2\sqrt{-g}}\cdot g\cdot g^{ij}\delta g_{ij}.$$

It results

$$\delta\sqrt{-g} = \frac{1}{2}\sqrt{-g}g^{ij}\delta g_{ij}.$$

From $g^{ks}g_{sl} = \delta_l^k$, it is $\delta g^{ks}g_{sl} + g^{ks}\delta g_{sl} = 0$, that is $g^{ks}\delta g_{sl} = -\delta g^{ks}g_{sl}$.
Multiplying by g_{mk}, we obtain

$$g_{mk}g^{ks}\delta g_{sl} = -g_{mk}g_{sl}\delta g^{ks}$$

and, after considering $s = m = i,\ l = j$, it is

$$\delta g_{ij} = -g_{ik}g_{ij}\delta g^{ik}.$$

Replacing in the formula of the variation of $\sqrt{-g}$ we obtain

$$\delta\sqrt{-g} = -\frac{1}{2}\sqrt{-g}g^{ij}g_{ik}g_{ji}\delta g^{ik},$$

that is

$$\delta\sqrt{-g} = -\frac{1}{2}\sqrt{-g}g_{ik}\delta g^{ik}.$$

$\square$

Theorem 9.3.3 (Palatini's Formula) $\delta R_{ij} = \delta\Gamma_{ij;s}^s - \delta\Gamma_{is;j}^s.$

Proof We start from

$$R_{ij} = R_{isj}^s = \frac{\partial\Gamma_{ij}^s}{\partial x^s} - \frac{\partial\Gamma_{is}^s}{\partial x^j} + \Gamma_{su}^s\Gamma_{ij}^u - \Gamma_{ju}^s\Gamma_{is}^u.$$

Then the variation of the Ricci tensor is

$$\delta R_{ij} = \frac{\partial(\delta\Gamma_{ij}^s)}{\partial x^s} - \frac{\partial\left(\delta\Gamma_{is}^s\right)}{\partial x^j} + \delta\Gamma_{su}^s\Gamma_{ij}^u + \Gamma_{su}^s\delta\Gamma_{ij}^u - \delta\Gamma_{ju}^s\Gamma_{is}^u - \Gamma_{ju}^s\delta\Gamma_{is}^u.$$

The variation $\delta\Gamma_{ij}^s$ is a $(1, 2)$ tensor type. Its covariant derivative is

$$\delta\Gamma_{ij;s}^s = \frac{\partial(\delta\Gamma_{ij}^s)}{\partial x^s} + \Gamma_{su}^s\delta\Gamma_{ij}^u - \delta\Gamma_{ju}^s\Gamma_{is}^u - \delta\Gamma_{iu}^s\Gamma_{js}^u.$$

In the same way the covariant derivative of $\delta\Gamma^s_{is}$ is

$$\delta\Gamma^s_{is;j} = \frac{\partial(\delta\Gamma^s_{is})}{\partial x^j} + \delta\Gamma^u_{is}\Gamma^s_{ju} - \delta\Gamma^s_{su}\Gamma^u_{ij} - \delta\Gamma^s_{iu}\Gamma^u_{js}.$$

Subtracting the second relation from the first we obtain the Palatini formula.   □

Theorem 9.3.4 *If V is a compact region of the Universe whose volume element is* dV, *such that on its boundary* ∂V, *the variations* $\delta\Gamma^i_{jk}$ *vanish, then*

$$\int_V g^{ij}\,\delta R_{ij}\,dV = 0.$$

Proof (*Palatini's Formula Consequence*) Since the volume element dV is expressed with respect to the given metric by

$$dV = \sqrt{-g}\,dx_0 dx_1 dx_2 dx_3,$$

our integral becomes

$$\int_V g^{ij}\,\delta R_{ij}\,\sqrt{-g}\,d^4x,$$

where we denoted $dx_0 dx_1 dx_2 dx_3$ by d^4x.

It exists a corresponding 3D-surface element do on ∂V, $do = \sqrt{-g'}\,d^3x$.

At each point of ∂V, it exists a normal outward vector n of components n_s, i.e. $n = n_s$.

All these results help us to express the divergence formula which, in the classical form, looks like

$$\int_V \operatorname{div} B\,dV = \int_{\partial V} B \cdot n\,do,$$

here, in its covariant form, being

$$\int_V B^s_{;s}\sqrt{-g}\,d^4x = \int_{\partial V} B^s n_s\sqrt{-g'}\,d^3x.$$

Now, Palatini's formula leads to

$$\int_V g^{ij}\,\delta R_{ij}\,\sqrt{-g}\,d^4x = \int_V g^{ij}\left(\delta\Gamma^s_{ij;s} - \delta\Gamma^s_{is;j}\right)\sqrt{-g}\,d^4x.$$

Taking into account that $g^{ij}_{;s} = 0$ and changing the dummy indexes, we can write

$$\int_V g^{ij}\,\delta R_{ij}\,\sqrt{-g}\,d^4x = \int_V \left[\left(g^{ij}\delta\Gamma^s_{ij}\right)_{;s} - \left(g^{ij}\delta\Gamma^s_{is}\right)_{;j}\right]\sqrt{-g}\,d^4x =$$

$$= \int_V \left[\left(g^{ij}\delta\Gamma^s_{ij}\right)_{;s} - \left(g^{is}\delta\Gamma^j_{ij}\right)_{;s}\right]\sqrt{-g}\,d^4x = \int_V \left[g^{ij}\delta\Gamma^s_{ij} - g^{is}\delta\Gamma^j_{ij}\right]_{;s}\sqrt{-g}\,d^4x.$$

Let us denote the contravariant vector $g^{ij}\delta\Gamma^s_{ij} - g^{is}\delta\Gamma^j_{ij}$ by B^s. Our initial integral

$$\int_V g^{ij}\,\delta R_{ij}\,\sqrt{-g}\,d^4x,$$

according to the covariant above form, becomes

$$\int_{\partial V} B^s n_s\,\sqrt{-g'}\,d^3x.$$

Since $B^s = g^{ij}\delta\Gamma^s_{ij} - g^{is}\delta\Gamma^j_{ij}$ vanishes on ∂V, the last integral is 0, that is

$$\int_V g^{ij}\,\delta R_{ij}\,\sqrt{-g}\,d^4x = 0.$$

$\square$

These considerations lead us to the following

Theorem 9.3.5 (Einstein's Field Equations in vacuum) *If V is a compact region of the Universe without matter and energy inside it, such that, on its boundary ∂V, the variations $\delta\Gamma^i_{jk}$ vanish, then*

$$R_{ij} - \frac{1}{2}R\,g_{ij} = 0.$$

Proof (*Hilbert*) We have proved both

$$\delta\sqrt{-g} = -\frac{1}{2}\sqrt{-g}\,g_{ij}\delta g^{ij}$$

and

$$\int_V g^{ij}\,\delta R_{ij}\,\sqrt{-g}\,d^4x = 0.$$

To derive Einstein's field equations, we have to choose an appropriate Lagrangian.

Hilbert's idea was to consider the Lagrangian expressed through the Ricci curvature scalar R, that is the *Einstein–Hilbert action* is

$$S_{EH} = \int_V R\sqrt{-g}\,d^4x.$$

Let us compute the first variation of S_{EH}. It is

$$\delta S_{EH} = \delta \int_V R\sqrt{-g}\, d^4x = \delta \int_V g^{ij} R_{ij} \sqrt{-g}\, d^4x = \int_V \delta\left(g^{ij} R_{ij} \sqrt{-g}\right) d^4x =$$

$$= \int_V \left(\delta g^{ij}\right) R_{ij} \sqrt{-g}\, d^4x + \int_V g^{ij} \left(\delta R_{ij}\right) \sqrt{-g}\, d^4x + \int_V R\, \delta\left(\sqrt{-g}\right) d^4x =$$

$$= \int_V \delta g^{ij} R_{ij} \sqrt{-g}\, d^4x + \int_V g^{ij} \delta R_{ij} \sqrt{-g}\, d^4x - \frac{1}{2} \int_V R\sqrt{-g}\, g_{ij} \delta g^{ij} d^4x$$

After rearranging the right side we have

$$\delta S_{EH} = \int_V \left[R_{ij} - \frac{1}{2} R\, g_{ij}\right] \sqrt{-g}\, \delta g^{ij}\, d^4x + \int_V g^{ij}\, \delta R_{ij}\, \sqrt{-g}\, d^4x.$$

Since

$$\int_V g^{ij}\, \delta R_{ij}\, \sqrt{-g}\, d^4x = 0,$$

the condition $\delta S_{EH} = 0$ for g^{ij} arbitrary, leads to the *Einstein field equations in vacuum*. Therefore in a region of space as the one described by the previous statement, without matter and energy, the Einstein field equations are

$$R_{ij} - \frac{1}{2} R\, g_{ij} = 0.$$

□

Theorem 9.3.6 (Einstein's field equations in presence of matter) *If V is a region of the Universe containing matter and energy, such that, on its boundary ∂V, the variations $\delta \Gamma^i_{jk}$ vanish, then there exists a $(2, 0)$ covariant tensor T_{ij} such that*

$$R_{ij} - \frac{1}{2} R\, g_{ij} = K\, T_{ij},$$

where K is a coupling constant.

Proof (*Hilbert*) In the previous theorem, we have used an action which describes the Geometry of space without matter and energy. If we want to describe the Geometry of a space with matter and energy inside it, the Einstein–Hilbert action has to contain a further term, denoted by S_M, depending on the matter-energy distribution in spacetime.

So, the *general Einstein–Hilbert action* S_{GEH} has the form $kS_{EH} + S_M$, where k is a constant. This can be written in the form

$$S_{GEH} = \int_V (kR + S)\sqrt{-g}\, d^4x,$$

that is

$$\delta S_{GEH} = k \int_V \left[R_{ij} - \frac{1}{2} R \, g_{ij} \right] \sqrt{-g} \, \delta g^{ij} d^4 x + \int_V \left[\frac{\delta S}{\delta g^{ij}} \right] \sqrt{-g} \, \delta g^{ij} d^4 x,$$

because we have already computed the variation

$$\delta S_{EH} = \int_V \left[R_{ij} - \frac{1}{2} R \, g_{ij} \right] \sqrt{-g} \, \delta g^{ij} d^4 x.$$

If the first variation of S_{GEH} vanishes, it results

$$R_{ij} - \frac{1}{2} R \, g_{ij} = K \, T_{ij},$$

where $T_{ij} := -\dfrac{\delta S}{\delta g^{ij}}$ and $K := \dfrac{1}{k}$.

The *general Einstein field equations* are obtained.   □

From

$$R_{ij} - \frac{1}{2} R \, g_{ij} = K \, T_{ij} ,$$

it results

$$g^{mi} R_{ij} - \frac{1}{2} R \, g^{mi} g_{ij} = K \, g^{mi} T_{ij},$$

i.e.

$$R_m^m - \frac{1}{2} \delta_m^m R = K \, T_m^m .$$

Denoting by $T := T_m^m$ the *Laue's scalar* and taking into account that the dimension is 4, we have $\delta_m^m = 4$, therefore $R - 2R = KT$, that is $R = -KT$.

We have obtained the following

Theorem 9.3.7 *The equivalent of the Einstein field equations*

$$R_{ij} - \frac{1}{2} R \, g_{ij} = K \, T_{ij},$$

written with respect to the Laue scalar T, are

$$R_{ij} = K \left(T_{ij} - \frac{1}{2} T \, g_{ij} \right).$$

Let us understand why this discussion was important even if, at this moment, we have not the exact value of the constant K.

If $T_{ij} = 0$, it results $T^i_j = 0$, therefore the Laue's scalar T is 0. Finally

Corollary 9.3.8 *Einstein's vacuum field equations are equivalent to*

$$R_{ij} = 0.$$

These results are particularly relevant because point out the symmetric role of matter-energy and curvature.In some sense they realize the two-way nature of the above Wheeler sentence that we report here again: *"Spacetime tells matter how to move; matter tells spacetime how to curve"*.

9.4 A Short Introduction to $f(R)$ Gravity

It is very interesting to observe that the previous theorems can be generalized if instead R we use any smooth function $f(R)$ in the Einstein–Hilbert action. In this way we obtain the field equations of the so-called $f(R)$ *gravity*. They are the straightforward generalization of Einstein field equations and have recently acquired a lot of interest in view of solving several problems in cosmology and astrophysics (for a comprehensive discussion, see, for example, [18]). For example, the model of $f(R) = R + \dfrac{R^2}{6M^2}$, where M has the dimension of mass, gives rise to the so-called Starobinski inflation [53] which gives rise to the accelerated expansion of the early Universe capable of addressing several issues of Cosmological Standard Model based on General Relativity and Standard Model of Particles [54]. This kind of theories can be useful also to address issues related to the late Universe, like recent accelerated expansion, often dubbed dark energy epoch [55–57] or astrophysical issues like dark matter [58].

A detailed discussion of these problems is out of the scope of this book but it is worth pointing out that they are very active research areas. We refer the interested reader to the cited bibliography.

In view of the present discussion, it is interesting to develop how the Einstein field equations can be generalized in the $f(R)$ gravity framework. In particular, it is interesting to point out that metric and Palatini's formalisms give different field equations that, however, can be related each other, see [59].

Taking into account the previous results related to the Einstein field equations, let us derive here the $f(R)$ gravity field equations We shall use the following facts proven in the previous section, that is:

$$\delta\sqrt{-g} = -\frac{1}{2}\sqrt{-g}\,g_{ij}\delta g^{ij},$$

$$\int_V g^{ij}\,\delta R_{ij}\,\sqrt{-g}\,d^4x = 0.$$

The second formula is a consequence of Palatini's identity

$$\delta R_{ij} = \delta \Gamma^s_{ij;s} - \delta \Gamma^s_{is;j}.$$

To begin, let us consider the basic objects introduced in the Differential Geometry chapter: g_{ij}, g^{ij}, $\Gamma_{ij,k}$, Γ^k_{ij}, R^i_{jkl}, R_{ij}, R_{ijkl}, R^i_j, R. They are smooth functions, that is functions having derivatives of all order everywhere in the domain, here V. Therefore, if we assume $f(R)$ as a smooth function of $R(V)$, then $f(R)$ is a smooth function for $x \in V$. If V is compact and connected region in the Euclidean n dimensional space, then $R(V)$ is a compact interval in $\mathbb{R}$. In particular f, f', $\ldots$ are at least continuous functions on a real compact interval, here $R(V)$. The values of $f(R)$, $f'(R), \ldots$ are in real compact intervals, too. The prime indicates derivative with respect to the Ricci scalar R.

Theorem 9.4.1 *If V is a compact and connected region of the universe without matter and energy inside it, such that on its boundary ∂V the variations $\delta \Gamma^i_{jk}$ vanish and f is a smooth real valued arbitrary function on $R(V)$, then*

$$f'(R)R_{ij} - \frac{1}{2}f(R)g_{ij} = 0.$$

Proof The line of the proof is similar to Theorem 9.3.5. The appropriate Lagrangian is

$$S_f = \int_V f(R)\sqrt{-g}\, d^4x.$$

Let us compute the first variation of S_f.

$$\delta S_f = \delta \int_V f(R)\sqrt{-g}\, d^4x = \int_V \delta[f(R)\sqrt{-g}]\, d^4x =$$

$$= \int_V f'(R)\delta R\sqrt{-g}\, d^4x + \int_V f(R)\delta\left(\sqrt{-g}\right)d^4x =$$

$$= \int_V f'(R)\left(\delta g^{ij}\right)R_{ij}\sqrt{-g}\, d^4x + \int_V f'(R)g^{ij}\left(\delta R_{ij}\right)\sqrt{-g}\, d^4x + \int_V f(R)\,\delta\left(\sqrt{-g}\right)d^4x =$$

$$= \int_V f'(R)\delta g^{ij} R_{ij}\sqrt{-g}\, d^4x + \int_V f'(R)g^{ij}\,\delta R_{ij}\sqrt{-g}\, d^4x - \frac{1}{2}\int_V f(R)\sqrt{-g}\, g_{ij}\delta g^{ij}\, d^4x$$

After rearranging the right-hand side, we have

$$\delta S_f = \int_V \left[f'(R)R_{ij} - \frac{1}{2}f(R)\, g_{ij}\right]\sqrt{-g}\,\delta g^{ij}\, d^4x + \int_V f'(R)\, g^{ij}\,\delta R_{ij}\,\sqrt{-g}\, d^4x.$$

Now, the mean value theorem implies the existence of a point $x \in V$ such that

$$\int_V f'(R)g^{ij}\,\delta R_{ij}\,\sqrt{-g}\,d^4x = f'(R(x))\int_V g^{ij}\,\delta R_{ij}\,\sqrt{-g}\,d^4x.$$

The last integral is 0, therefore the condition $\delta S_f = 0$ for g^{ij} arbitrary, leads to $f(R)$ *field equations in vacuum*. Therefore, in the condition of the above statement, in a region of space without matter and energy, the $f(R)$ field equations in vacuum are

$$f'(R)R_{ij} - \frac{1}{2}f(R)\,g_{ij} = 0.$$

Let us observe that, for $f(R) = R$, we obtain the Einstein field equations in vacuum.

□

Theorem 9.4.2 *If V is a compact and connected region of universe containing matter and energy, such that on its boundary ∂V the variations $\delta\Gamma^i_{jk}$ vanish and f is a smooth real valued arbitrary function on $R(V)$, then there exists a $(2, 0)$ covariant tensor T_{ij} such that*

$$f'(R)R_{ij} - \frac{1}{2}f(R)g_{ij} = KT_{ij},$$

where K is a constant.

Proof If we choose the action

$$S_{Gf} = \int_V (kf(R) + S)\sqrt{-g}\,d^4x,$$

that is

$$\delta S_{Gf} = k\int_V \left[f'(R)R_{ij} - \frac{1}{2}f(R)\,g_{ij} \right]\sqrt{-g}\,\delta g^{ij}\,d^4x + \int_V \left[\frac{\delta S}{\delta g^{ij}} \right]\sqrt{-g}\,\delta g^{ij}\,d^4x,$$

in the same way as in Theorem 9.3.6, we obtain the $f(R)$ *generalized field equations*

$$f'(R)R_{ij} - \frac{1}{2}f(R)\,g_{ij} = KT_{ij}.$$

□

Exercise 9.4.3 If the reader is interested in the differences between the Palatini and metric formalisms in $f(R)$ gravity, we propose the following exercise whose notation can be found in [18]. Starting from the action of $f(R)$, show that

$$f'(R)R_{ij} - \frac{1}{2}f(R)g_{ij} - f'(R)_{;ij} + g_{ij}\Box f'(R) = KT_{ij},$$

with the trace

$$3\Box f'(R) + f'(R)R - 2f(R) = KT,$$

are the field equations obtained by varying with respect to g_{ij} without using the above Palatini identity. Demonstrate that they are equivalent to

$$f'(R)R_{ij} - \frac{1}{2}f(R)\,g_{ij} = KT_{ij}\,,$$

unless a divergence free current. Here $\Box$ is the d'Alembert operator defined as $\Box :=$ $\nabla_i \nabla^i$ with ∇_i the covariant derivative.

Hint. Use results in [59].

We will consider again $f(R)$ gravity in view of the discussion of the de Sitter spacetime which is a solution of this theory.

9.5   The Energy-Momentum Tensor and Another Proof for Einstein's Field Equations

In the previous sections, considering an action S_M dependent on matter and energy, we have obtained a symmetric tensor acting as a source into the Einstein field equation. In order to discuss the properties of the tensor T_{ij}, we can start from the covariant divergence related to the flow of an incompressible fluid in a region where the parallelism is not the Euclidean one. The problem appears when we consider the difference

$$F_x\left(x + \frac{\Delta x}{2}, y, z\right)\Delta y \Delta z - F_x\left(x - \frac{\Delta x}{2}, y, z\right)\Delta y \Delta z$$

related to the motion by parallel transport of the vector

$$\left(-F_x\left(x - \frac{\Delta x}{2}\right), 0, 0\right)$$

to the other face at the point $\left(x + \frac{\Delta x}{2}, y, z\right)$.

Therefore we parallel transport the contravariant vector $\left(-F_x\left(x - \frac{\Delta x}{2}\right), 0, 0\right)$ along the infinitesimal vector $A^k = (\Delta x, 0, 0)$. Since, in general, $\Gamma_{ij}^k \neq 0$, the parallel transport along $A^1 = (\Delta x, 0, 0)$ for a contravariant vector $V = (V^1, 0, 0)$ leads to a vector whose first coordinate is

$$V^1\left(x - \frac{\Delta x}{2}, y, z\right) + \Delta V^1,$$

where
$$\Delta V^1 = -\Gamma_{ij}^1 V^j \Delta x^i = -\Gamma_{1j}^1 V^j \Delta x = -\Gamma_{11}^1 V^1 \Delta x.$$

The difference

$$\left[V^1\left(x + \frac{\Delta x}{2}, y, z\right) - V^1\left(x - \frac{\Delta x}{2}, y, z\right) + \Gamma^1_{11} V^1 \Delta x \right] \Delta y \Delta z$$

is

$$\left(\frac{\partial V^1}{\partial x} + \Gamma^1_{11} V^1 \right) \Delta x \Delta y \Delta z,$$

i.e. the covariant derivative with respect the first variable

$$V^1_{;1} \Delta x \Delta y \Delta z.$$

Taking into account the three pairs of opposite faces corresponding to the three directions, we can obtain the net outflow

$$(V^1_{;1} + V^2_{;2} + V^3_{;3}) \Delta x \Delta y \Delta z$$

for a parallelepiped in a region where the Euclidean parallel transport is replaced by the general parallel transport. The quantity $V^1_{;1} + V^2_{;2} + V^3_{;3}$, which is expressed with respect of the covariant derivatives, can be thought as a covariant divergence of a contravariant vector (V^1, V^2, V^3). Now let us use this idea to construct the main property of T_{ij}.

9.5.1 The Covariant Derivative of the Energy-Momentum Tensor

Now, let us assume a symmetric contravariant tensor (T^{ij}), expressed as a 4×4 matrix and a $4D$-space of coordinates (x^0, x^1, x^3, x^4) endowed with a metric

$$ds^2 = g_{ij} dx^i dx^j, \ i, j \in \{1, 2, 3, 4\}$$

such that the parallel transport depends on Γ^i_{jk} not all zero. The tensor (T^{ij}) looks like

$$(T^{ij}) = \begin{pmatrix} T^{00} & T^{01} & T^{02} & T^{03} \\ T^{10} & T^{11} & T^{12} & T^{13} \\ T^{20} & T^{21} & T^{22} & T^{23} \\ T^{30} & T^{31} & T^{32} & T^{33} \end{pmatrix}$$

Because of symmetry, the first line $T^{1i} = (T^{10}, T^{11}, T^{12}, T^{13})$ coincides with the first column $T^{i1} = (T^{01}, T^{11}, T^{21}, T^{31})$, and can be seen as the representation a contravariant 4-vector denoted by T^1, the same for the other rows and corresponding columns.

Before discussing something the physical aspects involved in the components of the energy-momentum tensor, let us assert that we can analyze the tensor like the vector F used to represent the flow of an incompressible fluid.

We can suppose the existence of a flow associated to the tensor above. For each line of the given tensor, we have the corresponding force T^k described above. We can also consider a 4-parallelepiped centered at a given point (x^0, x^1, x^3, x^4) with sides parallel to the axes of coordinates and having the small dimensions $\Delta x_0, \Delta x_1, \Delta x_2, \Delta x_3$. Small enough to suppose the vector T^k having the same components at each point of each considered face.

The difference of the total outflow, determined by F through the parallel faces corresponding to x direction, is

$$F_x\left(x + \frac{\Delta x}{2}, y, z\right) \Delta y \Delta z - F_x\left(x - \frac{\Delta x}{2}, y, z\right) \Delta y \Delta z.$$

In the case of T_{ij}, it can be replaced by the total outflow determined by each T^k through the parallel faces corresponding to x^i direction. Considering $i = 1$, we have the differences

$$T^{k1}\left(x_0, x_1 + \frac{\Delta x_1}{2}, x_2, x_3\right) \Delta x_0 \Delta x_2 \Delta x_3 - T^{k1}\left(x_0, x_1 - \frac{\Delta x_1}{2}, x_2, x_3\right) \Delta x_0 \Delta x_2 \Delta x_3,$$

where $k \in \{0, 1, 2, 3\}$.

Furthermore, we have to consider the differences with respect to the parallel transport of the given vectors $\left(0, T^{k1}\left(x_0, x_1 - \frac{\Delta x_1}{2}, x_2, x_3\right), 0, 0\right)$, $k \in \{0, 1, 2, 3\}$, and then we have

$$\left[T^{k1}\left(x_0, x_1 + \frac{\Delta x_1}{2}, x_2, x_3\right) - T^{k1}\left(x_0, x_1 - \frac{\Delta x_1}{2}, x_2, x_3\right) + \Gamma^1_{k1} T^{k1} \Delta x_1\right] \Delta x_0 \Delta x_2 \Delta x_3$$

with the approximation

$$\left[\frac{\partial T^{k1}}{\partial x^1} \Delta x_1 + \Gamma^1_{k1} T^{k1} \Delta x_1\right] \Delta x_0 \Delta x_2 \Delta x_3, \quad k \in \{0, 1, 2, 3\},$$

that is

$$T^{k1}_{;1} \Delta x_0 \Delta x_1 \Delta x_2 \Delta x_3, \quad k \in \{0, 1, 2, 3\}.$$

The total outflow for the 4-parallelepiped results equal to

$$T^{kl}_{;l} \Delta x_0 \Delta x_1 \Delta x_2 \Delta x_3.$$

It means that the quantity of matter and energy entering the 4-parallelepiped leaves completely the interior. Therefore, at each moment of time, the quantity of matter inside can be considered a constant. The energy-momentum tensor is conserved, and since it is conserved as $\Delta x^k \to 0$, the limiting net flow approaches 0, that is $T^{kl}_{;l} = 0$. We know how to lowering indexes using the metric tensor: $T^l_i = g_{ik}T^{kl}$. Since the covariant derivative of the metric tensor is 0,

$$T^l_{i;l} = g_{ik;l}T^{kl} + g_{ik}T^{kl}_{;l} = 0$$

Lowering again the indexes, we have $T_{ji} = g_{jl}T^l_i$. In the same way

$$T_{ji;l} = g_{jl;l}T^l_i + g_{jl}T^l_{i;l} = 0.$$

According to all the previous assumptions, we have proved an important property of the energy-momentum tensor, i.e.

$$T_{ij;l} = 0.$$

From a mathematical point of view, it is the divergence of T_{ij} that, as we will see below, corresponds to the contracted Bianchi identities.

9.5.2   Another Proof for Einstein's Field Equations

Let us remember, when starting from Bianchi's formula

$$R^s_{ijk;l} + R^s_{ikl;j} + R^s_{ilj;k} = 0,$$

we proved that the covariant derivative of the Einstein tensor is null, i.e.

$$\left(R_{ij} - \frac{1}{2}R \cdot g_{ij} \right)_{;l} = 0.$$

The two tensors, T_{ij} and $R_{ij} - \dfrac{1}{2}R \cdot g_{ij}$ are divergence-free, therefore they are proportional, that is

$$R_{ij} - \frac{1}{2}R \cdot g_{ij} = K \cdot T_{ij},$$

where K is a constant. This is another proof for the Einstein field equations in presence of matter. It means that both the Einstein tensor and the energy-momentum tensor satisfy the same contracted Bianchi identities, i.e. both tensors, being divergence-free, satisfy the same conservation laws. Below, we will give specific forms for T_{ij}.

9.6   Introducing the Cosmological Constant

The equation

$$\nabla^2 \Phi + \Lambda \Phi = 4\pi G \rho_0$$

has the particular solution $\Phi_0 = \dfrac{4\pi G \rho_0}{\Lambda}$.

Denote by Φ_1 the solution of the equation

$$\nabla^2 \Phi + \Lambda \Phi = 4\pi G (\rho + \rho_0).$$

A simple computation leads to the solution

$$\Phi := \Phi_1 - \Phi_0$$

for

$$\nabla^2 \Phi + \Lambda \Phi = 4\pi G \rho.$$

Observing this feature and knowing that $\nabla^2 \Phi$ can be generalized as $R_{ij} - \dfrac{1}{2} R \cdot g_{ij}$ and Φ is a component of g_{ij}, Einstein proposed to modify the left member of his equations in the form

$$R_{ij} - \frac{1}{2} R \cdot g_{ij} + \Lambda \cdot g_{ij},$$

where the constant Λ is called *cosmological constant*.

Since $g_{ij;l} = 0$, it results

$$\left(R_{ij} - \frac{1}{2} R \cdot g_{ij} + \Lambda \cdot g_{ij} \right)_{;l} = 0,$$

that is the new tensor proposed by Einstein is again divergence-free. Therefore the new Einstein field equations become

$$R_{ij} - \frac{1}{2} R \cdot g_{ij} + \Lambda \cdot g_{ij} = K \cdot T_{ij}.$$

Λ was called cosmological constant. Einstein introduced this ingredient to improve the total amount of matter-energy of the Universe in view of making it static. Λ can be supposed related to a uniform spatial density ρ. After the discovery of cosmic expansion by Edwin Hubble, he said that: *It was the biggest blunder of my life.* However, in '90 of last century, this concept has been revitalized and now represents one of the main question of modern Physics [60]. We will discuss this issue later in the book.

Going back to our formal discussion, we proved that Einstein's field equations, in the form

$$R_{ij} - \frac{1}{2} R \cdot g_{ij} = K \cdot T_{ij}$$

can be written with respect to Laue's scalar T in the equivalent form

$$R_{ij} = K \cdot \left(T_{ij} - \frac{1}{2} T \cdot g_{ij} \right).$$

The natural question is: There is an equivalent of the Einstein's field equations with cosmological constant

$$R_{ij} - \frac{1}{2} R \cdot g_{ij} + \Lambda \cdot g_{ij} = K \cdot T_{ij}$$

written with respect to the Laue scalar?

Theorem 9.6.1 *The equivalent of Einstein's field equations with cosmological constant are*

$$R_{ij} - \Lambda \cdot g_{ij} = K \left(T_{ij} - \frac{1}{2} T \cdot g_{ij} \right),$$

where T is the Laue scalar.

Proof We start from

$$g^{hi} R_{ij} - \frac{1}{2} R \cdot g^{hi} g_{ij} + \Lambda \cdot g^{hi} g_{ij} = K g^{hi} T_{ij}$$

which means

$$R_j^h - \frac{1}{2} R \cdot \delta_j^h + \Lambda \cdot \delta_j^h = K \cdot T_j^h,$$

that is

$$R - 2R + 4\Lambda = K \cdot T.$$

If we replace $R = 4\Lambda - K \cdot T$ in the original Einstein's field equations it results

$$R_{ij} - \Lambda \cdot g_{ij} = K \left(T_{ij} - \frac{1}{2} T \cdot g_{ij} \right).$$

$\square$

The nature of this cosmological constant can be easily understood in a space without standard matter-energy, that is in a space having $T_{ij} = 0$. Einstein equations reduces to

$$R_{ij} - \Lambda \cdot g_{ij} = 0.$$

Therefore, in this case, the nature of cosmological constant is geometrical one.

Specifically, let us consider $R_{ij} = R^h_{ihj}$. It contains partial derivatives of Γ^i_{jk} and terms as $\Gamma^s_{jk}\Gamma^i_{sl}$. Now, let us observe that the unit of measure for Γ^i_{jk} is given by the unit of measure for $\Gamma_{ij,k}$ which contains partial derivatives as $\dfrac{\partial g_{ik}}{\partial x^j}$. Since there is no unit for g_{ij} which is considered only as a geometric object, the unit for $\dfrac{\partial g_{ik}}{\partial x^j}$ is $\dfrac{1}{l}$, where l is a length. It is clear that R_{ij} has physical dimensions $\dfrac{1}{l^2}$. The cosmological constant Λ is measured in $\dfrac{1}{l^2}$.

9.7 The Schwarzschild Solution of Vacuum Field Equations

We intend to solve the Einstein field equations in vacuum, i.e. $R_{ij} = 0$ obtained previously assuming the spherical symmetry of spacetime. The Schwarzschild solution is an exact solution for the vacuum field equations. Another way to find Schwarzschild solution is presented in [23, 35].

Theorem 9.7.1 *Consider the vacuum field equations $R_{ij} = 0$. Then*

$$ds^2 = c^2 \left(1 + \frac{B}{r}\right) dt^2 - \frac{1}{1 + \dfrac{B}{r}} dr^2 - r^2 d\varphi^2 - r^2 \sin^2 \varphi d\theta^2$$

is the Schwarzschild solution for an arbitrary constant B.

Proof Karl Schwarzschild had the intuition to look for a spherically symmetric solution which describes the relativistic field outside of a non-rotating, massive body. This was the first exact solution of the Einstein field equations. Instead of the ordinary Cartesian coordinates $(x^0 = ct, x^1, x^2, x^3)$, Schwarzschild used spherical coordinates for the spatial part The new coordinate system $(x^0 = ct, r, \varphi, \theta)$ is related to the old one by the formulas

$$x^1 = r \sin \varphi \cos \theta, \quad x^2 = r \sin \varphi \sin \theta, \quad x^3 = r \cos \varphi,$$

so then, for the spatial part, it is

$$(dx^1)^2 + (dx^2)^2 + (dx^3)^2 = dr^2 + r^2 d\varphi^2 + r^2 \sin^2 \varphi d\theta^2.$$

Far from the source, the solution has to approximate the Minkowski metric

$$ds^2 = (dx^0)^2 - (dx^1)^2 - (dx^2)^2 - (dx^3)^2.$$

In fact, the solution has to approximate

$$ds^2 = c^2 dt^2 - dr^2 - r^2 d\varphi^2 - r^2 \sin^2 \varphi d\theta^2,$$

which is the *Minkowski metric in spherically spatial coordinates*.[1]

Therefore, it is natural to think the Schwarzschild metric in the form

$$ds^2 = c^2 \cdot e^T dt^2 - (e^Q - 1)dr^2 - dr^2 - r^2 d\varphi^2 - r^2 \sin^2 \varphi d\theta^2$$

where $T := T(r)$, $Q := Q(r)$ are two real functions that we need to determine from the vacuum field equations $R_{ij} = 0$. As we previously discussed, both $e^T \to 1$ and $e^Q \to 1$ have to go as $r \to \infty$. For the metric

$$ds^2 = c^2 \cdot e^T dt^2 - e^Q dr^2 - r^2 d\varphi^2 - r^2 \sin^2 \varphi d\theta^2$$

the coefficients are

$$g_{00} = e^T, \quad g_{11} = -e^Q, \quad g_{22} = -r^2, \quad g_{33} = -r^2 \sin^2 \varphi.$$

The inverse matrix coefficients are

$$g^{00} = e^{-T}, \quad g^{11} = -e^{-Q}, \quad g^{22} = -\frac{1}{r^2}, \quad g^{33} = -\frac{1}{r^2 \sin^2 \varphi}.$$

Let us observe that

$$\frac{\partial g_{ij}}{\partial x^0} = 0, \quad i, j = 0, \ldots, 3; \quad \Gamma^i_{jk} = 0, \quad i \neq j \neq k.$$

The non-zero Christoffel symbols are

$$\Gamma^0_{01} = \Gamma^0_{10} = \frac{T'}{2}, \quad \Gamma^1_{00} = -\frac{T'}{2} e^{T-Q}, \quad \Gamma^1_{11} = \frac{Q'}{2}, \quad \Gamma^1_{22} = -re^{-Q}, \quad \Gamma^1_{33} = -re^{-Q} \sin^2 \varphi,$$

$$\Gamma^2_{21} = \Gamma^2_{12} = \frac{1}{r}, \quad \Gamma^2_{33} = -\sin \varphi \cos \varphi, \quad \Gamma^3_{31} = \Gamma^3_{13} = \frac{1}{r}, \quad \Gamma^3_{23} = \Gamma^3_{32} = \cot \varphi.$$

---

[1] This request is an important property that any physical solution has to posses. In fact, very far from the source, a gravitational field has to go to zero. This means that Minkowski spacetime has to be recovered. This property is called "asymptotic flatness" and characterizes any physical gravitational field. It is worth noticing that this feature is fundamental for black hole solutions having physical meaning.

The only non-zero components of the Ricci tensor are

$$R_{00} = e^{T-Q}\left(\frac{T''}{2} + \frac{(T')^2}{4} - \frac{T'Q'}{4} + \frac{T'}{r}\right), \quad R_{11} = -\left(\frac{T''}{2} + \frac{(T')^2}{4} - \frac{T'Q'}{4} - \frac{Q'}{r}\right)$$

$$R_{22} = 1 - e^Q + re^{-Q}\left(\frac{Q'}{2} - \frac{T'}{2}\right), \quad R_{33} = \sin^2\varphi\, R_{22}.$$

The conditions $R_{00} = 0$ and $R_{11} = 0$ determine both T and Q.

Indeed, $e^{Q-T}R_{00} + R_{11} = 0$ implies $T' + Q' = 0$, that is $T + Q$=constant=k. Thus $e^T = e^{-Q}e^k$, i.e. the metric is

$$ds^2 = c^2 \cdot e^{-Q}e^k dt^2 - e^Q dr^2 - r^2 d\varphi^2 - r^2\sin^2\varphi d\theta^2.$$

If we let $t = e^{k/2}u$, then $dt^2 = e^k du^2$, and the metric becomes

$$ds^2 = c^2 \cdot e^{-Q} du^2 - e^Q dr^2 - r^2 d\varphi^2 - r^2\sin^2\varphi d\theta^2.$$

So, we may choose $T + Q = 0$, i.e. $Q = -T$. Replacing in the second equation, we have:

$$rT'' + r(T')^2 + 2T' = \left(re^T\right)'' = 0.$$

It results $\left(re^T\right)' = A$, that is $re^T = Ar + B$, i.e. $e^T = A + \dfrac{B}{r}$. We impose that $e^T \to 1$ as $r \to \infty$; it results $A = 1$. Therefore $e^T = 1 + \dfrac{B}{r}$ and $e^Q = e^{-T} = \dfrac{1}{1 + \dfrac{B}{r}}$.

Let us observe that for T and Q so far determined, $R_{22} = R_{33} = 0$.

The *Schwarzschild metric* is exactly the solution in the theorem statement, that is

$$ds^2 = c^2\left(1 + \frac{B}{r}\right)dt^2 - \frac{1}{1 + \dfrac{B}{r}}dr^2 - r^2 d\varphi^2 - r^2\sin^2\varphi d\theta^2.$$

$\square$

It is important to note that the Schwarzschild solution is independent of time. According to this property, the solution is not only stationary but also static. This is the statement of the Birkhoff theorem. See [61] for a detailed proof.

9.7.1 Orbit of a Planet in the Schwarzschild Metric

The above result can be immediately applied to celestial mechanics. Let us recall the classical orbit from first Kepler's law. The differential equation which describes the gravitational attraction between a planet and a star is

$$\vec{\ddot{X}} = -\frac{GM}{r^3} \cdot \vec{X},$$

where $\vec{X}$ is the position vector, $G = 6.67 \cdot 10^{-11} (\text{m})^3/(\text{kg}) \cdot (\text{s})^2$ is the gravitational constant, M is the mass of the star and $r = \| \vec{X} \|$. Let J be the magnitude of the angular moment of the planet. If we consider polar coordinates and $r = r(\theta) = \frac{1}{u(\theta)}$, then the previous equation becomes

$$\frac{d^2 u}{d\theta^2} + u = \frac{\mu}{J^2}, \quad \mu = GM.$$

The classical solution is:

$$u(\theta) = \frac{\mu}{J^2} + A\cos(\theta - \theta_0),$$

where A is an arbitrary constant which can be obtained from the initial condition and θ_0 is a phase shift. Since the phase shift alters the position of the planet at time $t = 0$ and we are interested only in the orbit itself, we may consider $\theta_0 = 0$. Denoting by e the eccentricity $e := \frac{AJ^2}{\mu}$, the orbit described by the solution

$$u(\theta) = \frac{\mu}{J^2}(1 + e\cos\theta) \quad \text{is the conic} \quad r(\theta) = \frac{\frac{J^2}{\mu}}{1 + e\cos\theta}.$$

The next result provides the differential equation which predicts the orbit of a planet in its movement around the Sun in the new context of Schwarzschild metric.

Theorem 9.7.2 *The orbit of a planet in the Schwarzschild metric is described by the equation*

$$\frac{d^2 u}{d\theta^2} + u = -\frac{c^2 B}{2J^2} - \frac{3B}{2}u^2.$$

Proof In the same way as before, we denote $x^0 := ct$. The worldcurve of the planet is the geodesic $\zeta(\tau) := (t(\tau), r(\theta), \varphi(\tau), \theta(\tau))$ of the Schwarzschild metric. We are looking for a solution in the (x, y) plane, that is $\varphi = \frac{\pi}{2}$. The reduced metric is

$$ds^2 = \left(1 + \frac{B}{r}\right)(dx^0)^2 - \frac{1}{1 + \frac{B}{r}}dr^2 - r^2 d\theta.$$

Since $\Gamma^3_{23} = \Gamma^3_{32} = \dfrac{1}{r}$ and $\Gamma^3_{ij} = 0$, the equation corresponding to the variable θ is:

$$\ddot{\theta}\,(\tau) + \frac{2}{r(\tau)} \cdot \dot{r}\,(\tau) \cdot \dot{\theta}\,(\tau) = \frac{1}{r^2}\left(r^2(\tau)\,\dot{\theta}\,(\tau)\right)' = 0.$$

We denote $r^2\,\dot{\theta} = J$ and this J describes the magnitude of the angular momentum of the planet exactly as in the classical case. We cancel τ in the next computations.

Let us continue with the geodesic equation corresponding to the variable x^0. Since only $\Gamma^0_{01} = \Gamma^0_{10} = -\dfrac{1}{\left(1 + \dfrac{B}{r}\right)} \cdot \dfrac{B}{2r^2}$, the equation in x^0 is

$$\ddot{x}^0 - \frac{B}{r^2} \cdot \frac{1}{\left(1 + \frac{B}{r}\right)} \cdot \dot{x}^0 \cdot \dot{r} = 0.$$

By replacing x^0 with ct, it results

$$\ddot{t} - \frac{B}{r^2} \cdot \frac{1}{\left(1 + \frac{B}{r}\right)} \cdot \dot{t} \cdot \dot{r} = 0 \quad \text{i.e.} \quad \left(\left(1 + \frac{B}{r}\right) \cdot \dot{t}\right)' = 0,$$

that is

$$\dot{t} = \frac{E}{1 + \frac{B}{r}},$$

where E is a constant.

In the case of the equation corresponding to the variable r, we use directly the metric condition taking into account that $ds^2 = c^2 d\tau^2$. After canceling $d\tau^2$, it results

$$c^2 = c^2\left(1 + \frac{B}{r}\right)\dot{t}^2 - \frac{1}{1 + \frac{B}{r}}\,\dot{r}^2 - r^2\,\dot{\theta}^2.$$

Let us replace $\dot{t}$ and $\dot{\theta}$ in the previous equation. We have:

$$c^2\left(1 - E^2\right) + c^2 \cdot B \cdot \frac{1}{r} = -\dot{r}^2 - \frac{J^2}{r^2} - \frac{B \cdot J^2}{r^3}.$$

Consider $r = r(\theta)$. It results:

$$\dot{r} = \frac{dr}{d\theta} \cdot \dot{\theta} = \frac{dr}{d\theta} \cdot \frac{J}{r^2}.$$

If $r := \dfrac{1}{u}$, then $\dfrac{dr}{d\theta} = -\dfrac{1}{u^2}\dfrac{du}{d\theta}$, i.e.

$$\dot{r} = -J \cdot \frac{du}{d\theta}.$$

Since $\dot{r}^2 = J^2 \cdot \left(\frac{du}{d\theta}\right)^2$, the previous equation becomes

$$c^2(1 - E^2) + c^2 Bu = -J^2 \left(\frac{du}{d\theta}\right)^2 - J^2 u^2 - BJ^2 u^3.$$

If we differentiate with respect to θ, then we divide by $\frac{du}{d\theta}$, we obtain the equation

$$\frac{d^2 u}{d\theta^2} + u = -\frac{c^2 B}{2J^2} - \frac{3B}{2} u^2.$$

$\square$

9.7.2 Relativistic Solution of the Mercury Perihelion Drift Problem

Now we need to clarify who is B in the Schwarzschild metric. We have requested that, as $r \to \infty$, the Schwarzschild metric approaches the ordinary Minkowski metric. Let us continue by taking into account the following two equations.

1. The classical orbit is described by

$$\frac{d^2 u}{d\theta^2} + u = \frac{\mu}{J^2}, \quad \mu = GM$$

2. The relativistic orbit is described, in the Schwarzschild metric, by

$$\frac{d^2 u}{d\theta^2} + u = -\frac{c^2 B}{2J^2} - \frac{3B}{2} u^2.$$

From $g_{00} = c^2 \left(1 + \frac{B}{r}\right)$, we compute $\Gamma_{00}^1 = \frac{c^2 B}{2r^2}$ which is the only nonzero Γ_{00}^i.

So, the r component of the geodesic equation is

$$\frac{d^2 r}{d\tau^2} = \Gamma_{00}^1 \frac{d\tau}{d\tau} \frac{d\tau}{d\tau},$$

that is

$$\frac{d^2 r}{d\tau^2} = \frac{c^2 B}{2r^2}.$$

As r approaches the infinity, $d\tau$ becomes dt and the previous equation is the original Newton equation $\dfrac{d^2r}{dt^2} = -\dfrac{GM}{r^2}$ if and only if $B = -\dfrac{2GM}{c^2}$.

It results

$$1 + \frac{B}{r} = 1 - \frac{2GM}{c^2} \cdot \frac{1}{r}.$$

In this way, the gravitational Newtonian potential $\phi(x, y, z) = -\dfrac{GM}{r}$ is involved in the coefficients of the metric. The coefficient $\dfrac{1}{c^2}$ highlights the weak gravitational field which we will discuss later. See also [4].

The quantity $r_M := \dfrac{2GM}{c^2}$ has the dimension of a length and it is called *gravitational radius*, or *the Schwarzschild radius*, corresponding to the mass M. It is an intrinsic characteristic of any body with mass.

In General Relativity, we can define a *proper time interval* $\Delta\tau$ between two events along a timelike path l following the definition given in Special Relativity. Using constant space coordinates, the proper time satisfies the same equality

$$ds^2 = c^2(d\tau)^2$$

as in Special Relativity. Therefore using the same constant coordinates x^1, x^2, x^3, it results

$$\Delta\tau = \int_l ds = \int_l \frac{1}{c}\sqrt{g_{ij}dx^i dx^j} = \int_l \frac{1}{c}\sqrt{g_{00}}dx^0.$$

A discussion about how gravity influences the proper time is in [40].

Next result allows to make distinction between the proper time and the time coordinate in the case of Schwarzschild metric.

Theorem 9.7.3 *The gravitational field described by the Schwarzschild metric*

$$ds^2 = c^2\left(1 - \frac{2GM}{c^2 r}\right)dt^2 - \frac{1}{1 - \frac{2GM}{c^2 r}}dr^2 - r^2 d\varphi^2 - r^2 \sin^2\varphi d\theta^2$$

causes the slow down of clocks.

Proof Let us consider the Schwarzschild metric in a frame at rest R and apply the previous results in the following way. The source of the gravitational field is at the origin O of R. Consider two motionless observers, one close to the source O, denoted by O_1, and the other one far from the source, O_2. Each observer has a clock. For both observers the variation of the space coordinates is 0. We have $dr = d\varphi = d\theta = 0$ for the first observer, therefore, according to him

$$ds^2 = c^2\left(1 - \frac{2GM}{c^2 r}\right)^2 dt^2.$$

For the second motionless observer at rest, far from source, the influence of the gravitational field is almost not observable. There, for $r \to \infty$, we have $ds^2 = c^2 d\tau^2$. Therefore the proper time is affected by the gravity according to the rule

$$c^2 d\tau^2 = ds^2 = c^2 \left(1 - \frac{2GM}{c^2 r}\right) dt^2.$$

Considering the clocks, it results

$$c^2 \Delta\tau^2 = c^2 \left(1 - \frac{2GM}{c^2 r}\right) \Delta t^2,$$

that is, the time interval $\Delta\tau$ of O_2's clock appears to be less than Δt on O_1's clock. If you are close to a source, your clock will slow down and will continue to slow down if you come closer and closer to the source.   □

Let c be the speed of light in vacuum. If we write formally an expression Q as a Taylor series in powers of $\frac{1}{c}$

$$Q = a_0 + a_1 \cdot \frac{1}{c} + a_2 \cdot \frac{1}{c^2} + a_3 \cdot \frac{1}{c^3} + \cdots + a_k \cdot \frac{1}{c^k} + \cdots,$$

we say that the order of Q is $O\left(\frac{1}{c^m}\right)$ if $a_0 = a_1 = \cdots = a_{m-1} = 0$ and $a_m \neq 0$. How is working this formal definition in a given physical context? Let us write each relativistic expression (components of the gravitational field, metric tensor, equations) as a Taylor series in powers of $\frac{1}{c}$. The computations with these series can be truncated at the term that is appropriate for the physical context we are considering.

Theorem 9.7.4 *In the relativistic field described by the Schwarzschild metric*

$$ds^2 = c^2 \left(1 - \frac{2GM}{c^2 r}\right) dt^2 - \frac{1}{1 - \frac{2GM}{c^2 r}} dr^2 - r^2 d\varphi^2 - r^2 \sin^2 \varphi d\theta^2$$

the planet equation of motion

$$\frac{d^2 u}{d\theta^2} + u = \frac{\mu}{J^2} + \frac{3\mu}{c^2} u^2$$

has the solution

$$u(\theta) = \frac{\mu}{J^2} (1 + e\cos(\theta - F\theta)) + O\left(\frac{1}{c^2}\right),$$

where $F := \frac{3\mu^2}{c^2 J^2}$, *being* $\mu = GM$.

Proof We start from the classical equation of an orbit of a planet,

$$\frac{d^2a}{d\theta^2} + a = \frac{\mu}{J^2}$$

with the classical solution $a(\theta) = \frac{\mu}{J^2}(1 + e\cos\theta)$.

The new equation of the orbit

$$\frac{d^2u}{d\theta^2} + u = \frac{\mu}{J^2} + \frac{3\mu}{c^2} \cdot u^2$$

differs from the classical one by $\frac{3\mu}{c^2}u^2$. This "correction" of the classical orbit is due to the gravity related to the Schwarzschild metric. It is natural to search for the solution as

$$u(\theta) := a(\theta) + \frac{w(\theta)}{c^2}.$$

If we replace it in the new orbit equation

$$\frac{d^2u}{d\theta^2} + u = \frac{\mu}{J^2} + \frac{3\mu}{c^2} \cdot u^2,$$

we obtain

$$\frac{d^2a}{d\theta^2} + a + \frac{1}{c^2}\left(\frac{d^2w}{d\theta^2} + w\right) = \frac{\mu}{J^2} + \frac{3\mu}{c^2}\left(\frac{\mu}{J^2}(1 + e\cos\theta) + \frac{w}{c^2}\right)^2,$$

or equivalently

$$\frac{1}{c^2}\left(\frac{d^2w}{d\theta^2} + w\right) = \frac{3\mu^3}{c^2 J^4}(1 + e\cos\theta)^2 + O\left(\frac{1}{c^4}\right).$$

The term $O\left(\frac{1}{c^4}\right)$ has a small influence on w. It remains to solve

$$\frac{d^2w}{d\theta^2} + w = \frac{3\mu^3}{J^4}\left(1 + \frac{e^2}{2} + 2e\cos\theta + \frac{e^2}{2}\cos 2\theta\right).$$

The solutions of the following three equations

$$\frac{d^2w_1}{d\theta^2} + w_1 = \frac{3\mu^3}{J^4}\left(1 + \frac{e^2}{2}\right); \quad \frac{d^2w_2}{d\theta^2} + w_2 = \frac{6e\mu^3}{J^4}\cos\theta; \quad \frac{d^2w_3}{d\theta^2} + w_3 = \frac{3\mu^3 e^2}{2J^4}\cos 2\theta$$

are

$$w_1 = \frac{3\mu^3}{J^4}\left(1 + \frac{e^2}{2}\right), \quad w_2 = \frac{3e\mu^3}{J^4}\theta\cos\theta, \quad w_3 = -\frac{3\mu^3 e^2}{2J^4}\cos 2\theta.$$

Therefore, the solution of the new orbit equation is

$$u(\theta) = \frac{\mu}{J^2}(1 + e\cos\theta) + \frac{3\mu^3}{J^4 c^2}\left(1 + \frac{e^2}{2} + e\theta\sin\theta - \frac{e^2}{2}\cos 2\theta\right).$$

Einstein's idea was to use only the non-periodic term in the classical solution. Then

$$u(\theta) = \frac{\mu}{J^2}\left[1 + e\left(\cos\theta + \frac{3\mu^2}{c^2 J^2}\theta\sin\theta\right)\right] + O\left(\frac{1}{c^2}\right),$$

which can be written as

$$u(\theta) = \frac{\mu}{J^2}[1 + e\cos(\theta - F\theta)] + O\left(\frac{1}{c^2}\right),$$

where $F := \frac{3\mu^2}{c^2 J^2}$. Neglecting the term $O\left(\frac{1}{c^2}\right)$ which adds only a small contribution, the trajectory is still the old conic. $\qquad\square$

The correction to the classical trajectory, described in the Schwarzschild metric, reaches the perihelion for $\cos(\theta - F\theta) = 1$, therefore, it is $\theta = \theta_n = \frac{2n\pi}{1 + F}$ for an integer n. It results $\theta \approx 2n\pi\left(1 - F + O\left(F^2\right)\right)$; that is $2\pi F$ is the perihelion drift for each revolution.

If N is the number of orbits for a given period of time T, then the perihelion drift P_d is

$$P_d = \frac{6\pi G^2 M^2}{c^2 J^2} \cdot N.$$

For Mercury, if we replace the constants, we obtain 43 arcseconds per century which was observed by astronomers, without explanation, in the context of Classical Mechanics. This was considered one of the first confirmations of General Relativity. See Sect. 9.8 and [40] for a detailed discussion also in the historical context.

9.7.3 Speed of Light in a Given Metric

Consider a Minkowski spacetime and suppose the worldcurve $\mathbb{X}(t) = (ct, x^1(t), x^2(t), x^3(t))$ of a spatial object parameterized by the time t.

Then, its relativistic speed is

$$\left\|\frac{d\mathbb{X}}{dt}\right\|^2 = c^2 - \left(\frac{dx^1}{dt}\right)^2 - \left(\frac{dx^2}{dt}\right)^2 - \left(\frac{dx^3}{dt}\right)^2 = c^2 - v^2,$$

where

$$v = \sqrt{\left(\frac{dx^1}{dt}\right)^2 + \left(\frac{dx^2}{dt}\right)^2 + \left(\frac{dx^3}{dt}\right)^2}$$

is the ordinary velocity of the object.

If the object is a photon,

$$\left\|\frac{d\mathbb{X}}{dt}\right\|^2 = c^2 - c^2 = 0$$

as we expected.

If we consider the same worldcurve $\mathbb{X}(t) = (ct, x^1(t), x^2(t), x^3(t))$ in the metric

$$ds^2 = g_{00}(dx^0)^2 + g_{\alpha\beta}dx^\alpha dx^\beta,$$

where α, β are spatial indexes according to the above formalism, we have

$$ds^2 = g_{00}(dx^0)^2 + g_{11}(dx^1)^2 + g_{22}(dx^2)^2 + g_{33}(dx^3)^2 + \sum_{\alpha \neq \beta = 1}^{3} g_{\alpha\beta}dx^\alpha dx^\beta.$$

Then

$$ds^2 = (dx^0)^2 - (dx^1)^2 - (dx^2)^2 - (dx^3)^2 + (g_{00} - 1)(dx^0)^2 + (g_{11} + 1)(dx^1)^2+$$

$$+(g_{22} + 1)(dx^2)^2 + (g_{33} + 1)(dx^3)^2 + \sum_{\alpha \neq \beta = 1}^{3} g_{\alpha\beta}dx^\alpha dx^\beta,$$

i.e.

$$ds^2 = (dx^0)^2 - (dx^1)^2 - (dx^2)^2 - (dx^3)^2 + (g_{00} - 1)(dx^0)^2 + \sum_{\alpha,\beta = 1}^{3} \bar{g}_{\alpha\beta}dx^\alpha dx^\beta$$

where $\bar{g}_{\alpha\beta} = g_{\alpha\beta}$, if $\alpha \neq \beta$ and $\bar{g}_{\alpha\beta} = 1 + g_{\alpha\beta}$, if $\alpha = \beta$.

This means that, for a photon, we obtain

$$0 = \left(\frac{d\mathbb{X}}{dt}\right)^2 = c^2 - \gamma^2 + (g_{00} - 1)\left(\frac{dx^0}{dt}\right)^2 + \sum_{\alpha,\beta = 1}^{3} \bar{g}_{\alpha\beta}\frac{dx^\alpha}{dt}\frac{dx^\beta}{dt}$$

where

$$\gamma = \sqrt{\left(\frac{dx^1}{dt}\right)^2 + \left(\frac{dx^2}{dt}\right)^2 + \left(\frac{dx^3}{dt}\right)^2}.$$

The quantity γ, given by the formula

$$\gamma = \sqrt{c^2 + (g_{00} - 1)\left(\frac{dx^0}{dt}\right)^2 + \sum_{\alpha,\beta=1}^{3} \bar{g}_{\alpha\beta}\frac{dx^\alpha}{dt}\frac{dx^\beta}{dt}},$$

is called *the speed of light in a gravitational field* derived by the above metric. We say that γ does not *violate the speed of light limit* if $\gamma \leq c$.

9.7.4 Bending of Light in Schwarzschild Metric

Let us consider now the light traveling in the spacetime described by the Schwarzschild metric. First of all, we need to compute the speed γ of light in the gravitational field induced by the metric above.

Theorem 9.7.5 *Consider the Schwarzschild metric*

$$ds^2 = c^2\left(1 - \frac{2\mu}{c^2 r}\right)dt^2 - \frac{1}{1 - \frac{2\mu}{c^2 r}}dr^2 - r^2 d\varphi^2 - r^2\sin^2\varphi d\theta^2.$$

(i) In Cartesian coordinates this metric has the form

$$ds^2 = \sum_{i=0}^{3}(dx^i)^2 - \frac{2\mu}{c^2 r}\left((dx^0)^2 + \frac{1}{1 - \frac{2\mu}{c^2 r}}\sum_{\alpha,\beta=1}^{3}\frac{x^\alpha x^\beta}{r^2}dx^\alpha dx^\beta\right).$$

(ii) A deflected photon in the (x^1, x^2) plane, which comes from the undeflected photon $X(t) = (ct, h, ct, 0)$, has the speed

$$\gamma = c - \frac{\mu}{cr} - \frac{\mu}{cr}\cdot\frac{(x^2)^2}{r^2}\cdot\frac{1}{1 - \frac{2\mu}{c^2 r}}.$$

(iii) The deflected photon does not violate the speed of light limit c and γ can be written in the equivalent form

$$\gamma = c - \frac{\mu}{cr} - \frac{\mu(x^2)^2}{cr^3} + O\left(\frac{1}{c^3}\right).$$

Proof (i) Let $(x^0, x^1, x^2, x^3) = (ct, x, y, z)$ and $r^2 = x^2 + y^2 + z^2$.

It results $rdr = xdx + ydy + zdz$ which gives

$$dr = \sum_{\alpha=1}^{3} \frac{x^{\alpha}}{r} dx^{\alpha}, \quad dr^2 = \sum_{\alpha,\beta=1}^{3} \frac{x^{\alpha}x^{\beta}}{r^2} dx^{\alpha} dx^{\beta}.$$

Taking into account that

$$dx^2 + dy^2 + dz^2 = dr^2 + r^2 d\varphi^2 + r^2 \sin^2 \varphi d\theta^2 = \sum_{\alpha=1}^{3} (dx^{\alpha})^2 ,$$

it results

$$ds^2 = c^2 dt^2 - \frac{2\mu}{c^2 r} c^2 dt^2 - \frac{1 - \frac{2\mu}{c^2 r} + \frac{2\mu}{c^2 r}}{1 - \frac{2\mu}{c^2 r}} dr^2 - r^2 d\varphi^2 - r^2 \sin^2 \varphi d\theta^2$$

$$= c^2 dt^2 - dr^2 - r^2 d\varphi^2 - r^2 \sin^2 \varphi d\theta^2 - \frac{2\mu}{c^2 r} \left(c^2 dt^2 + \frac{1}{1 - \frac{2\mu}{c^2 r}} dr^2 \right)$$

$$= (dx^0)^2 - \sum_{\alpha=0}^{3} (dx^{\alpha})^2 - \frac{2\mu}{c^2 r} \left((dx^0)^2 + \frac{1}{1 - \frac{2\mu}{c^2 r}} \sum_{\alpha,\beta=1}^{3} \frac{x^{\alpha}x^{\beta}}{r^2} dx^{\alpha} dx^{\beta} \right) .$$

(ii) According to the technique previously described, suppose that $X(t) = \left(ct, x^1(t), x^2(t), x^3(t)\right)$ is the worldcurve of an object parameterized by the time t. In the Minkowski metric, it is

$$\left(\frac{ds}{dt} \right)^2 = \| \dot{X}(t) \|^2 = c^2 - \left((\dot{x}^1(t))^2 + (\dot{x}^2(t))^2 + (\dot{x}^3(t))^2 \right) = c^2 - v^2,$$

where v is the usual spatial speed of the object.

If the object is a photon, then $\left(\dfrac{ds}{dt} \right)^2 = c^2 - c^2 = 0$ and so

$$0 = \left(\frac{ds}{dt} \right)^2 = c^2 - \gamma^2 - \frac{2\mu}{c^2 r} \left(c^2 + \frac{1}{1 - \frac{2\mu}{c^2 r}} \sum_{\alpha,\beta=1}^{3} \frac{x^{\alpha}x^{\beta}}{r^2} \frac{dx^{\alpha}}{dt} \frac{dx^{\beta}}{dt} \right) ,$$

where

$$\gamma(t) = \sqrt{(\dot{x}^1(t))^2 + (\dot{x}^2(t))^2 + (\dot{x}^3(t))^2}$$

is the speed of the photon in the gravitational field described by the above metric. In fact,

$$\gamma = \sqrt{c^2 - \frac{2\mu}{c^2 r}\left(c^2 + \frac{1}{1 - \frac{2\mu}{c^2 r}}\sum_{\alpha,\beta=1}^{3}\frac{x^\alpha x^\beta}{r^2}\frac{dx^\alpha}{dt}\frac{dx^\beta}{dt}\right)}.$$

We determine γ along the worldcurve X of an undeflected photon in the (x^1, x^2) plane at the fixed distance h from the x^2-axis.

The undeflected worldcurve of the photon is $X(t) := (ct, h, ct, 0)$. The deflection will add only lower order terms, therefore the deflected photon has, in the same plane, a worldcurve which components have extra terms of order $O\left(\frac{1}{c}\right)$. The deflected photon is parameterized by

$$X_d(t) := \left(ct, h + O\left(\frac{1}{c}\right), ct + O(1), 0\right).$$

Since $\dot{X}_d(t) := \left(c, O\left(\frac{1}{c}\right), c + O(1), 0\right)$, we have

$$\frac{dx^1}{dt} = O\left(\frac{1}{c}\right), \quad \frac{dx^2}{dt} = c + O(1), \quad \frac{dx^3}{dt} = 0.$$

It results the approximation

$$\gamma^2 = c^2 - \frac{2\mu}{c^2 r}\left(c^2 + \frac{1}{1 - \frac{2\mu}{c^2 r}}\frac{(x^2)^2}{r^2}\cdot c^2\right),$$

equivalent to

$$\gamma^2 = c^2\left(1 - \frac{2\mu}{c^2 r} - \frac{2\mu(x^2)^2}{c^2 r^3}\frac{1}{1 - \frac{2\mu}{c^2 r}}\right),$$

i.e.

$$\gamma = c \cdot \sqrt{1 - \frac{2\mu}{c^2 r} - \frac{2\mu(x^2)^2}{c^2 r^3}\frac{1}{1 - \frac{2\mu}{c^2 r}}}.$$

Taking into account that $\sqrt{1 + 2A} \approx 1 + A$, the result is

$$\gamma = c - \frac{\mu}{cr} - \frac{\mu}{cr}\cdot\frac{(x^2)^2}{r^2}\cdot\frac{1}{1 - \frac{2\mu}{c^2 r}}.$$

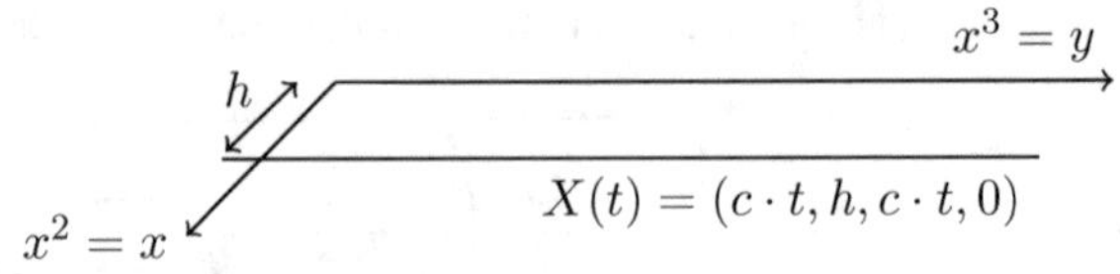

Fig. 9.1 Trajectory of an
undeflected photon

(iii) Since $\dfrac{1}{1-\frac{2\mu}{c^2 r}} \approx 1 + \dfrac{2\mu}{c^2 r} + O\left(\dfrac{1}{c^3}\right)$ it results both the formula

$$\gamma = c - \frac{\mu}{cr} - \frac{\mu(x^2)^2}{cr^3} + O\left(\frac{1}{c^3}\right)$$

and the fact that the deflected photon does not violate the light limit speed
(Fig. 9.1).   □

Theorem 9.7.6 *The total deflection of the trajectory $X_d(t)$ of a deflected photon in
the gravitational field described by the Schwarzschild metric*

$$ds^2 = c^2\left(1 - \frac{2\mu}{c^2 r}\right) dt^2 - \frac{1}{1 - \frac{2\mu}{c^2 r}} dr^2 - r^2 d\varphi^2 - r^2 \sin^2 \varphi d\theta^2$$

is $T_D = \dfrac{4GM}{c^2 h}$.

Proof Let us recall that the trajectory of the deflected photon is

$$X_d(t) = \left(ct, h + O\left(\frac{1}{c}\right), ct + O(1), 0\right)$$

and it comes from the undeflected photon trajectory $X(t) = (ct, h, ct, 0)$, in which
the deflection added small contribution terms.

The previous theorem proves that the speed of a deflected photon in (x, y) plane
is

$$\gamma = c - \frac{\mu}{cr} - \frac{\mu(x^2)^2}{cr^3} + O\left(\frac{1}{c^3}\right).$$

Let us imagine a line l and two points on it having coordinates x and $x + dx$
respectively. Two parallel lines constructed through the given points make the same
$\Delta\theta$ angle with the perpendicular to the l direction. This lines can be imagined as
trajectories of photons, the first one traveling with the speed $\gamma(x)$, the second one
traveling with the speed $\gamma(x + \Delta x)$.

After Δt seconds, the last two parallel lines change the trajectories into other
two parallel lines, etc. The first photon traveled $\gamma(x)\Delta t$, the second one traveled
$\gamma(x + \Delta x)\Delta t$.

Let us suppose now that $\gamma(x)\Delta t > \gamma(x + \Delta x)\Delta t$. It is easy to see that there is a rectangle triangle which leads to the relation

$$\Delta\theta \approx \sin \Delta\theta = \frac{\gamma(x)\Delta t - \gamma(x + \Delta x)\Delta t}{dx},$$

that is

$$\frac{\Delta\theta}{dt} \approx \frac{\gamma(x) - \gamma(x + \Delta x)}{dx}.$$

As $\Delta t \to 0$ and $\Delta x \to 0$, the last relation becomes

$$\frac{\Delta\theta}{dt} = -\frac{\partial\gamma}{\partial x}.$$

If $s = ct$,

$$\frac{\Delta\theta}{ds} = -\frac{1}{c}\frac{\partial\gamma}{\partial x}.$$

At the same time, $\dfrac{\Delta\theta}{ds}$ is the geometric curvature determined for the photons trajectories when the parameter is s. If we denote x by x^1, the perpendicular direction coordinate by x^2, the *total deflection* is related to the integral of the geometric curvature $\dfrac{\Delta\theta}{ds}$, that is

$$T_D := -\frac{1}{c}\int_{-\infty}^{\infty}\frac{\partial\gamma}{\partial x^1}dx^2.$$

To perform the computation, we start from canceling the $O\left(\dfrac{1}{c^4}\right)$ term. We have

$$\frac{\partial\gamma}{\partial x^1}\bigg|_{X_d} = \frac{GMx^1}{cr^3}\bigg|_{X_d} + \frac{3GM(x^2)^2x^1}{cr^5}\bigg|_{X_d} = \frac{GMh}{c(h^2 + (x^2)^2)^{\frac{3}{2}}} + \frac{3GMh(x^2)^2}{c(h^2 + (x^2)^2)^{\frac{5}{2}}}$$

and elementary computations lead to

$$\frac{1}{c}\int_{-\infty}^{\infty}\frac{\partial\gamma}{\partial x^1}dx^2 = \frac{GMh}{c^2}\left(\int_{-\infty}^{\infty}\frac{1}{(h^2 + (x^2)^2)^{\frac{3}{2}}}dx^2 + \int_{-\infty}^{\infty}\frac{3(x^2)^2}{(h^2 + (x^2)^2)^{\frac{5}{2}}}dx^2\right) =$$

$$= \left(\frac{2}{h^2} + \frac{2}{h^2}\right)\frac{GMh}{c^2}.$$

The total deflection is then $T_D = \dfrac{4GM}{c^2h}$. $\qquad\qquad\square$

At the surface of the Sun, we have

$$h = \text{radius of the Sun} = 7 \times 10^8 (\text{m}); G = 6,67 \times 10^{-11}(\text{m}^3)/(\text{kg}) \cdot (\text{s}^2),$$
$$M = \text{mass of the Sun} = 2 \times 10^{30}(\text{kg}); c = 3 \times 10^8(\text{m})/(\text{s}^2).$$

It results for $T_{DS} \approx 1.75''$. This was another sensational confirmation of General Relativity due to Dyson and Eddington in 1919. See [40] for details.

9.8 About Einstein's Metric: Einstein's Computations Related to Perihelion's Drift and Bending of the Light Rays

Even if Einstein was the one who discovered the vacuum field equations, he did not solve them. In order to make computations possible, he choose a spherically symmetric metric, independent of time, metric who approximates the Minkowski metric as $r \to \infty$. He took care to involve the gravitational potential $\varPhi = -\dfrac{GM}{r}$ in the first two coefficients. Therefore, the chosen metric was

$$ds^2 = c^2 \left(1 - \frac{2\mu}{c^2 r}\right) dt^2 - \left(1 + \frac{2\mu}{c^2 r}\right) dr^2 - r^2 d\varphi^2 - r^2 \sin^2 \varphi d\theta^2,$$

being always $\mu = GM$. Obviously, this metric does not satisfy the field equations $R_{ij} = 0$.

Einstein's computations on perihelion drift and bending of light were performed with this metric.

Theorem 9.8.1 (Einstein's First Theorem) *In the relativistic field described by Einstein's metric*

$$ds^2 = c^2 \left(1 - \frac{2\mu}{c^2 r}\right) dt^2 - \left(1 + \frac{2\mu}{c^2 r}\right) dr^2 - r^2 d\varphi^2 - r^2 \sin^2 \varphi d\theta^2$$

the planet equation of motion

$$\frac{d^2 u}{d\theta^2} + u = \frac{\mu}{J^2} + \frac{3\mu}{c^2} u^2$$

has the solution

$$u(\theta) = \frac{\mu}{J^2}(1 + e\cos(\theta - F\theta)) + O\left(\frac{1}{c^2}\right),$$

where $F := \dfrac{3\mu^2}{c^2 J^2}.$

Proof In the same way as before, we denote $x^0 := ct$. The worldcurve of the planet is the geodesic $\zeta(\tau) := (t(\tau), r(\theta), \varphi(\tau), \theta(\tau))$ of Einstein's metric. We are looking for a solution in the (x, y) plane, that is $\varphi = \dfrac{\pi}{2}$. The reduced metric is

$$ds^2 = c^2 \left(1 - \frac{2\mu}{c^2 r}\right) dt^2 - \left(1 + \frac{2\mu}{c^2 r}\right) dr^2 - r^2 d\theta^2 .$$

We cancel out τ in the next computations. Since $\Gamma_{23}^3 = \Gamma_{32}^3 = \dfrac{1}{r}$ and $\Gamma_{ij}^3 = 0$ in the other cases, the equation corresponding to the variable θ is

$$\ddot{\theta} + \frac{2}{r} \cdot \dot{r} \cdot \dot{\theta} = \frac{1}{r^2} \left(r^2 \dot{\theta}\right)' = 0.$$

It results $r^2 \dot{\theta} =$ constant. We denote $J := r^2 \dot{\theta}$.

The constant J describes the magnitude of the angular momentum of the planet exactly as in the classical case.

Let us continue with the geodesic equation corresponding to the variable x^0.

Since only $\Gamma_{01}^0 = \Gamma_{10}^0 = \dfrac{1}{1 - \dfrac{2\mu}{c^2 r}} \cdot \dfrac{\mu}{c^2 r^2}$, the equation in x^0 is

$$\ddot{x}^0 + \frac{2\mu}{c^2 r^2} \cdot \frac{1}{1 - \frac{2\mu}{c^2 r}} \cdot \dot{x}^0 \cdot \dot{r} = 0.$$

Replacing x^0 by ct, it results

$$\ddot{t} + \frac{2\mu}{c^2 r^2} \cdot \frac{1}{1 - \frac{2\mu}{c^2 r}} \cdot \dot{t} \cdot \dot{r} = 0.$$

that is

$$\dot{t} = \frac{E}{1 - \frac{2\mu}{c^2 r}},$$

where E is a constant.

In the case of the equation corresponding to the variable r, we use directly the metric condition. Taking into account that $ds^2 = c^2 d\tau^2$, after we cancel $d\tau^2$, it results

$$c^2 = c^2 \left(1 - \frac{2\mu}{c^2 r}\right) \dot{t}^2 - \left(1 + \frac{2\mu}{c^2 r}\right) \dot{r}^2 - r^2 \dot{\theta}^2 .$$

Let us replace $\dot{t} = \dfrac{E}{1 - \frac{2\mu}{c^2 r}}$, and $\dot{\theta} = \dfrac{J}{r^2}$ in the previous equation. We have:

$$c^2\left(1-E^2\right)-\frac{2\mu}{r}=\left(\frac{4\mu^2}{c^4r^2}-1\right)\dot{r}^2-r^2\dot{\theta}^2\left(1-\frac{2\mu}{c^2r}\right)$$

Consider $r=r(\theta)$. It results:

$$\dot{r}=\frac{dr}{d\theta}\cdot\dot{\theta}=\frac{dr}{d\theta}\cdot\frac{J}{r^2}.$$

If $r:=\dfrac{1}{u}$, then $\dfrac{dr}{d\theta}=-\dfrac{1}{u^2}\dfrac{du}{d\theta}$, i.e. $\dot{r}=-J\cdot\dfrac{du}{d\theta}$.

Since $\dot{r}^2=J^2\cdot\left(\dfrac{du}{d\theta}\right)^2$, the previous equation becomes

$$c^2(1-E^2)-2\mu\cdot u=-J^2\left(\frac{du}{d\theta}\right)^2-J^2u^2\left(1-\frac{2\mu}{c^2r}u\right)+O\left(\frac{1}{c^4}\right).$$

We can neglect the $O\left(\dfrac{1}{c^4}\right)$ terms which add only a small contribution to the trajectory. If we differentiate with respect to θ and then we divide by $\dfrac{du}{d\theta}$, we obtain exactly the equation derived in the Schwarzschild metric case, that is

$$\frac{d^2u}{d\theta^2}+u=\frac{\mu}{J^2}+\frac{3\mu}{c^2}u^2.$$

Of course, Einstein found the same solution and the same perihelion drift as in the case of Schwarzschild metric.   □

Now, let us compute the Einstein metric

$$ds^2=c^2\left(1-\frac{2\mu}{c^2r}\right)dt^2-\left(1+\frac{2\mu}{c^2r}\right)dr^2-r^2d\varphi^2-r^2\sin^2\varphi d\theta^2$$

in Cartesian coordinates. As in the case of Schwarzschild metric, let $(x^0,x^1,x^2,x^3)=(ct,x,y,z)$ and $r^2=x^2+y^2+z^2$. It results $rdr=xdx+ydy+zdz$ which gives

$$dr=\sum_{\alpha=1}^{3}\frac{x^\alpha}{r}dx^\alpha,\quad dr^2=\sum_{\alpha,\beta=1}^{3}\frac{x^\alpha x^\beta}{r^2}dx^\alpha dx^\beta.$$

Taking into account that

$$dx^2+dy^2+dz^2=dr^2+r^2d\varphi^2+r^2\sin^2\varphi d\theta^2=\sum_{\alpha=1}^{3}(dx^\alpha)^2,$$

it results

$$ds^2 = c^2 dt^2 - dr^2 - r^2 d\varphi^2 - r^2 \sin^2 \varphi d\theta^2 - \frac{2\mu}{c^2 r} \left(c^2 dt^2 + dr^2 \right).$$

Therefore, *Einstein's metric in Cartesian coordinates* is

$$ds^2 = (dx^0)^2 - \sum_{\alpha=0}^{3} (dx^\alpha)^2 - \frac{2\mu}{c^2 r} \left((dx^0)^2 + \sum_{\alpha,\beta=1}^{3} \frac{x^\alpha x^\beta}{r^2} dx^\alpha dx^\beta \right).$$

We determine the speed of light in the gravitational field described by Einstein's metric. We use the same technique as in the case of Schwarzschild metric.

If $X(t) = \left(ct, x^1(t), x^2(t), x^3(t) \right)$ is the worldcurve of an object parameterized by the time t, then, in the Minkowski metric, it is

$$\left(\frac{ds}{dt} \right)^2 = \| \dot{X}(t) \|^2 = c^2 - \left((\dot{x}^1(t))^2 + (\dot{x}^2(t))^2 + (\dot{x}^3(t))^2 \right) = c^2 - v^2,$$

where v is the usual spatial speed of the object.

If the object is a photon, then $\left(\dfrac{ds}{dt} \right)^2 = c^2 - c^2 = 0$ and so

$$0 = \left(\frac{ds}{dt} \right)^2 = c^2 - \gamma^2 - \frac{2\mu}{c^2 r} \left(c^2 + \sum_{\alpha,\beta=1}^{3} \frac{x^\alpha x^\beta}{r^2} \frac{dx^\alpha}{dt} \frac{dx^\beta}{dt} \right)$$

where $\gamma(t) = \sqrt{ (\dot{x}^1(t))^2 + (\dot{x}^2(t))^2 + (\dot{x}^3(t))^2 }$ is the speed of the photon in the gravitational field described by the metric above. In fact

$$\gamma = c \cdot \sqrt{ 1 - \frac{2\mu}{c^2 r} \left(1 + \frac{1}{c^2} \sum_{\alpha,\beta=1}^{3} \frac{x^\alpha x^\beta}{r^2} \frac{dx^\alpha}{dt} \frac{dx^\beta}{dt} \right) }.$$

It remains, as an exercise, to determine the speed of a deflected photon using the same technique as in the case of Schwarzschild metric. We highlight the quick answer.

We determine γ along the worldcurve X of an undeflected photon in the (x^1, x^2) plane at the fixed distance h from x^2-axis.

So, the undeflected worldcurve of the photon is $X(t) := (ct, h, ct, 0)$. The deflection will add only lower order terms, therefore the deflected photon has in the same plane extra terms of order $O\left(\dfrac{1}{c} \right)$. Then, the deflected photon is parameterized by

$$X_d(t) := \left(ct, h + O\left(\frac{1}{c}\right), ct + O(1), 0 \right).$$

Since $\dot{X}_d(t) := \left(c, O\left(\frac{1}{c}\right), c + O(1), 0 \right)$ we have

$$\frac{dx^1}{dt} = O\left(\frac{1}{c}\right), \quad \frac{dx^2}{dt} = c + O(1), \quad \frac{dx^3}{dt} = 0.$$

It results the approximation

$$\gamma^2 = c^2 - \frac{2\mu}{c^2 r}\left(c^2 + \frac{(x^2)^2}{r^2} \cdot c^2 \right),$$

equivalent to

$$\gamma = c \cdot \sqrt{1 - \frac{2\mu}{c^2 r} - \frac{2\mu(x^2)^2}{c^2 r^3}}.$$

Taking into account

$$\sqrt{1 + 2A} \approx 1 + A,$$

the final result is

$$\gamma = c - \frac{\mu}{cr} - \frac{\mu}{cr} \cdot \frac{(x^2)^2}{r^2}.$$

The total deflection is computed as in the Schwarzschild case. Therefore, we succeeded to prove

Theorem 9.8.2 (Einstein's Second Theorem) *Consider the Einstein metric*

$$ds^2 = c^2 \left(1 - \frac{2\mu}{c^2 r} \right) dt^2 - \left(1 + \frac{2\mu}{c^2 r} \right) dr^2 - r^2 d\varphi^2 - r^2 \sin^2 \varphi d\theta^2.$$

(i) In Carthesian coordinates, the above metric has the form

$$ds^2 = \sum_{i=0}^{3} \left(dx^i \right)^2 - \frac{2\mu}{c^2 r}\left((dx^0)^2 + \sum_{\alpha,\beta=1}^{3} \frac{x^\alpha x^\beta}{r^2} dx^\alpha dx^\beta \right).$$

(ii) A deflected photon in the (x^1, x^2) plane, which comes from the undeflected photon $X(t) = (ct, h, ct, 0)$, has the speed

$$\gamma = c - \frac{\mu}{cr} - \frac{\mu}{cr} \cdot \frac{(x^2)^2}{r^2}$$

and does not violate the speed of light limit.

(iii) The total deflection of the trajectory $X_d(t)$ of a deflected photon in the grav-itational field, described by the Einstein metric, is $\dfrac{4GM}{c^2 h}$.

All the computations made in Einstein's metric lead to the same results in Schwarzschild's metric. According to these considerations, we can say that the Schwarzschild metric reduces to the Einstein metric in the weak field limit.

9.9   Solutions of General Einstein's Field Equations: The Friedmann–Lemaître–Robertson–Walker Models of Universe

If we intend to find a metric for the general Einstein field equations describing the Universe, we have to consider the fact that the observed Universe appears homogeneous and isotropic beyond a given scale according to the Cosmological Principle, therefore we have to consider, at the beginning, a spherical symmetry for the cosmic spacetime.[2]

The spatial part has to be as

$$dr^2 + q^2(r)\left(d\theta^2 + \sin^2\theta d\phi^2\right),$$

where $q(r)$ will be determined, so, we can try with the metric

$$ds^2 = dt^2 - a^2(t)\left[dr^2 + q^2(r)d\theta^2 + q^2(r)\sin^2\theta d\phi^2\right]$$

which introduces a new function $a(t)$ necessary to preserve the spherical symmetry of the spatial part of the metric which can, eventually, expand under a homothetic transformation In this way, $a(t)$ becomes an expansion factor of the Universe. We will discuss this fact a little bit later. Observe that we are working in geometric coordinates, that is $c = 1$.

Let us search for $a(t)$ and $q(r)$ such that the previous metric satisfies the Einstein field equations. To address the answer, there are three possible forms for $q(r)$ depending on a constant of integration, while $a(t)$ is determined from Einstein's field equations. We prove

---

[2]From observational surveys, the Universe can be considered homogeneous and isotropic beyond scales of the order 100–120 Megaparsecs. See [62] for details. This means that, over these scales, no large scale structure, like clusters or super-clusters of galaxies are detected. According to these data, matter density can be considered homogeneously distributed in all directions.

Theorem 9.9.1 *The following three metrics*

$$ds^2 = dt^2 - a^2(t)\left[dr^2 + R^2 \sinh^2 \frac{r}{R}d\theta^2 + R^2 \sinh^2 \frac{r}{R} \sin^2 \theta d\phi^2\right],$$

$$ds^2 = dt^2 - a^2(t)\left[dr^2 + R^2 \sin^2 \frac{r}{R}d\theta^2 + R^2 \sin^2 \frac{r}{R} \sin^2 \theta d\phi^2\right],$$

$$ds^2 = dt^2 - a^2(t)\left[dr^2 + r^2 d\theta^2 + r^2 \sin^2 \theta d\phi^2\right],$$

satisfy the Einstein field equations

$$R_{ij} - \frac{1}{2}Rg_{ij} = KT_{ij}$$

in the case when the contravariant energy-momentum tensor is describing a perfect fluid with components

$$T^{ij} = (\rho_0 + p_0)u^i u^j - p_0 g^{ij},$$

where g^{ij} are the inverse components of the metric tensor matrix which satisfies Einstein's field equations, ρ_0 is the density of the fluid, p_0 is the pressure of the fluid and u^i are the components $(u^t, v_x u^t, v_y u^t, v_z u^t)$ of the fluid 4-velocity. For the moment, ρ_0 and p_0 are assumed constant.

Proof We start by calculating the Ricci symbols.

$$g_{00} = 1, \ g_{11} = -a^2(t), \ g_{22} = -a^2(t)q^2(r), \ g_{33} = -a^2(t)q^2(r)\sin^2\theta$$

$$g^{00} = 1, \ g^{11} = -\frac{1}{a^2(t)}, \ g^{22} = -\frac{1}{a^2(t)q^2(r)}, \ g^{33} = -\frac{1}{a^2(t)q^2(r)\sin^2\theta}.$$

We observe

$$\Gamma^i_{jk} = g^{is}\Gamma_{jk,s} = g^{ii}\Gamma_{jk,i}; \ \Gamma^i_{jk} = 0, \ i \neq j \neq k.$$

Therefore

$$\Gamma^0_{11} = a \cdot \dot{a}, \ \Gamma^0_{22} = a \cdot \dot{a} \cdot q^2, \ \Gamma^0_{33} = a \cdot \dot{a} \cdot q^2 \cdot \sin^2\theta, \text{ otherwise } \Gamma^0_{ij} = 0,$$

$$\Gamma^1_{01} = \Gamma^1_{10} = \frac{\dot{a}}{a}, \ \Gamma^1_{22} = -q \cdot q', \ \Gamma^1_{33} = -q \cdot q' \cdot \sin^2\theta, \text{ otherwise } \Gamma^1_{ij} = 0,$$

$$\Gamma^2_{02} = \Gamma^2_{20} = \frac{\dot{a}}{a}, \ \Gamma^2_{12} = \Gamma^2_{21} = \frac{q'}{q}, \ \Gamma^2_{33} = -\sin\theta\cos\theta, \text{ otherwise } \Gamma^2_{ij} = 0,$$

$$\Gamma^3_{03} = \Gamma^3_{30} = \frac{\dot{a}}{a}, \ \Gamma^3_{13} = \Gamma^3_{31} = \frac{q'}{q}, \ \Gamma^3_{32} = \Gamma^3_{23} = -\cot\theta, \text{ otherwise } \Gamma^3_{ij} = 0.$$

If we compute

$$R_{00} = R^s_{0s0} = -\frac{\partial \Gamma^1_{01}}{\partial t} - \frac{\partial \Gamma^2_{02}}{\partial t} - \frac{\partial \Gamma^3_{03}}{\partial t} - \Gamma^1_{01}\Gamma^1_{10} - \Gamma^2_{02}\Gamma^2_{20} - \Gamma^3_{03}\Gamma^3_{30},$$

it results

$$R_{tt} = R_{00} = -3\frac{\ddot{a}}{a}.$$

We obtain

$$R_{rr} = R_{11} = 2\dot{a}^2 + \ddot{a} \cdot a - 2\frac{q''}{q},$$

$$R_{\theta\theta} = R_{22} = 2q^2 \cdot \dot{a}^2 + q^2 \cdot a \cdot \ddot{a} - q \cdot q'' + 1 - (q')^2,$$

$$R_{\phi\phi} = R_{33} = R_{22} \sin^2 \theta.$$

Using $R^i_j = g^{is} R_{sj}$, we rise an index, therefore

$$R^t_t = -3\frac{\ddot{a}}{a},$$

$$R^r_r = -2\frac{\dot{a}^2}{a^2} - \frac{\ddot{a}}{a} + 2\frac{q''}{a^2 \cdot q},$$

$$R^\theta_\theta = -2\frac{\dot{a}^2}{a^2} - \frac{\ddot{a}}{a} - \frac{1}{a^2 \cdot q^2}\left(1 - q \cdot q'' - (q')^2\right),$$

$$R^\phi_\phi = -2\frac{\dot{a}^2}{a^2} - \frac{\ddot{a}}{a} - \frac{1}{a^2 \cdot q^2}\left(1 - q \cdot q'' - (q')^2\right) = R^\theta_\theta.$$

The key of finding the metric is related to the way the physicists describe the energy-momentum tensor. They look at the galaxies in the Universe such that they are imagined as the molecules of an ideal gas which move arbitrarily. In this case, the gas is described as in the statement of the theorem, by the contravariant energy-momentum tensor

$$T^{ij} = (\rho_0 + p_0)u^i u^j - p_0 g^{ij},$$

where

- g^{ij} are the inverse components of the metric tensor matrix which satisfies Einstein's field equations $R_{ij} - \frac{1}{2}Rg_{ij} = KT_{ij}$,
- ρ_0 is the density of the gas,
- p_0 is the constant pressure of the gas and
- u^i are the components $(u^t, v_x u^t, v_y u^t, v_z u^t)$ of the gas 4-velocity.

It is convenient to use the $(1, 1)$ tensor T^i_j by lowering the second index, so

$$T^i_j = (\rho_0 + p_0)u^i g_{jk}u^k - p_0\delta^i_j.$$

Our chosen metric has $g_{tt} = 1$.

This ideal fluid is, by definition, at rest in these comoving coordinates, therefore the conditions $u^r = u^\theta = u^\phi = 0$ for every t and $u^i g_{ij} u^j = 1$ lead to $1 = g_{tt}(u^t)^2$, that is $u^t = 1$.

It follows that $T_t^t = \rho_0$ and $T_r^r = T_\theta^\theta = T_\phi^\phi = -p_0$, that is $T = T_i^i = \rho_0 - 3p_0$. If we arrange the Einstein's field equation in the form

$$R_j^i = K \cdot \left(T_j^i - \frac{1}{2}\delta_j^i T \right)$$

we obtain $R_r^r = R_\theta^\theta = R_\phi^\phi = -\dfrac{K}{2}(\rho_0 - p_0) \cdot$

The condition $R_r^r = R_\theta^\theta$ highlights the equality

$$-2\frac{\dot a^2}{a^2} - \frac{\ddot a}{a} + 2\frac{q''}{a^2 \cdot q} = -2\frac{\dot a^2}{a^2} - \frac{\ddot a}{a} - \frac{1}{a^2 \cdot q^2}\left(1 - q \cdot q'' - (q')^2\right)$$

and it remains to solve the differential equation

$$(q')^2 - q \cdot q'' = 1.$$

I. Determining $q(r)$.

From the beginning, we observe that $q(r) = r$ is a possible solution. We continue: for $p := q' = \dfrac{dq}{dr}$ we obtain $q'' = \dfrac{dp}{dr} = \dfrac{dp}{dq}\dfrac{dq}{dr} = \dfrac{dp}{dq}p$.

The differential equation transforms to $p^2 - qp\dfrac{dp}{dq} = 1$, that is

$$2\frac{dq}{q} = \frac{2p\,dp}{p^2 - 1}.$$

The solution written as $2\ln|q| = \ln q^2 = \ln|p^2 - 1| - \ln|k|$, leads first to $q^2 = \dfrac{p^2 - 1}{k}$, then, after replacing in $(q')^2 - q \cdot q'' = 1$, to $q'' = kq$. It results

$$\left(\frac{dq}{dr}\right)^2 = (q')^2 = 1 + q \cdot q'' = 1 + kq^2.$$

Since in the metric appears q^2, we are not interested in the solutions with minus. Without loosing the generality, we can suppose that $q(0) = 0$ and $q'(0) = 1$. Therefore, having these initial conditions, we have to solve

$$\frac{dq}{\sqrt{1 + kq^2}} = dr.$$

Case $k > 0$. We choose $k = \dfrac{1}{R^2}$. We have

$$r = \int \frac{1}{\sqrt{1 + \left(\dfrac{q}{R}\right)^2}} dq = R \sinh^{-1} \frac{q}{R},$$

that is $q = R \sinh \dfrac{r}{R}$. In this case the metric is

$$ds^2 = dt^2 - a^2(t) \left[dr^2 + R^2 \sinh^2 \frac{r}{R} d\theta^2 + R^2 \sinh^2 \frac{r}{R} \sin^2 \theta d\phi^2 \right].$$

Case $k < 0$. We choose $k = -\dfrac{1}{R^2}$. We have

$$r = \int \frac{1}{\sqrt{1 - \left(\dfrac{q}{R}\right)^2}} dq = R \arcsin \frac{q}{R},$$

that is $q = R \sin \dfrac{r}{R}$. In this case the metric is

$$ds^2 = dt^2 - a^2(t) \left[dr^2 + R^2 \sin^2 \frac{r}{R} d\theta^2 + R^2 \sin^2 \frac{r}{R} \sin^2 \theta d\phi^2 \right].$$

For $q(r) = r$ the metric is

$$ds^2 = dt^2 - a^2(t) \left[dr^2 + r^2 d\theta^2 + r^2 \sin^2 \theta d\phi^2 \right].$$

□

Let us now determine $a(t)$. To proceed, we consider again the above field equations:

$$R_t^t = -3\frac{\ddot{a}}{a} = K \cdot \left(T_t^t - \frac{1}{2} T \right) = \frac{K}{2} \cdot (\rho_0 + 3 p_0)$$

$$R_r^r = -2\frac{\dot{a}^2}{a^2} - \frac{\ddot{a}}{a} + 2\frac{q''}{a^2 \cdot q} = K \cdot \left(T_\theta^\theta - \frac{1}{2} T \right) = -\frac{K}{2} (\rho_0 - p_0)$$

Since $q'' = kq$, in the case when $k = \pm\dfrac{1}{R^2}$, the following two equations have to be considered:

$$\frac{\ddot{a}}{a} = -\frac{K}{6} \cdot (\rho_0 + 3 p_0),$$

$$2\frac{\dot{a}^2}{a^2} + \frac{\ddot{a}}{a} - 2\frac{k}{a^2} = \frac{K}{2}(\rho_0 - p_0).$$

It results the equation

$$\frac{\dot{a}^2}{a^2} - \frac{k}{a^2} = \frac{K}{3}\rho_0$$

that is

$$\dot{a}^2 - \frac{K}{3}\rho_0 \cdot a^2 = k,$$

which can be solved. The equation can be written as

$$\dot{a}^2 - B \cdot a^2 = k$$

where $B = \frac{K}{3}\rho_0 > 0$. Since in metric appears a^2, as in the case of $q(r)$, we are not interested in the solutions with minus. Furthermore, we are not interested in using constants which can be eliminated by a convenient change of coordinates.

In the case $k = \frac{1}{R^2} > 0$, if we arrange the equation in the form

$$\frac{1}{k}\dot{a}^2 - \frac{B}{k} \cdot a^2 = 1$$

the solution is $a(t) = \frac{1}{R\sqrt{B}}\sinh(t\sqrt{B})$. Replacing B, it results

$$a(t) = \frac{1}{R \cdot \sqrt{\frac{K}{3}\rho_0}}\sinh\left(t\sqrt{\frac{K}{3}\rho_0}\right).$$

In the case $k = -\frac{1}{R^2} < 0$, if we arrange the equation in the form

$$-R^2\dot{a}^2 + R^2 B \cdot a^2 = 1$$

the solution is $a(t) = \frac{1}{R\sqrt{B}}\cosh(t\sqrt{B})$. After replacing B,

$$a(t) = \frac{1}{R \cdot \sqrt{\frac{K}{3}\rho_0}}\cosh\left(t\sqrt{\frac{K}{3}\rho_0}\right).$$

If $q(r) = r$ it results $q''(r) = 0$. Let us consider the first two equations:

$$\frac{\ddot{a}}{a} = -\frac{K}{6} \cdot (\rho_0 + 3p_0),$$

$$2\frac{\dot{a}^2}{a^2} + \frac{\ddot{a}}{a} = \frac{K}{2}(\rho_0 - p_0).$$

This kind of differential equations in $a(t)$ are called the $(FLRW)$ equations. We obtain

$$\frac{\dot{a}^2}{a^2} = \frac{K}{3}\rho_0.$$

Taking into account our notation $B = \dfrac{K}{3}\rho_0$, two solutions are possible: $a_1(t) = e^{t\sqrt{B}}$ and $a_2(t) = e^{-t\sqrt{B}}$.

We may observe that as $t \to +\infty$, $a_2(t) \to 0$ which does not correspond to the known expansion of the Universe, related to the observational evidences. The other solution can be accepted. As we will see in the following subsection, it is related to the Hubble constant.

Let us stress again that these metrics have been obtained in the case when T_{ij} has the above special form. We may conceive other possible T_{ij} having the property $T_r^r = T_\theta^\theta = T_\phi^\phi$ and some other metrics can appear.   □

Denote $d\Omega^2 := d\theta^2 + \sin^2\theta d\phi^2$. In the process of finding the metric, we have used

$$\frac{dq^2}{1 + kq^2} = dr^2.$$

Replace this position in the metric

$$ds^2 = dt^2 - a^2(t)\left[dr^2 + q^2 d\theta^2 + q^2 \sin^2\theta d\phi^2\right],$$

we can write all possible metrics in the form

$$ds^2 = dt^2 - a^2(t)\left[\frac{dq^2}{1 + kq^2} + q^2 d\Omega^2\right].$$

This metric is known as the *Friedman–Lemaître–Robertson–Walker metric (or FLRW metric) of the Universe.*

Problem 9.9.2 Consider the case of the cosmological fluid such that the contravariant energy-momentum tensor is

$$T^{ij} = (\rho_0 + p_0)u^i u^j - p_0 g^{ij} + \frac{\Lambda}{K}g^{ij},$$

where Λ is the cosmological constant. Under the conditions of the previous theorem, let us find the coefficients if the metric for Universe is

$$ds^2 = c^2 dt^2 - a^2(t) \left[dr^2 + q^2(r) d\theta^2 + q^2(r) \sin^2 \theta d\phi^2 \right].$$

Hint. We obtain $T_t^t = \rho_0 + \dfrac{\Lambda}{K}$, $T_r^r = T_\theta^\theta = T_\phi^\phi = -\rho_0 + \dfrac{\Lambda}{K}$ and $T = \rho_0 - 3\rho_0 + \dfrac{4\Lambda}{K}$, but we have to complete the computations using

$$R_j^i - \Lambda \delta_j^i = K \cdot \left(T_j^i - \frac{1}{2} \delta_j^i T \right).$$

Problem 9.9.3 Consider the metric

$$ds^2 = \alpha(x + y + z) dt^2 - \frac{1}{2} \left(dx^2 + dy^2 + dz^2 \right),$$

where α is a constant. Compute $R_{ij} - \dfrac{1}{2} R\, g_{ij}$.

Hint. Denote $x^0 := t$, $x^1 := x$, $x^2 := y$, $x^3 := z$. It is easy to obtain $\Gamma_{10}^0 = \Gamma_{01}^0 = \Gamma_{20}^0 = \Gamma_{02}^0 = \Gamma_{30}^0 = \Gamma_{03}^0 = \dfrac{1}{2(x + y + z)}$ and $\Gamma_{00}^1 = \Gamma_{00}^2 = \Gamma_{00}^3 = \alpha$. Then

$$R_{ii} = -\frac{3\alpha}{2(x + y + z)}, \quad R_{00} = \frac{\alpha}{4(x + y + z)^2}, \quad i = 1, 2, 3,$$

that is

$$R = R_i^i = -3 \frac{\alpha}{(x + y + z)^2}.$$

It results

$$R_{ij} - \frac{1}{2} R\, g_{ij} = -\frac{1}{2} \begin{pmatrix} 0 & 0 & 0 & 0 \\ 0 & (x+y+z)^{-2} & 0 & 0 \\ 0 & 0 & (x+y+z)^{-2} & 0 \\ 0 & 0 & 0 & (x+y+z)^{-2} \end{pmatrix}.$$

Can you derive some conclusions about T_{ij} tensor?

The equations of geodesics are

$$\begin{cases} \dfrac{d^2t}{d\tau^2} + \dfrac{3}{x+y+z}\dfrac{dt}{d\tau}\left(\dfrac{dx}{d\tau} + \dfrac{dy}{d\tau} + \dfrac{dz}{d\tau}\right) = 0 \\[4mm] \dfrac{d^2x}{d\tau^2} + \alpha\left(\dfrac{dt}{d\tau}\right)^2 = 0 \\[4mm] \dfrac{d^2y}{d\tau^2} + \alpha\left(\dfrac{dt}{d\tau}\right)^2 = 0 \\[4mm] \dfrac{d^2z}{d\tau^2} + \alpha\left(\dfrac{dt}{d\tau}\right)^2 = 0 \end{cases}$$

A important question for the reader is:

Exercise 9.9.4 Can these equations be the geodesic equations of the classical constant gravitational field $(-\alpha, -\alpha, -\alpha)$?

Hint. Start by analyzing the necessary condition $\dfrac{dt}{d\tau} = 1$ and the norm (with respect the metric) of the tangent vector to the geodesic.

9.9.1 The Cosmological Expansion

This subsection is dedicated to the expansion of the Universe. We saw that Einstein Static Universe imposed the existence of a new term in the fields equations, because in a Universe in which the matter is constrained to interact only by gravity, all the matter sources will be concentrated in the same region, in contrast with the desired Einstein static structure.

The new term was proposed to establish a repulsive effect to counterweight the attractive effect of gravity. However, Einstein discarded the cosmological term when Hubble discovered evidences for cosmological expansion. In any case, Theorem 9.9.1 suggests that we can obtain an expanding universe even if the cosmological constant is not considered. Is Hubble's law related to the cosmological metrics obtained above? The answer is yes!

Let us describe the Hubble law for recession of galaxies.

First, we have to mention that Hubble used Doppler's effect to establish his result related to the redshift of distant galaxies. The light in the Universe is produced by stars. The hydrogen of stars, in thermonuclear fusion, produces primarily helium and energy that radiates in space, some of it in form of light. Hubble considered the four lines of the hydrogen light spectrum. For distant galaxies, the same four lines of hydrogen spectrum are seen shifted to the right in comparison to normal pattern of light decomposition detected in laboratory. Hubble realized that this is a Doppler effect and the observed redshift means that the distant galaxies are moving away from us. He stated that the redshifts in spectra of distant galaxies are proportional to the distance of galaxies from us. The mathematical form is

$$V = H \cdot D,$$

where D is the proper distance from us to the galaxy, $V := \dot{D}$ is the proper speed of the galaxy and H is a constant called the Hubble constant. The farther away the galaxy is, the faster it moves away from us. The entire space, the entire texture of the Universe is moving away from us carrying the galaxies in it.

Alternatively, let us suppose we have a ruler of coordinates marked 0, 1, 2, 3, 4, ... The distance between two consecutive coordinates is denoted by a. The distance measured with this ruler is denoted by D. $D = a \cdot \Delta x$, where Δx is the difference between the coordinates of the chosen points we wish to measure.

Now, suppose we have a rubber band marked in the same way as our ruler; we pin the origin and start to stretch. The coordinate points remain drawn on the rubber band but the distance between them increases. Therefore a depends on time, it is $a(t)$. The distance D becomes $D(t) = \Delta x \cdot a(t)$. We have

$$V := \dot{D} = \Delta x \cdot \dot{a}.$$

This relation can be written as

$$V \cdot a = a \cdot \Delta x \cdot \dot{a} = D \cdot \dot{a},$$

that is

$$V = \frac{\dot{a}(t)}{a(t)} D.$$

We define $H := \dfrac{\dot{a}(t)}{a(t)}$ and obtain Hubble's law

$$V = H \cdot D.$$

What is new in this approach is the fact that it is suggested the stretch of the texture of the universe. Such a stretch was seen in Sect. 9.9 when we discussed about a possible metric for the Cosmos. The metric proposed was

$$ds^2 = c^2 dt^2 - a^2(t) \left[dr^2 + q^2(r)d\theta^2 + q^2(r)\sin^2\theta d\phi^2 \right],$$

where $q(t)$ and $a(t)$ were determined from the general Einstein field equations under some conditions imposed by the energy-momentum tensor T_{ij}.

The differential equation for $a(t)$ will be found now under some physical conditions and important consequences will come out.

Consider two galaxies in the Universe such that the distance between them is D. Let us consider one of them and the sphere of radius D centered at the chosen galaxy.

Denote by M the total mass of galaxies inside the sphere and by m the mass of the second galaxy. This galaxy moves away from the galaxy at the center of the sphere with speed $V = H \cdot D$. The gravitational force which acts on the galaxy of mass m is

$$F = \frac{GMm}{D^2}.$$

The potential energy for that galaxy is

$$PE = -\frac{GMm}{D}$$

and the kinetic energy is

$$KE = \frac{mV^2}{2}.$$

The total energy acting on the second galaxy is a constant,

$$PE + KE = const = k_1.$$

Thanks to the Equivalence Principle, we can divide by m and then it results

$$-\frac{2GM}{D} + V^2 = k$$

But $D(t) = \Delta x \cdot a(t)$ and $V(t) = \Delta x \cdot \dot{a}(t)$, that is

$$(\Delta x \cdot \dot{a}(t))^2 - \frac{2GM}{\Delta x \cdot a(t)} = k.$$

Some remarks are in order now. $M = Vol \times density$. If the volume increases when the Universe is expanding but the number of galaxies does not change, the density decreases. Since

$$M = \frac{4}{3}\pi \cdot D^3 \cdot \rho(t) = \frac{4}{3}\pi \cdot (\Delta x \cdot a(t))^3 \cdot \rho(t)$$

we have

$$(\Delta x)^2 \cdot (\dot{a}(t))^2 - \frac{8\pi G}{3} \cdot (\Delta x)^2 \cdot (a(t))^2 \cdot \rho(t) = k.$$

We arrange in a dimensional way the previous formula replacing k by $-K \cdot \Delta x$. Finally, we obtain the differential equation

$$\left(\frac{\dot{a}(t)}{a(t)}\right)^2 - \frac{8\pi G}{3} \cdot \rho(t) = -\frac{K}{a^2(t)}.$$

This is a sort of Friedman–Lemaître–Robertson–Walker equation as obtained in the previous Sect. 9.9. The term $\dfrac{8\pi G}{3} \cdot \rho(t)$ is always positive. If K is negative, the equation written in the form

$$\left(\frac{\dot{a}(t)}{a(t)}\right)^2 = \frac{8\pi G}{3} \cdot \rho(t) - \frac{K}{a^2(t)}$$

can be solved. Such an equation describes a spatially open Universe. If, for some t, K is such that the right member becomes negative at a point, this Universe will increase until that point; then it can remain unchanged or even it can contract. This kind of Universe is called a spatially closed universe. If $K = 0$, the universe will be called a spatially flat Universe. This Universe expands too. In such a Universe there is a perfect balance between the kinetic and the potential energy.

The observational evidences show that our Universe is a flat one. So, it remains to solve the equation

$$\left(\frac{\dot{a}(t)}{a(t)}\right)^2 = \frac{8\pi G}{3} \cdot \rho(t).$$

In a flat, matter dominated Universe (mdu), in a cube of side $a(t)$, having inside galaxies whose total mass is M, the density is expressed by the formula $\rho_{mdu}(t) = \dfrac{M}{a^3(t)}$. The corresponding ($FLRW$) equations is

$$\left(\frac{\dot{a}(t)}{a(t)}\right)^2 = \frac{8\pi G}{3} \cdot \frac{M}{a^3(t)}.$$

The solution, expressing the expansion of a matter dominated Universe, is then

$$a(t) = B \cdot t^{2/3},$$

where B is a positive constant.

After the Big-Bang and inflation [62], there was a period when the universe was radiation dominated (rdu). To describe its expansion, we consider the same cube of side $a(t)$, now full of photons. Since the energy is expressed by the formula $E = h\nu = h\dfrac{c}{\lambda}$ and, when $a(t)$ is increasing, the wavelength λ is increasing too, we can suppose that $E = \dfrac{C}{a(t)}$ is describing the energy formula. Here C is a constant.
The density of such a Universe is given by

$$\rho_{rdu} = \frac{E}{a^3(t)} = \frac{C}{a^4(t)}.$$

The corresponding $(FLRW)$ equation is

$$\left(\frac{\dot{a}(t)}{a(t)}\right)^2 = \frac{8\pi G}{3} \cdot \frac{C}{a^4(t)}.$$

The solution, which expresses the expansion of a radiation dominated universe, is

$$a(t) = A \cdot t^{1/2},$$

where A is a positive constant.

Now, let us observe something crucial. We know two important physic formulas, Planck's one $E = h\nu = h\frac{c}{\lambda}$ and Boltzmann's one, $E = k_B T$. Using the same reasoning as before we deduce the direct proportionality between the temperature T and $\frac{1}{a(t)}$. As said, after Big Bang and inflation, our Universe was radiation dominated, and the temperature at which the atoms can form is less than 3×10^3 K degrees. Now, the today cosmic background radiation has approximatively 3 K degrees. Therefore, if we suppose that in the period in our Universe started to be matter dominated, the temperature decreases of the ratio $\sim \frac{3 \times 10^3}{3}$, it is easy to see the ratio $\frac{a_{today}}{a_{ionized}}$, where $a_{ionized}$ is the epoch in which ionized atoms appear. It is

$$\frac{T_{ionized}}{T_{today}} = \frac{a_{today}}{a_{ionized}} \simeq 10^3 \simeq \frac{t_{today}^{2/3}}{t_{ionized}^{2/3}}.$$

Since t_{today} is about 10^{10} years, i.e. the age of the observed Universe, $t_{ionized}$ becomes about 3×10^5 years after Big Bang. It means that the Universe was radiation dominated for almost 3×10^5 years. More precisely, it takes about 3×10^5 years for the Universe, expanding and cooling after Big Bang, to allow electrons and protons to couple and form neutral atoms. At this point, even the photons are free to move and get to us, providing us with the first "photograph" of the Universe that can be obtained, that is the Cosmic Microwave Background Radiation.[3] Clearly, this is only a rough calculation to derive the order of magnitudes. For a detailed discussion on primordial Universe phenomenology, see [62, 64].

We are now ready to understand some basic facts about dark energy and the pressure exerted to expand our Universe. Specifically, dark energy is the hypothetical fluid fueling the observed accelerated expansion revealed at the end of XXth century [65]. Let us begin by analyzing the pressure exerted on the faces of a cube imagined in our Universe. Obviously, there is no pressure in a matter dominated Universe because the galaxies inside the cube do not exert any pressure on the faces of the cube.

---

[3] Actually the recombination of hydrogen happened at a redshift $z = 1089$ corresponding to a period of 3.79×10^5 years after Big Bang. Here the redshift correspond to the above $a_{today}/a_{ionized}$. See [63].

In a radiation dominated universe, it is possible to study the pressure in the following way. Let us consider a photon which can move between "the extremities" of a segment line of length L. The small amount of time necessary to move between the extremities can be denoted as dt and we have the formula $dt = \dfrac{2L}{c}$. The force which produces the pressure on the extremities is

$$F = \frac{dp}{dt} = \frac{2p}{\dfrac{2L}{c}} = \frac{pc}{L} = \frac{E}{L}.$$

If we denote by L the length of the side of a cube in a radiation dominated Universe and by dA the infinitesimal area of a square drawn on a face (the sides parallel to the sides of the the face), now corresponding to the perpendicular direction on the given face, we have

$$P = \frac{F}{dA} = \frac{E}{L\,dA}.$$

Therefore the pressure P exerted can be though as the ratio between the energy and a volume corresponding to dA and the above mentioned perpendicular direction, that is an energy density ρ. In fact we have

$$P = w\rho,$$

where $w = 0$ in the case of matter dominated Universe and $w = \dfrac{1}{3}$ in an radiation dominated Universe; 3 appears because we have three perpendicular direction on faces.

These numbers represent two possibilities for the equation of state of a standard perfect fluid where $0 \leq w \leq 1$ is the so-called Zel'dovich interval [66]. Being $w = \left(\frac{c_s}{c}\right)^2$, with c_s the sound speed, the fluids in the Zel'dovich interval agree with the the causality condition implying that the speed of light has to be $c > c_s$. In other words, standard matter cannot be constituted by tachyons, that is particles faster than light.

Suppose now that the pressure expands the cosmic cube of a dV volume. Taking into account the work done by the force F, $F \cdot d = P \cdot A \cdot d = P \cdot dV$, and the variation of the energy E, we have

$$dE = -P \cdot dV.$$

At the same time,

$$E = \rho \cdot V.$$

It results

$$dE = d\rho \cdot V + \rho \cdot dV = -P \cdot dV,$$

i.e.

$$V \cdot d\rho = -(P + w)dV = -\rho(w + 1)dV.$$

We have obtained the differential equation

$$\frac{d\rho}{\rho} = -(w + 1)\frac{dV}{V}$$

with the solution

$$\rho = NV^{-(w+1)} = Na^{-3(w+1)},$$

where N is a constant.

For $w = 0$, we obtain the formula corresponding to a matter dominated Universe, while for $w = \frac{1}{3}$ we obtain the formula of a radiation dominated Universe.

Let us insert this last formula in the $(FLRW)$ equation, we get

$$\left(\frac{\dot{a}(t)}{a(t)}\right)^2 = \frac{8\pi G}{3} \cdot \frac{N}{(a(t))^{3(w+1)}}.$$

Clearly, in the above discussion, the functions P and ρ are functions of time and the above definition of the energy-momentum tensor can be generalized to describe a perfect fluid of the form

$$T^{ij} = (\rho + P)u^i u^j - pg^{ij}.$$

What happens if $w = -1$? The $(FLRW)$ equation becomes

$$\left(\frac{\dot{a}(t)}{a(t)}\right)^2 = \frac{8\pi G}{3} \cdot \rho_0,$$

where ρ_0 is a constant. Now we are in the case of a Universe expanding according to the law

$$\frac{\dot{a}(t)}{a(t)} = H_0 = \sqrt{\frac{8\pi G \rho_0}{3}}.$$

The solution of the expansion is exponential,[4] that is

$$a(t) = a_0 e^{H_0 t},$$

where a_0 is a constant related to the initial value of the scale factor $a(t)$. The value $w = -1$ is clearly out of the above Zel'dovich interval, i.e. it is not a standard perfect fluid, and corresponds to "something" which determines the exponential accelerated

---

[4] H_0 is assumed constant because ρ_0 is constant.

expansion. Such an expansion is in agreement with the existence of a possible cosmological constant Λ (that is ρ_0). This "something" manifests itself as a pressure implying an energy density. As said, this energy is neither produced by the ordinary matter nor by the radiation.[5] This is a simple example of *dark energy* that gives rise to accelerated expansion. The mechanism can work both in early Universe, giving rise to inflation, and in late Universe, giving the observed accelerated expansion of the Hubble flow. Clearly the scales of energy are completely different and between inflation and recent accelerated epoch there are radiation and matter dominated eras. It is worth noticing that, according to data, the dark energy constitutes $\sim$70% of the total amount of matter-energy content of the Universe [65]. Understanding nature and dynamics of dark energy is one of the main challenges of modern cosmology.

9.10 The Fermi Coordinates

After the above summary on cosmological expansion, let us define a system of coordinates very useful to describe the geodesic motion. From a mathematical point of view, Fermi's coordinates are *local coordinates adapted to a geodesic*, that is, at a given point P on a geodesic $c(\tau)$, there exists a local system of coordinates around P such that:

- the geodesic locally becomes $(x^0, 0, 0, \ldots, 0)$;
- the metric tensor along geodesic is the Minkowski metric (or the Euclidean metric; it depends on the context);
- all the Christoffel symbols vanish along geodesic.

A nice treatment of this subject[6] and its applications can be seen in [4, 33]. In our context, we intend to describe the topic in a simplified way.

Consider a coordinate frame at rest denoted by $R : (y^0, y^1, y^2, y^3)$ together with a given metric $ds^2 = \bar{g}_{ij} dy^i dy^j$.

We intend to describe the free fall of an observer F in the gravitational field induced by $\bar{g}_{ij}$.

1. In the coordinate frame at rest, R, the freely falling observer F is moving on a geodesic of the metric $ds^2 = \bar{g}_{ij} dy^i dy^j$, say $c(\tau)$.

The geodesic equations of $c(\tau)$ are
$$\frac{d^2 y^i}{d\tau^2} + \bar{\Gamma}^i_{jk} \frac{dy^j}{d\tau} \frac{dy^k}{d\tau} = 0.$$
This geodesic is the world line of F in R.

2. From F point of view, there is no field. Consider F in a spacecraft, somewhere in an almost empty region of the space. That is, to describe the free falling, means to create a coordinate frame $F : (x^0, x^1, x^2, x^3)$ such that, along the world line of F

---

[5]It is important to note that any form of standard matter, in the interval $0 \le w \le 1$, gives rise to decelerated expansion.

[6]It is interesting saying that the paper reporting these results was the first one written by Enrico Fermi when he was student at Scuola Normale Superiore di Pisa [67].

in R in these coordinates, we have $\Gamma^i_{jk} = 0$. For F, the geodesic equations become $\dfrac{d^2 x^i}{d\tau^2} = 0$, that is F should move on a straight line.

We make the assumption: Let x^0 axis be the world line of F in R.

3. Now, more clearly, we have to construct a map $M : F \to R$ which transfers x^0 axis into the geodesic $c(\tau)$, in such a way that the x^0 axis becomes a geodesic in F endowed with the metric $g_{ij} = dM^t_x \cdot \bar{g}_{ij} \cdot dM_x$.

Therefore, M maps the x^0 axis into the the image of the geodesic $c(\tau)$.

If τ is the geodesic parameter for the curve $c(\tau)$, we can consider the same parameter for the x^0 axis, i.e. $x^0 = \tau$ is the current coordinate of this axis.

At each point τ, we have

$$c(\tau) = (y^0(\tau), y^1(\tau), y^2(\tau), y^3(\tau)) \in R,$$

therefore the map $M : F \to R$ gives rise to

$$(x^0, 0, 0, 0) \to (y^0(x^0), y^1(x^0), y^2(x^0), y^3(x^0)),$$

where $(y^0(x^0), y^1(x^0), y^2(x^0), y^3(x^0))$ are the coordinates of the points of the geodesic in R. Therefore our transformation $M : F \to R$ can be thought

$$(\tau, 0, 0, 0) \to c(\tau),$$

with some considerations on the functions y^k we need to describe.

Let us keep in our mind that we are interested in transferring the property *"c is a geodesic in R"* to the x^0 axis in F. So, we have

Lemma 9.10.1 *Along the geodesic c in R, it can be highlighted an orthonormal frame with respect the metric $\bar{g}_{ij}$ whose time-like vector is the tangent vector $\dfrac{dc}{d\tau}$.*

Proof We know that at each point $c(\tau)$, i.e. along the geodesic c, the tangent vector $\dfrac{dc}{d\tau} = \left(\dfrac{dy^0}{d\tau}, \dfrac{dy^1}{d\tau}, \dfrac{dy^2}{d\tau}, \dfrac{dy^3}{d\tau} \right)$ is a time-like unit vector. We denote it by $e_0(\tau)$. We know that $e_0(\tau)$ is parallel transported along the geodesic c in R preserving all its properties.

Consider the point corresponding to $\tau = 0$, that is the point $c(0)$ on the geodesic. We choose the spatial vectors $e_1(0)$, $e_2(0), e_3(0)$ such that the frame $\{e_0(0), e_1(0), e_2(0), e_3(0)\}$ is orthonormal with respect to the metric $\bar{g}_{ij}$ and we parallel transport it along the geodesic c.

At each point $c(\tau)$, the vectors $\{e_0(\tau), e_1(\tau), e_2(\tau), e_3(\tau)\}$ form an orthonormal frame with respect to the metric $\bar{g}_{ij}$. $\qquad \square$

Lemma 9.10.2 *Every point (x^0, x^1, x^2, x^3) in F can be uniquely described in the form $(\tau, l \vec{v})$, where $l \vec{v}$ is an appropriate Euclidean description of its spatial part.*

Proof Consider a point (x^0, x^1, x^2, x^3) in F which does not belong to x^0 axis. This point is (τ, x^1, x^2, x^3) and at least one spatial component is nonzero.

Denote

$$l := \sqrt{(x^1)^2 + (x^2)^2 + (x^3)^2}$$

and construct the vector

$$\vec{v} := \left(\frac{x^1}{l}, \frac{x^2}{l}, \frac{x^3}{l} \right) := (v^1, v^2, v^3).$$

It appears the possibility to describe the point (x^0, x^1, x^2, x^3) by (τ, lv^1, lv^2, lv^3) or simply, by $(\tau, l\,\vec{v})$. $\qquad\square$

9.10.1   Determining the Fermi Coordinates

Consider in $T_{c(\tau)}R$ the vector

$$\vec{V}(\tau) := v^1 e_1(\tau) + v^2 e_2(\tau) + v^3 e_3(\tau).$$

Observe

$$ds^2\left(\vec{V}(\tau), \vec{V}(\tau)\right) = \bar{g}_{\alpha\beta} v^\alpha v^\beta = -(v^1)^2 - (v^2)^2 - (v^3)^2.$$

We may impose $ds^2\left(\vec{V}(\tau), \vec{V}(\tau)\right) = -1$, that is $\vec{V}$ is a spatial unit vector. Let us observe that this spatial part has the same property $(v^1)^2 + (v^2)^2 + (v^3)^2 = 1$ as the vector $\vec{v}$ from the above lemma.

According to the equations of geodesics, we may conclude that it exists a unique geodesic of R, denoted by $y_{\vec{V}}(s)$ passing through the point $c(\tau)$ at $s = 0$, such that its tangent vector at origin is $\vec{V}$, that is $\dfrac{dy_{\vec{V}}}{ds}(0) = \vec{V}$.

According to the above notations, the local map $M : F \to R$, describing the Fermi coordinates, is

$$M(x^0, x^1, x^2, x^3) = M(\tau, s\,\vec{v}) := y_{\vec{V}}(s).$$

Observe that the tangent vector along the spatial geodesic $y_{\vec{V}}(s)$ is a unit vector.

The immediate consequence is: For a given point $(\tau, s_0\,\vec{v})$, the spatial distance to $(\tau, 0, 0, 0)$ is s_0. The length of the spatial geodesic between its initial point $c(\tau) = y_{\vec{V}}(0)$ and $y_{\vec{V}}(s_0)$ is also s_0, because the length formula is

$$\int_0^{s_0} \left\| \frac{d y_{\overrightarrow{V}}}{ds}(s) \right\| ds = \int_0^{s_0} ds = s_0.$$

The coordinates induced in F by M are called *Fermi's coordinates*.

It remains to prove that in F, in Fermi's coordinates, with respect to the induced metric g_{ij}, the x^0 axis $(\tau, 0, 0, 0)$ is a geodesic and $\Gamma^i_{jk}(\tau, 0, 0, 0) = 0$.

Let us discuss the consequences on the map M.

Theorem 9.10.3 *The map M in invertible in the neighborhood of each point $P(\tau, 0, 0, 0)$ of the x^0 axis.*

Proof According to the inverse function theorem, it is enough to prove that the matrix dM_P transforms a basis of the tangent space $T_P F$ into linear independent vectors of $T_{M(P)} R$.

We know

$$M(P) = M(\tau, 0, 0, 0) = c(\tau).$$

Consider the standard basis of $T_P F$, denoted by ε_i, $i \in \{0, 1, 2, 3\}$, ε_i having 1 on the ith row, 0 elsewhere. Therefore, by the way we defined M, $dM_P(\varepsilon_i) = e_i(\tau)$, $i \in \{0, 1, 2, 3\}$, i.e. M is locally invertible. □

The meaning of the word "neighborhood" in this context is "tube around the geodesic".

Now, it makes sense the metric $g_{ij} = dM^t_x \cdot \bar{g}_{ij} \cdot dM_x$ as a metric of F.

Theorem 9.10.4 *The x^0 axis is a geodesic of F with respect to the metric g_{ij}.*

Proof The previous theorem allows us to observe that the tangent vector ε_0 is parallel transported along the x^0 axis, therefore x^0 axis is a geodesic of F. □

Exercise 9.10.5 All orthogonal lines to x^0 axis are geodesics of F with respect to the g_{ij} metric.

Hint. M maps these orthogonal lines into geodesics $y_{\overrightarrow{V}}$.

Proposition 9.10.6 *At each point $(\tau, 0, 0, 0)$ of the x^0 axis, it is*

$$g_{ij}(P) = \begin{pmatrix} 1 & 0 & 0 & 0 \\ 0 & -1 & 0 & 0 \\ 0 & 0 & -1 & 0 \\ 0 & 0 & 0 & -1 \end{pmatrix}.$$

Proof We have

$$ds_F^2(\varepsilon_i, \varepsilon_j) = ds_R^2(dM_P(\varepsilon_i), dM_P(\varepsilon_j)) = ds_R^2(e_i(\tau), e_j(\tau)) = \delta_{ij},$$

where δ_{ij} is the Kronecker symbol. □

According to the local definition of M, we can prove the main

Theorem 9.10.7 *In Fermi coordinates, at every point P belonging of the x^0 axis, the gravitational field is null, that is $\Gamma^i_{jk}(P) = 0$.*

Proof From the above exercise, we know that the line $\gamma(s) := (\tau, sv^1, sv^2, sv^3)$ is a geodesic. The geodesic equations are

$$\frac{d^2x^i}{ds^2} + \Gamma^i_{jk}(\tau, s\,\vec{v})\frac{dx^j}{ds}\frac{dx^k}{ds} = 0, \; i, j, k \in \{0, 1, 2, 3\}.$$

Since $\dfrac{dx^\alpha}{ds} = v^\alpha$, $\alpha \in \{1, 2, 3\}$, it results $\dfrac{d^2x^\alpha}{ds^2} = 0$. Then, being $\dfrac{dx^0}{ds} = 0$ (because x^0 is parameterized by τ), it is $\dfrac{d^2x^i}{ds^2} = 0$. From the geodesic equations, it remains only

$$\Gamma^i_{\alpha\beta}(\tau, s\,\vec{v})v^\alpha v^\beta = 0, \; i \in \{0, 1, 2, 3\}, \; \alpha, \beta \in \{1, 2, 3\}.$$

Now, for $s = 0$, we have

$$\Gamma^i_{\alpha\beta}(\tau, 0, 0, 0)v^\alpha v^\beta = 0, \; i \in \{0, 1, 2, 3\}, \; \alpha, \beta \in \{1, 2, 3\}$$

for any given vector $\vec{v}$, therefore

$$\Gamma^i_{\alpha\beta}(\tau, 0, 0, 0) = 0, \; i \in \{0, 1, 2, 3\}, \; \alpha, \beta \in \{1, 2, 3\}.$$

It remains to prove

$$\Gamma^i_{j0}(\tau, 0, 0, 0) = 0, \; i, j \in \{0, 1, 2, 3\}.$$

We know that the vectors ε_i, are parallel transported along x^0 axis. Let us write the parallel transport equations for these vectors with all components constant, $\varepsilon_k = \delta^i_k$. It is

$$\frac{d\delta^i_k}{d\tau} + \Gamma^i_{jl}(\tau, 0, 0, 0)\delta^j_k\frac{dx^l}{d\tau} = 0.$$

The only non null terms are obtained when $j = k$ and $l = 0$, which ends the proof. $\qquad\square$

Three consequences can immediately be proved:

1. From $\Gamma_{ij,k} = g_{kr}\Gamma^r_{ij}$ we have

$$\Gamma_{ij,k}(\tau, 0, 0, 0) = 0.$$

2. From $\dfrac{\partial g_{ij}}{\partial x^k} = \Gamma_{ik,j} + \Gamma_{jk,i}$ it results

$$\frac{\partial g_{ij}}{\partial x^k}(\tau, 0, 0, 0) = 0.$$

3. From

$$\lim_{h \to 0} \frac{\dfrac{\partial g_{ij}}{\partial x^k}(\tau + h, 0, 0, 0) - \dfrac{\partial g_{ij}}{\partial x^k}(\tau, 0, 0, 0)}{h} = 0$$

we obtain

$$\frac{\partial^2 g_{ij}}{\partial x^0 \partial x^k}(\tau, 0, 0, 0) = 0.$$

Starting from these considerations, the Fermi coordinates offer another view, more physical than geometrical, about the field equations in vacuum.

9.10.2 The Fermi Viewpoint on Einstein's Field Equations in Vacuum

Consider the tidal acceleration equations, written in Fermi's coordinates, with respect to a freely falling observer whose world line has the equation $a^h(\tau) = (\tau, 0, 0, 0, 0)$. Suppose this world line is part of a family of geodesic $x^h(\tau, q)$ such that $x^h(\tau, 0) = a^h(\tau)$.

Therefore

$$\frac{\nabla^2}{d\tau^2} \frac{\partial x^h}{\partial q} = -K^h_j \frac{\partial x^j}{\partial q},$$

where

$$K^h_j = R^h_{ijk} \frac{dx^i}{d\tau} \frac{dx^k}{d\tau}.$$

For the components of our curve $a^h(\tau) = (\tau, 0, 0, 0)$, there is only the term R^h_{0j0} for K^h_j.

So, the relativistic equations of tidal acceleration vector along the curve $a^h(\tau) = (\tau, 0, 0, 0)$ are

$$\frac{\nabla^2}{d\tau^2} \frac{\partial x^h}{\partial q} = -R^h_{0j0} \frac{\partial x^j}{\partial q}, \quad h, j \in \{0, 1, 2, 3\}.$$

Theorem 9.10.8 *In Fermi's coordinates, the tidal acceleration equations along the curve $a^h(\tau) = (\tau, 0, 0, 0)$ have the form*

$$\frac{\nabla^2}{d\tau^2}\frac{\partial x^h}{\partial q} = -\frac{\partial \Gamma^h_{00}}{\partial x^j}\frac{\partial x^j}{\partial q}.$$

Proof Denote by A a point belonging to the curve a^h. Since $\Gamma^i_{jk}(A) = 0$, it results

$$R^h_{0j0}(A) = \frac{\partial \Gamma^h_{00}}{\partial x^j}(A) - \frac{\partial \Gamma^h_{0j}}{\partial x^0}(A).$$

Now, $\Gamma^h_{0j}(\tau, 0, 0, 0) = 0$ for every τ, that is $\frac{\partial \Gamma^h_{0j}}{\partial x^0}(A) = 0$, therefore

$$K^h_j(A) = R^h_{0j0}(A) = \frac{\partial \Gamma^h_{00}}{\partial x^j}(A).$$

The tidal acceleration equations for all points $(\tau, 0, 0, 0)$ become

$$\frac{\nabla^2}{d\tau^2}\frac{\partial x^h}{\partial q} = -\frac{\partial \Gamma^h_{00}}{\partial x^j}\frac{\partial x^j}{\partial q}.$$

$\square$

It remains to compute $\frac{\partial \Gamma^h_{00}}{\partial x^j}(A)$.

Theorem 9.10.9 *It is*

$$\frac{\partial \Gamma^h_{00}}{\partial x^j}(A) = \pm\frac{\partial^2 g_{00}}{\partial x^h \partial x^j}$$

Proof $\Gamma^h_{00} = g^{hs}\Gamma_{00,s} = \frac{g^{hs}}{2}\left(2\frac{\partial g_{0s}}{\partial x^0} - \frac{\partial g_{00}}{\partial x^s}\right)$

$$\frac{\partial \Gamma^h_{00}}{\partial x^j} = \frac{1}{2}\frac{\partial g^{hs}}{\partial x^j}\left(2\frac{\partial g_{0s}}{\partial x^0} - \frac{\partial g_{00}}{\partial x^s}\right) + \frac{g^{hs}}{2}\left(2\frac{\partial^2 g_{0s}}{\partial x^j \partial x^0} - \frac{\partial^2 g_{00}}{\partial x^j \partial x^s}\right)$$

We know $2\frac{\partial g_{0s}}{\partial x^0}(A) - \frac{\partial g_{00}}{\partial x^s}(A) = 0$ and $\frac{\partial^2 g_{0s}}{\partial x^j \partial x^0}(A) = 0$, therefore

$$\frac{\partial \Gamma^h_{00}}{\partial x^j}(A) = -\frac{g^{hs}}{2}\frac{\partial^2 g_{00}}{\partial x^j \partial x^s}(A)$$

Since $g^{00}(A) = 1$, $g^{\alpha\alpha}(A) = -1$, $\alpha \in \{1, 2, 3\}$, $g^{hs} = 0$ when $h \neq s$, we obtain

$$\frac{\partial \Gamma^h_{00}}{\partial x^j}(A) = \pm\frac{1}{2}\frac{\partial^2 g_{00}}{\partial x^h \partial x^j}.$$

$\square$

Let us construct now the matrix K_j^h. It results $\dfrac{\partial \Gamma_{00}^\alpha}{\partial x^\beta}(A) = \dfrac{1}{2}\dfrac{\partial^2 g_{00}}{\partial x^\alpha \partial x^\beta}(A)$, $\alpha, \beta \in \{1, 2, 3\}$, i.e.

$$K_\beta^\alpha(A) = K_\alpha^\beta(A) = \frac{1}{2}\frac{\partial^2 g_{00}}{\partial x^\alpha \partial x^\beta}(A), \quad \alpha, \beta \in \{1, 2, 3\}.$$

Using $\dfrac{\partial^2 g_{00}}{\partial x^k \partial x^0}(A) = 0$, $k \in 0, 1, 2, 3$, it is

$$K_j^0 = \frac{\partial \Gamma_{00}^0}{\partial x^j}(A) = \frac{1}{2}\frac{\partial^2 g_{00}}{\partial x^j \partial x^0}(A) = 0, \quad j \in \{0, 1, 2, 3\}$$

and

$$K_0^h = \frac{\partial \Gamma_{00}^h}{\partial x^0}(A) = \frac{1}{2}\frac{\partial^2 g_{00}}{\partial x^0 \partial x^h}(A) = 0, \quad h \in \{0, 1, 2, 3\}.$$

Therefore, in Fermi's coordinates, the tidal acceleration equations

$$\frac{\nabla^2}{d\tau^2}\frac{\partial x^h}{\partial q} = -K_j^h \frac{\partial x^j}{\partial q},$$

along the world line $a^h(\tau) = (\tau, 0, 0, 0)$ highlight the symmetric matrix

$$K_j^h(A) = \frac{1}{2}\begin{pmatrix} 0 & 0 & 0 & 0 \\ 0 & \dfrac{\partial^2 g_{00}}{(\partial x^1)^2} & \dfrac{\partial^2 g_{00}}{\partial x^1 \partial x^2} & \dfrac{\partial^2 g_{00}}{\partial x^1 \partial x^3} \\ 0 & \dfrac{\partial^2 g_{00}}{\partial x^2 \partial x^1} & \dfrac{\partial^2 g_{00}}{(\partial x^2)^2} & \dfrac{\partial^2 g_{00}}{\partial x^2 \partial x^3} \\ 0 & \dfrac{\partial^2 g_{00}}{\partial x^3 \partial x^1} & \dfrac{\partial^2 g_{00}}{\partial x^3 \partial x^2} & \dfrac{\partial^2 g_{00}}{(\partial x^3)^2} \end{pmatrix}$$

having, as components, second order partial derivatives.

The information about the gravitational field depends on the gravitational potential.

If in these Fermi's coordinates, we identify the classical gravitational potential Φ as $\dfrac{1}{2}g_{00}$, the Hessian matrix of the gravitational potential Φ,

$$d^2\Phi_{\bar{x}} = \left(\frac{\partial^2 \Phi(\bar{x})}{\partial x^i \partial x^k}\right)_{i,k}$$

can be identified with the "spatial part" $\dfrac{1}{2}\left(\dfrac{\partial^2 g_{00}}{\partial x^\alpha \partial x^\beta}\right)_{\alpha,\beta \in \{1,2,3\}}$ of the matrix K_j^h.

The information encapsulates in the trace of the Hessian of the gravitational field, that is the vacuum field equation $\nabla^2 \Phi = 0$, appears when we consider the trace of entire matrix K^h_j in the form $Tr\,K^h_j = K^h_h = 0$. This means that $K^h_h = R^h_{ihk} = 0$, i.e. $R_{ik} = 0$.

Now, we apply the Principle of General Covariance.

The equations

$$R_{ij} = 0$$

represent, in any system of coordinates, the *relativistic field equations in vacuum*.

9.10.3 The Gravitational Coupling in Einstein's Field Equations: $K = \frac{8\pi G}{c^4}$

Let us considered the energy-momentum tensor as a perfect fluid. We can choose such a tensor as a 4×4 symmetric matrix (T^{ij})

$$(T^{ij}) = \begin{pmatrix} T^{00} & T^{01} & T^{02} & T^{03} \\ T^{10} & T^{11} & T^{12} & T^{13} \\ T^{20} & T^{21} & T^{22} & T^{23} \\ T^{30} & T^{31} & T^{32} & T^{33} \end{pmatrix}$$

whose most important property is its null divergence expressed in terms of covariant derivative,

$$T^{kl}_{;l} = 0.$$

This property means that "at each moment, the quantity of matter and energy in the interior of a given infinitesimal parallelepiped is constant."

A first example of energy-momentum tensor was related to the Friedman–Lemaître–Robertson–Walker metric of the Universe. In fact, the key point in the computations of the metric in geometric coordinates was related to the chosen form of energy-momentum tensor. Physicists proposed to look at galaxies as molecules of an ideal gas. In this case, the contravariant energy-momentum tensor was

$$T^{ij} = (\rho + p)u^i u^j - pg^{ij},$$

where g^{ij} are the inverse components of the metric tensor matrix which satisfies Einstein's field equations $R_{ij} - \frac{1}{2}Rg_{ij} = KT_{ij}$,

- ρ is the density;
- p is the pressure;
- u^i are the components $(u^t, v_x u^t, v_y u^t, v_z u^t)$ of the gas 4-velocity. The previous null divergence property is obviously recovered.

Energy and matter can be seen in different ways according to physics models.

The next description, known as the *energy-momentum tensor of a swarm of particles*, is useful to determine the constant K in Einstein's field equations

$$R_{ij} - \frac{1}{2} R g_{ij} = K T_{ij}.$$

How can we describe a *swarm of particles*?

They have to be identical, they have to be uniformly distributed in space and they have to be non-interacting. Each particle has the rest mass m_0 and we suppose that, in an unit of volume of a given spacetime, if the swarm is at rest, there are exactly n_0 particles (see [4]).

The mass can be incorporated in a 4-*momentum vector*

$$\mathbb{P} := (m, m\, \vec{v}),$$

where m is the mass of each non-interacting particle which moves at speed v. If the swarm is at rest,

$$\mathbb{P}_0 = (m_0, \vec{0}).$$

The proper 4-*velocity of a particle moving at speed v* is

$$\mathbb{V} := \left(\frac{1}{\sqrt{1 - v^2}}, \frac{\vec{v}}{\sqrt{1 - v^2}} \right),$$

that is $\mathbb{P} = m_0 \mathbb{V}$.

Another 4-vector can related to the number of particles, denoted by n, in the unit of volume of the previous spacetime which move at speed v,

$$\mathbb{N} := (n, n\, \vec{v}).$$

At rest, we choose

$$\mathbb{N}_0 = (n_0, \vec{0}).$$

It results $\mathbb{N} = n_0 \mathbb{V}$.

If we define the *density of mass for the swarm* by the product between the mass of a particle and the number of particles in a unit volume of the spacetime, we have:

$$\rho_v := mn$$

if the swarm is moving at speed v and

$$\rho_0 := m_0 n_0$$

if the swarm is at rest. Even if

$$mn = \frac{m_0}{\sqrt{1-v^2}}\frac{n_0}{\sqrt{1-v^2}} = \frac{m_0 n_0}{1-v^2},$$

that is, the product of first components is not a covariant quantity, the $(1,0)$ contravariant vectors $\mathbb{P} = (p^0, p^1, p^2, p^3)$ and $\mathbb{N} = (n^0, n^1, n^2, n^3)$ produce a $(2,0)$ contravariant tensor,

$$T^{ij} := p^i n^j = \begin{pmatrix} p^0 n^0 & p^0 n^1 & p^0 n^2 & p^0 n^3 \\ p^1 n^0 & p^1 n^1 & p^1 n^2 & p^1 n^3 \\ p^2 n^0 & p^2 n^1 & p^2 n^2 & p^2 n^3 \\ p^3 n^0 & p^3 n^1 & p^3 n^2 & p^3 n^3 \end{pmatrix},$$

such that the mass-density is incorporated in the T^{00} component.

Now let us cancel the geometric coordinates which helped us to find a possible energy-momentum tensor and consider the dimensional coordinates $(x^0, x^1, x^2, x^3) = (ct, x, y, z)$. It results

$$\mathbb{V} = (c, \dot{x}, \dot{y}, \dot{z}) = (c, v^1, v^2, v^3) = (c, \vec{v}),$$

$$\mathbb{P} = \left(\frac{E}{c}, p^1, p^2, p^3\right),$$

where $E = mc^2$ is the relativistic energy of a particle of the swarm and

$$\mathbb{N} = (nc, nv^1, nv^2, nv^3).$$

The *energy-momentum tensor* becomes

$$T^{ij} := \begin{pmatrix} En & Env^1/c & Env^2/c & Env^3/c \\ cp^1 n & p^1 nv^1 & p^1 nv^2 & p^1 nv^3 \\ cp^2 n & p^2 nv^1 & p^2 nv^2 & p^2 nv^3 \\ cp^3 n & p^3 nv^1 & p^3 nv^2 & p^3 nv^3 \end{pmatrix}.$$

We have

$$T^{00} = En = mnc^2 = \rho c^2,$$

therefore we can call T^{00} the *density of the relativistic energy of the swarm*.

One may describes all the components of the energy-momentum tensor according to the physic units. However only T^{00} is used to determine K.

Suppose we are working in Fermi's coordinates with a swarm of non-interacting particles which move together such that the world line of a particle is the x^0 axis. Therefore

$$\mathbb{V} = (c, 0, 0, 0)$$

and the energy-momentum T^{ij} has only one term,

$$T^{00} = T_0^0 = T_{00} = T = \rho c^2.$$

Along the x^0 axis, the Einstein equations, written in the form

$$R_{ij} = K \left(T_{ij} - \frac{1}{2} T g_{ij} \right),$$

become the only equation

$$R_{00} = K \left(T_{00} - \frac{1}{2} T g_{00} \right) = K \left(\rho c^2 - \frac{1}{2} \rho c^2 \right),$$

that is

$$R_{00} = K \frac{1}{2} \rho c^2.$$

Since dimensionally we have

$$R_{00} = K_s^s = \frac{1}{2} \sum_{\alpha=1}^{3} \frac{\partial^2 g_{00}}{\partial (x^\alpha)^2} = \frac{1}{c^2} \nabla^2 \Phi = \frac{1}{c^2} 4\pi G \rho,$$

it results

$$\frac{1}{c^2} 4\pi G \phi = K \frac{1}{2} \rho c^2,$$

that is

$$K = \frac{8\pi G}{c^4}.$$

Therefore *Einstein's field equations* are

$$R_{ij} - \frac{1}{2} R g_{ij} = \frac{8\pi G}{c^4} T_{ij},$$

where the gravitational coupling is written in physical constants.

9.11 Weak Gravitational Field and the Classical Counterparts of the Relativistic Equations

We are interested in seeing under which conditions it is possible to recover the Classical Mechanics basic formulas involving gravity from the relativistic formulas seen in the present chapter.

Let us discuss this point in a mathematical language: In this section we show that, in the case of a "week gravitational field", for "particles with slow motion", the classical field equations emerge from their *relativistic counterparts*, that is

$$\frac{d^2x^i}{d\tau^2} = -\Gamma^i_{jk}\frac{dx^j}{d\tau}\frac{dx^k}{d\tau} \longrightarrow \frac{d^2x^\alpha}{dt^2} = -\frac{\partial\Phi}{\partial x^\alpha},$$

$$R_{ij} - \frac{1}{2}g_{ij}R = \frac{8\pi G}{c^4}T_{ij} \longrightarrow \nabla^2\Phi = 4\pi\rho,$$

$$R_{ij} = 0 \longrightarrow \nabla^2\Phi = 0.$$

A complete treatment of these results can be found in [4]. Of course, the basic facts were presented by Einstein himself in [3].

Consider the Minkowski metric which describes a frame with no gravity

$$J := J_{ij} = J^{ij} := \begin{pmatrix} 1 & 0 & 0 & 0 \\ 0 & -1 & 0 & 0 \\ 0 & 0 & -1 & 0 \\ 0 & 0 & 0 & -1 \end{pmatrix}.$$

Adding small variations of order $\frac{1}{c^k}$, $k \geq 2$ we introduce gravitational effects. Therefore, the following definition is necessary to introduce our working frame:

Definition 9.11.1 A weak gravitational field is described by a metric

$$g_{ij} = J_{ij} + \frac{1}{c^2}g^{(2)}_{ij} + \frac{1}{c^3}g^{(3)}_{ij} + O\left(\frac{1}{c^4}\right)$$

with the supplementary properties

$$g^{(m)}_{ij} = O(1), \quad \frac{\partial g^{(m)}_{ij}}{\partial t} = O(1), \quad \frac{\partial g^{(m)}_{ij}}{\partial x^\alpha} = O(1), \quad m \in \{2, 3\}, \ \alpha \in \{1, 2, 3\}.$$

Here, $g^{(2)}_{ij}$ are coefficients of the metric g_{ij} related to the factor $\frac{1}{c^2}$, etc. In other words, $\frac{1}{c^2}$ is our expansion parameter related to the strength of the field.

Let us first observe that, for a weak gravitational field, it is

$$g_{ij} = J_{ij} + O\left(\frac{1}{c^2}\right)$$

and

$$g^{ij} = J^{ij} + O\left(\frac{1}{c^2}\right).$$

Therefore, for a weak gravitational field, we have the following consequences of the previous definition:

$$g_{0j} = O\left(\frac{1}{c^2}\right), \; g_{\alpha\alpha} = O(1), \; \frac{\partial g_{ij}}{\partial x^\alpha} = O\left(\frac{1}{c^2}\right) \text{ and } det(g_{ij}) = -1 + O\left(\frac{1}{c^2}\right).$$

Then, it is easy to see that

$$\frac{\partial g_{ij}^{(m)}}{\partial x^0} = \frac{\partial g_{ij}^{(m)}}{\partial t}\frac{\partial t}{\partial x^0} = O(1)\frac{1}{c} = O\left(\frac{1}{c}\right).$$

In the same way,

$$\frac{\partial g_{ij}}{\partial x^0} = \frac{\partial g_{ij}}{\partial t}\frac{\partial t}{\partial x^0} = O\left(\frac{1}{c^2}\right)\frac{1}{c} = O\left(\frac{1}{c^3}\right).$$

Theorem 9.11.2 *The Christoffel symbols of a weak gravitational field have the properties*

$$\Gamma_{k0,k} = O\left(\frac{1}{c^3}\right), \; \Gamma_{ij,k} = O\left(\frac{1}{c^2}\right), \; \Gamma_{00}^0 = \Gamma_{h0}^h = O\left(\frac{1}{c^3}\right),$$

$$\Gamma_{00}^\alpha = \frac{1}{2c^2}\frac{\partial g_{00}^{(2)}}{\partial x^\alpha} + O\left(\frac{1}{c^3}\right), \; \Gamma_{\alpha\beta}^i = O\left(\frac{1}{c^2}\right).$$

Proof We present two computations and we leave to the reader the details.
The first one:

$$\Gamma_{00}^0 = g^{0i}\Gamma_{00,i} = g^{00}\Gamma_{00,0} + g^{0\alpha}\Gamma_{00,\alpha} =$$

$$\left(1 + O\left(\frac{1}{c^2}\right)\right)O\left(\frac{1}{c^2}\right) + O\left(\frac{1}{c^2}\right)\left(\frac{1}{c^3}\right) = O\left(\frac{1}{c^3}\right).$$

The second one:

$$\Gamma_{00}^\alpha = g^{\alpha i}\Gamma_{00,i} = g^{\alpha\alpha}\Gamma_{00,\alpha} + g^{\alpha0}\Gamma_{00,0} + g^{\alpha\beta}\Gamma_{00,\beta} =$$

$$= \left(-1 + O\left(\frac{1}{c^2}\right)\right)\Gamma_{00,\alpha} + O\left(\frac{1}{c^2}\right)\left(\Gamma_{00,0} + \Gamma_{00,\beta}\right).$$

Replacing $\Gamma_{00,\alpha}$, it results

$$\Gamma_{00}^\alpha = \left(-1 + O\left(\frac{1}{c^2}\right)\right)\Gamma_{00,\alpha} + O\left(\frac{1}{c^3}\right) =$$

$$= \left(-1 + O\left(\frac{1}{c^2}\right)\right)\left(\frac{\partial g_{0\alpha}}{\partial x^0} - \frac{1}{2}\frac{\partial g_{00}}{\partial x^\alpha}\right) + O\left(\frac{1}{c^3}\right),$$

that is

$$\Gamma^\alpha_{00} = \frac{1}{2}\frac{\partial g_{00}}{\partial x^\alpha} + O\left(\frac{1}{c^3}\right) = \frac{1}{2c^2}\frac{\partial g^{(2)}_{00}}{\partial x^\alpha} + O\left(\frac{1}{c^3}\right).$$

$\square$

Denote by $X(t) := (x^1(t), x^2(t), x^3(t))$ the trajectory of a classical particle; its classical speed is $\dot{X} := (\dot{x}^1(t), \dot{x}^2(t), \dot{x}^3(t))$.

Definition 9.11.3 The particle is "slow" if $\dot{x}^\alpha(t) = O(1)$, $\alpha \in \{1, 2, 3\}$.

The corresponding worldcurve is $\overrightarrow{X} = (ct, X(t))$ and its relativistic speed is $\overrightarrow{V} = (c, \dot{X}(t))$. Observe that

$$L(t) = \int_{t_0}^t \| \overrightarrow{V}(s)\|_g ds = \int_{t_0}^t \sqrt{g_{ij}\dot{x}^i(s)\dot{x}^j(s)}ds$$

has length dimension.

Parameterizing $\overrightarrow{X}$ by proper time means to consider $\tau(t) := \frac{1}{c}L(t)$. Let us observe that $\tau(t)$ has time dimension.

Theorem 9.11.4 *In a Minkowski metric, if a particle is moving "slow" uniformly along a curve parameterized by proper time, then*

$$\frac{d\tau}{dt} = 1 + O\left(\frac{1}{c^2}\right).$$

Proof From

$$d\tau = \frac{\| \overrightarrow{V} \|_M}{c}dt = \sqrt{1 - \frac{1}{c^2}\sum_{\alpha=1}^3 (\dot{x}^\alpha(t))^2}dt = \sqrt{1 - \frac{1}{c^2}O(1)},$$

it results

$$\frac{d\tau}{dt} = \sqrt{1 - O\left(\frac{1}{c^2}\right)}dt.$$

Since $\sqrt{1 - A} \approx 1 + \frac{A}{2}$, we have

$$\frac{d\tau}{dt} = 1 + O\left(\frac{1}{c^2}\right).$$

$\square$

Now, parameterizing with respect to τ in a metric g_{ij}, we have

$$d\tau = \frac{\| \vec{V} \|_g}{c} dt = \sqrt{\frac{g_{ij} \dot{x}^i \dot{x}^j}{c^2}} dt = \sqrt{g_{ij} \frac{\dot{x}^i}{c} \frac{\dot{x}^j}{c}} dt.$$

Theorem 9.11.5 *If a particle is moving "slow" in a weak gravitational field along a curve parameterized by proper time, then*

$$\frac{d\tau}{dt} = 1 + O\left(\frac{1}{c^2}\right).$$

Proof We have $\dot{x}^0 = \dfrac{d}{dt}(ct) = c$. The particle is "slow" and, by definition, this means $\dot{x}^\alpha = O(1)$. Therefore

$$g_{ij} \frac{\dot{x}^i}{c} \frac{\dot{x}^j}{c} = g_{00} + 2g_{0\alpha} \frac{\dot{x}^\alpha}{c} + g_{\alpha\beta} \frac{\dot{x}^\alpha}{c} \frac{\dot{x}^\beta}{c} = 1 + O\left(\frac{1}{c^2}\right).$$

We have used $g_{00} = 1$, $g_{\alpha\alpha} = O(1)$, $g_{\alpha\beta} = O\left(\dfrac{1}{c^2}\right)$, $\alpha \neq \beta$, $g_{0\alpha} = O\left(\dfrac{1}{c^2}\right)$.
Finally,

$$\frac{d\tau}{dt} = \sqrt{g_{ij} \frac{\dot{x}^i}{c} \frac{\dot{x}^j}{c}} dt = \sqrt{1 + O\left(\frac{1}{c^2}\right)} dt,$$

i.e.

$$\frac{d\tau}{dt} = 1 + O\left(\frac{1}{c^2}\right).$$

Observe we can also obtain

$$\frac{dt}{d\tau} = 1 + O\left(\frac{1}{c^2}\right).$$

$\square$

Theorem 9.11.6 *If a particle is moving "slow" along a curve parameterized by proper time, then it is* $\dfrac{d^2 x^0}{d\tau^2} = O\left(\dfrac{1}{c}\right)$ *and* $\dfrac{d^2 x^\alpha}{d\tau^2} = \ddot{x}^\alpha + O\left(\dfrac{1}{c^2}\right)$.

Proof From $\dfrac{dx^0}{d\tau} = \dfrac{dx^0}{dt} \dfrac{dt}{d\tau} = c\left(1 + O\left(\dfrac{1}{c^2}\right)\right) = c + O\left(\dfrac{1}{c}\right)$, we obtain

$$\frac{d^2 x^0}{d\tau^2} = \frac{d}{dt}\left(c + O\left(\frac{1}{c}\right)\right) = O\left(\frac{1}{c}\right),$$

and, from

$$\frac{dx^\alpha}{d\tau} = \frac{dx^\alpha}{dt}\frac{dt}{d\tau} = \dot{x}^\alpha\left(1 + O\left(\frac{1}{c^2}\right)\right) = \dot{x}^\alpha + O\left(\frac{1}{c^2}\right),$$

it results

$$\frac{d^2 x^\alpha}{d\tau^2} = \frac{d}{dt}\left(\dot{x}^\alpha + O\left(\frac{1}{c^2}\right)\right) = \ddot{x}^\alpha + O\left(\frac{1}{c^2}\right).$$

$\square$

Theorem 9.11.7 *In a weak gravitational field, the four geodesic equations for "slow" particles reduce to the three classical equations of motion, that is*

$$\frac{d^2 x^i}{d\tau^2} = -\Gamma^i_{jk}\frac{dx^j}{d\tau}\frac{dx^k}{d\tau}, \ i, j, k \in \{0, 1, 2, 3\} \quad\longrightarrow\quad \frac{d^2 x^\alpha}{dt^2} = -\frac{\partial\Phi}{\partial x^\alpha}, \ \alpha \in \{1, 2, 3\}.$$

Proof We already proved that, if a particle is moving "slow," then $\dfrac{d^2 x^0}{d\tau^2} = O\left(\dfrac{1}{c}\right)$, so, the left hand-side part is a $O\left(\dfrac{1}{c}\right)$ quantity.

We consider

$$\frac{d^2 x^0}{d\tau^2} = -\Gamma^0_{jk}\frac{dx^j}{d\tau}\frac{dx^k}{d\tau} = -\Gamma^0_{00}\left(\frac{dx^0}{d\tau}\right)^2 - 2\Gamma^0_{0\alpha}\frac{dx^0}{d\tau}\frac{dx^\alpha}{d\tau} - \Gamma^0_{\alpha\beta}\frac{dx^\alpha}{d\tau}\frac{dx^\beta}{d\tau}$$

and we observe

$$\Gamma^0_{00}\left(\frac{dx^0}{d\tau}\right)^2 = O\left(\frac{1}{c^3}\right)\left(c + O\left(\frac{1}{c}\right)\right)^2 = O\left(\frac{1}{c}\right),$$

$$\Gamma^0_{0\alpha}\frac{dx^0}{d\tau}\frac{dx^\alpha}{d\tau} = O\left(\frac{1}{c^2}\right)\left(c + O\left(\frac{1}{c}\right)\right)\left(\dot{x}^\alpha + O\left(\frac{1}{c^2}\right)\right) = O\left(\frac{1}{c}\right)$$

$$\Gamma^0_{\alpha\beta}\frac{dx^\alpha}{d\tau}\frac{dx^\beta}{d\tau} = O\left(\frac{1}{c^2}\right)\left(\dot{x}^\alpha + O\left(\frac{1}{c^2}\right)\right)\left(\dot{x}^\beta + O\left(\frac{1}{c^2}\right)\right) = O\left(\frac{1}{c^2}\right)$$

The right hand-side part of the equation of geodesic is $O\left(\dfrac{1}{c}\right)$, therefore, for a "slow" particle, the first geodesic equation is an equality between "very small" quantities. As a consequence, we can neglect it.

We already proved that the left hand-side part of the geodesic equations is

$$\frac{d^2x^\alpha}{d\tau^2} = \ddot{x}^\alpha + O\left(\frac{1}{c^2}\right).$$

Now, for the right hand-side part, we proceed as above.

$$\frac{d^2x^\alpha}{d\tau^2} = -\Gamma^\alpha_{jk}\frac{dx^j}{d\tau}\frac{dx^k}{d\tau} = -\Gamma^\alpha_{00}\left(\frac{dx^0}{d\tau}\right)^2 - 2\Gamma^\alpha_{0\beta}\frac{dx^0}{d\tau}\frac{dx^\beta}{d\tau} - \Gamma^\alpha_{\beta\gamma}\frac{dx^\beta}{d\tau}\frac{dx^\gamma}{d\tau}$$

and we observe

$$\Gamma^\alpha_{00}\left(\frac{dx^0}{d\tau}\right)^2 = \left(\frac{1}{2c^2}\frac{\partial g^{(2)}_{00}}{\partial x^\alpha} + O\left(\frac{1}{c^3}\right)\right)\left(c^2 + O(1)\right)^2 = \frac{1}{2}\frac{\partial g^{(2)}_{00}}{\partial x^\alpha} + O\left(\frac{1}{c}\right),$$

$$\Gamma^\alpha_{0\beta}\frac{dx^0}{d\tau}\frac{dx^\beta}{d\tau} = O\left(\frac{1}{c^3}\right)\left(c + O\left(\frac{1}{c}\right)\right)\left(\dot{x}^\alpha + O\left(\frac{1}{c^2}\right)\right) = O\left(\frac{1}{c^2}\right),$$

$$\Gamma^\alpha_{\beta\gamma}\frac{dx^\beta}{d\tau}\frac{dx^\gamma}{d\tau} = O\left(\frac{1}{c^2}\right)\left(\dot{x}^\beta + O\left(\frac{1}{c^2}\right)\right)\left(\dot{x}^\gamma + O\left(\frac{1}{c^2}\right)\right) = O\left(\frac{1}{c^2}\right),$$

that is the right hand-side of the geodesic equations is

$$-\frac{1}{2}\frac{\partial g^{(2)}_{00}}{\partial x^\alpha} + O\left(\frac{1}{c}\right), \alpha \in \{1, 2, 3\}.$$

Neglecting the "small" quantities, the geodesic equations

$$\frac{d^2x^i}{d\tau^2} = -\Gamma^i_{jk}\frac{dx^j}{d\tau}\frac{dx^k}{d\tau}, \ i, j, k \in \{0, 1, 2, 3\}$$

reduce to

$$\frac{d^2x^\alpha}{dt^2} = -\frac{\partial \Phi}{\partial x^\alpha}, \ \alpha \in \{1, 2, 3\}, \ \Phi = \frac{1}{2}\frac{\partial g^{(2)}_{00}}{\partial x^\alpha}.$$

$\square$

Theorem 9.11.8 *The relativistic equations of the weak gravitational field reduce to the classical Poisson field equation:*

$$R_{ij} - \frac{1}{2}g_{ij}R = \frac{8\pi G}{c^4}T_{ij} \quad \longrightarrow \quad \nabla^2\Phi = 4\pi G\rho,$$

Proof We consider the relativistic equation written with respect to the Laue scalar

$$R_{ij} = \frac{8\pi G}{c^4}\left(T_{ij} - \frac{1}{2}g_{ij}T\right).$$

Suppose the matter-energy tensor written in the previous form consisting of a swarm of identical non-interacting particles having density ρ. The only non-zero component is $T_{00} = \rho c^2$. The right member is

$$\frac{8\pi G}{c^4}\left(T_{ij} - \frac{1}{2}g_{ij}T\right) = \frac{8\pi G}{c^4}\left[\rho c^2 - \frac{1}{2}\left(1 + O\left(\frac{1}{c^2}\right)\right)\rho c^2\right] = \frac{8\pi G}{c^4}\left[\frac{\rho c^2}{2} + O(1)\right] =$$

$$= \frac{1}{c^2}\cdot 4\pi G\rho + O\left(\frac{1}{c^4}\right).$$

Now, let us look at the left member.

$$R_{00} = R^s_{0s0} = R^0_{000} + R^\alpha_{0\alpha 0} = R^\alpha_{0\alpha 0} = \frac{\partial \Gamma^\alpha_{00}}{\partial x^\alpha} - \frac{\partial \Gamma^\alpha_{0\alpha}}{\partial x^0} + \Gamma^m_{00}\Gamma^\alpha_{m\alpha} - \Gamma^m_{0\alpha}\Gamma^\alpha_{m0}.$$

The two products of Christoffel symbols are at least $O\left(\dfrac{1}{c^4}\right)$.

Then, since $\Gamma^\alpha_{0\alpha} = O\left(\dfrac{1}{c^3}\right)$, it results $\dfrac{\partial \Gamma^\alpha_{0\alpha}}{\partial x^0} = \dfrac{\partial \Gamma^\alpha_{0\alpha}}{\partial t}\dfrac{\partial t}{\partial x^0} = O\left(\dfrac{1}{c^3}\right)\dfrac{1}{c} = O\left(\dfrac{1}{c^4}\right).$

If we consider the derivative with respect x^α of the equality $\Gamma^\alpha_{00} = \dfrac{1}{2c^2}\dfrac{\partial g^{(2)}_{00}}{\partial x^\alpha} + O\left(\dfrac{1}{c^3}\right)$, it is

$$\frac{\partial \Gamma^\alpha_{00}}{\partial x^\alpha} = \frac{1}{2c^2}\frac{\partial^2 g^{(2)}_{00}}{(\partial x^\alpha)^2} + O\left(\frac{1}{c^3}\right), \quad \alpha \in \{1, 2, 3\},$$

that is

$$R_{00} = \frac{1}{2c^2}\sum_{\alpha=1}^{3}\frac{\partial^2 g^{(2)}_{00}}{(\partial x^\alpha)^2} + O\left(\frac{1}{c^3}\right).$$

Since $\Phi = \dfrac{1}{2}g^{(2)}_{00}$, we finally obtain the right member as

$$R_{00} = \frac{1}{c^2}\nabla^2\Phi + O\left(\frac{1}{c^3}\right).$$

Neglecting the small quantities, the relativistic weak field equations reduce to the classical Poisson field equation

$$\nabla^2\Phi = 4\pi G\rho.$$

$\square$

Corollary 9.11.9 *The relativistic equations of the weak gravitational field in vacuum reduce to the classical Laplace field equation in vacuum:*

$$R_{ij} = 0 \quad \longrightarrow \quad \nabla^2 \Phi = 0.$$

9.12 The Einstein Static Universe and its Cosmological Constant

On February 8, 1917, the Prussian Academy of Science in Berlin published a paper by Albert Einstein where the first application of his theory, published on November 25, 1915, was presented [68]. The paper discussed a dynamical system describing a static spacetime representing the Universe. It can be considered the birth of modern Cosmology. The model proved wrong after the discovery or recession of galaxies by Hubble, however it is important because several concepts presented in it were used in the further developments of this science. Let us give now a quick presentation of it.

Consider before some mathematical preliminaries. Let be $U := (0, 2\pi) \times (-\frac{\pi}{2}, \frac{\pi}{2}) \times (-\frac{\pi}{2}, \frac{\pi}{2})$, $(\alpha, \beta, \theta) \in U$ and the map $f : U \to \mathbb{R}^4$,

$$f(\alpha, \beta, \theta) := \begin{cases} u_1 = r \cos \alpha \cos \beta \cos \theta \\ u_2 = r \sin \alpha \cos \beta \cos \theta \\ u_3 = r \sin \beta \cos \theta \\ u_4 = r \sin \theta \end{cases}$$

It is easy to see that $u_1^2 + u_2^2 + u_3^2 + u_4^2 = r^2$.

The image of f in $\mathbb{R}^4$, $f(U)$, is the 3-sphere centered at the origin having radius r. In classical notation, it is $S^3(O, r)$.

The coefficients of the metric are computed with the Euclidean inner product of the partial derivatives of f. The only nonzero coefficients of the metric are

$$g_{\alpha\alpha} = r^2 \cos^2 \beta \cos^2 \theta, \ g_{\beta\beta} = r^2 \cos^2 \theta, \ g_{\theta\theta} = r^2,$$

therefore the metric induced by the Euclidean 4-space in the tangent 3-planes of this surface is

$$ds^2 = g_{\alpha\alpha}(dx^\alpha)^2 + g_{\beta\beta}(dx^\beta)^2 + g_{\theta\theta}(dx^\theta)^2.$$

The volume of $S^3(O, r)$ is

$$Vol[S^3(O, r)] = \int_0^{2\pi} \int_{-\pi/2}^{\pi/2} \int_{-\pi/2}^{\pi/2} \sqrt{det(g_{ij})} d\alpha d\beta d\theta =$$

$$= r^3 \int_0^{2\pi} \int_{-\pi/2}^{\pi/2} \int_{-\pi/2}^{\pi/2} \cos \beta \cos^2 \theta d\alpha d\beta d\theta = 2\pi^2 r^3.$$

Einstein's static universe is $\mathbb{E} := \mathbb{R} \times S^3(O, r)$ and it does not evolve. If we accept the Einstein point of view, the Universe is static. Let us study the mathematical formalism to find its properties.

Let us choose the homogeneous coordinates $(x^0, x^1, x^2, x^3) := (ct, \alpha, \beta, \theta)$. Now, we use the Minkowski product of partial derivatives to determine the metric of $\mathbb{E}$. It results a simple form for the metric of Einstein's static universe

$$ds^2 = (dx^0)^2 - r^2 \cos^2 x^1 \cos^2 x^3 (dx^1)^2 - r^2 \cos^2 x^3 (dx^2)^2 - r^2 (dx^3)^2.$$

The only nonzero Christoffel symbols are

$$\Gamma^1_{12} = \Gamma^1_{21} = -\tan x^2; \ \Gamma^1_{13} = \Gamma^1_{31} = \Gamma^2_{23} = \Gamma^2_{32} = -\tan x^3;$$

$$\Gamma^2_{11} = \sin x^2 \cos x^2; \ \Gamma^3_{11} = \cos^2 x^2 \sin x^3 \cos x^3; \ \Gamma^3_{22} = \sin x^3 \cos x^3.$$

Using

$$R_{jl} = R^h_{jhl} = \frac{\partial \Gamma^h_{jl}}{\partial x^h} - \frac{\partial \Gamma^h_{jh}}{\partial x^l} + \Gamma^s_{jl} \Gamma^h_{sh} - \Gamma^s_{jh} \Gamma^h_{sl}$$

it results

$$R_{00} = \frac{\partial \Gamma^h_{00}}{\partial x^h} - \frac{\partial \Gamma^h_{0h}}{\partial x^0} + \Gamma^s_{00} \Gamma^h_{sh} - \Gamma^s_{0h} \Gamma^h_{s0} = 0.$$

In the same way, we compute $R_{ii} = -\dfrac{2}{r^2} g_{ii}$, $i \in \{1, 2, 3\}$.

If we consider the Einstein field equations with the cosmological constant included, it is

$$R_{ij} - \frac{1}{2} R g_{ij} + \Lambda g_{ij} = \frac{8\pi G}{c^4} T_{ij},$$

in the equivalent form

$$R_{ij} - \Lambda g_{ij} = \frac{8\pi G}{c^4} \left(T_{ij} - \frac{1}{2} T g_{ij} \right).$$

If $i = j = 0$, the Einstein field equations reduced to:

$$R_{00} - \Lambda g_{00} = \frac{8\pi G}{c^4} \left(T_{00} - \frac{1}{2} T g_{00} \right).$$

If $i = j \neq 0$,

$$R_{ii} - \Lambda g_{ii} = \frac{8\pi G}{c^4} \left(T_{ii} - \frac{1}{2} T g_{ii} \right).$$

Einstein assumed that the matter-energy appears only in the form of a swarm of non-interacting particles of uniform density ρ. Only $T^{00} = \rho c^2$, all the other components are 0. The first equation becomes

$$\Lambda = -\frac{4\pi G}{c^2}\rho.$$

The other three equations lead to a single equation,

$$-\frac{2}{r^2} - \Lambda \cdot (-1) = \frac{8\pi G}{c^4}\left(0 - \frac{1}{2}\rho c^2 \cdot (-1)\right),$$

i.e.

$$-\frac{1}{r^2} = \Lambda$$

Combining the two equations, it results

$$r = \frac{c}{2\sqrt{\pi G\rho}}.$$

The total amount of matter, denoted as $\mathbb{M}_{\mathbb{E}}$, in Einstein's universe is finite. It is computed as the product between the spatial density ρ and the volume of $S^3(O, r)$, that is

$$\mathbb{M}_{\mathbb{E}} = \rho \cdot 2\pi^2 r^3 = \rho \cdot 2\pi^2 \cdot r \cdot r^2 = \rho 2\pi^2 \frac{1}{\sqrt{-\Lambda}}\frac{c^2}{4\pi G\rho} = \frac{\pi c^2}{2G\sqrt{-\Lambda}}.$$

Of course, the radius of this universe is the constant computed before, that is

$$r = \frac{c}{2\sqrt{\pi G\rho}}.$$

Einstein did not consider any more this model after Hubble discovered the evidence of cosmological expansion, however the concept of cosmological constant, used here, was considered later in view of the issues of cosmological inflation and dark energy discussed above.

9.13   Cosmic Strings

Cosmic strings are 1-dimensional hypothetical structures emerged as topological defects of spacetime in some phase transition after the Big Bang. They should have acted like seeds for cosmological large scale structure formation [64]. A nice presentation of this topic is in [35]. Here, we adapted it for a metric with signature $(+ - - -)$.

In a system of geometric coordinates (t, r, ϕ, z), the metric which describes a static string around and along the z axis is

$$ds^2 = dt^2 - dr^2 - f^2(r)d\phi^2 - dz^2,$$

where $f(r)$ has to be determined. The only non-zero Christoffel symbols are

$$\Gamma^1_{22} = f(r)f'(r); \quad \Gamma^2_{12} = \Gamma^2_{21} = \frac{f'(r)}{f(r)}.$$

We have

$$\frac{\partial \Gamma^2_{12}}{\partial r} = \frac{\partial \Gamma^2_{21}}{\partial r} = \frac{f''(r)f(r) - (f'(r))^2}{f^2(r)}; \quad \frac{\partial \Gamma^1_{22}}{\partial r} = (f'(r))^2 + f''(r)f(r);$$

Then

$$R^2_{121} = \frac{\partial \Gamma^2_{11}}{\partial \phi} - \frac{\partial \Gamma^1_{12}}{\partial r} + \Gamma^2_{s2}\Gamma^s_{11} - \Gamma^2_{s1}\Gamma^s_{12} = -\frac{f''(r)f(r) - (f'(r))^2}{f^2(r)} - \frac{(f'(r))^2}{f^2(r)} = \frac{-f''(r)}{f(r)},$$

that is

$$R_{11} = R^s_{1s1} = \frac{-f''(r)}{f(r)}.$$

From

$$R^1_{212} = \frac{\partial \Gamma^1_{22}}{\partial r} - \frac{\partial \Gamma^1_{21}}{\partial \phi} + \Gamma^1_{s1}\Gamma^s_{22} - \Gamma^1_{s2}\Gamma^s_{21} = (f'(r))^2 + f''(r)f(r) - f(r)f'(r)\frac{f'(r)}{f(r)},$$

we deduce

$$R_{22} = R^s_{2s2} = f''(r)f(r);$$

Since

$$R = R^i_i = g^{11}R_{11} + g^{22}R_{22} = \frac{f''(r)}{f(r)} + \left(\frac{-1}{f^2(r)}\right)f''(r)f(r) = 0,$$

we obtain

$$\begin{cases} R_{11} + \dfrac{1}{2}R\,g_{11} = \dfrac{-f''(r)}{f(r)} \\[3mm] R_{22} + \dfrac{1}{2}R\,g_{22} = f''(r)f(r). \end{cases}$$

Einstein's field equations in geometric coordinates are

$$R_{ij} - \frac{1}{2}Rg_{ij} = 8\pi GT_{ij};$$

Now using the previous result, that is $R_1^1 = -R_2^2 = \dfrac{f''(r)}{f(r)}$, to choose the tensor T_j^i: $T_j^i = 0$ except $T_1^1 = -T_2^2 = \sigma(r)$, where σ is a given positive smooth function expressing the unit energy density of the string. It remains to find f which satisfies the equation

$$f''(r) = -8\pi G\sigma(r)f(r).$$

In order to avoid singularities, the metric has to reduce to the flat Minkowski metric at the origin. This means that $f(r)$ approaches r for small r. Therefore, two conditions for f have to be given: $f(0) = 0$ and $f'(0) = 1$.

Denote by r_s the value of r such that $\sigma(r) = 0$ if $r \geq r_s$. The physical area of a ring of radius r and width dr in the given metric is

$$\int_0^{2\pi} \int_0^{dr} \sqrt{-\det g_{ij}}\,d\phi dz = 2\pi f(r)dr.$$

It implies that the string energy per unit of length is

$$E = \int_0^{r_s} \sigma(r)2\pi f(r)dr,$$

therefore, integrating the equation $R_{11} = -8\pi G\sigma(r)$, we obtain

$$f'(r_s) - f'(0) = \int_0^{r_s} f''(r)dr = -4G \int_0^{r_s} \sigma(r)2\pi f(r)dr = -4GE.$$

If $r > r_s$, it is

$$f'(r) = 1 - 4GE.$$

Integrating again, it results

$$f(r) = (1 - 4GE)r + K,$$

where K is a constant that should be 0 because $f(0) = 0$.

The metric obtained outside the string is

$$ds^2 = dt^2 - dr^2 - (1 - 4GE)^2 r^2 d\phi^2 - dz^2,$$

which a flat metric. In this way, we have the simplest expression of a metric describing an infinite, straight and independent of time string lying along the z axis of our chosen coordinate system.

9.14 The Case of Planar Gravitational Waves

Also the issue of gravitational waves can be dealt under the standard of our geometric approach. Here, we shall give just a short summary of this important topic. For a detailed discussion on the history, the theoretical foundation and the discovery, we refer the reader to specialized texts and papers [69–71].

In order to deal with gravitational waves, we have to obtain metrics in a geometric coordinate system as

$$ds^2 = (\alpha_{ij} + \epsilon h_{ij})dx^i dx^j,$$

such that both Einstein's field equations and $\Box h_{ij} = 0$ are satisfied.

Here,

$$\Box := (\partial^0)^2 - (\partial^1)^2 - (\partial^2)^2 - (\partial^3)^2,$$

where

$$(\partial^k)^2 := \frac{\partial^2}{(\partial x^k)^2}.$$

If μ_{ij} are the coefficients of the classical Minkowski metric, the previous *d'Alembert operator* definition can be written in a simpler form as

$$\Box := \mu_{ij}\partial^i \partial^j.$$

In any case it is difficult to find such kind of metrics because we have to develop the whole theory of tensor perturbations in General Relativity [69]. Instead of trying to find out general gravitational wave solutions, let us focus on planar gravitational waves which are easier to obtain. See [44] for details. We follow this last reference to offer a first glance on this subject. Consider the metric

$$ds^2 = (1 + \cos(t - x)[2 + \cos(t - x)])dt^2 - (1 - \cos^2(t - x))dx^2 - dy^2 - dz^2 -$$

$$-2\cos(t - x)(1 + \cos(t - x))dtdx.$$

The previous metric can be seen as a slightly perturbation of the Minkowski metric μ_{ij} because

$$g_{ij} = \mu_{ij} + h_{ij}.$$

A metric which such coefficients is called a *linearized metric*. If it satisfies both Einstein's vacuum field equations and the conditions $\Box h_{ij} = 0$, such a linearized metric describes *gravitational planar waves*.

Theorem 9.14.1 *The previous metric having the following nonzero coefficients*

$$g_{00} = 1 + \cos(t - x)(2 + \cos(t - x)); \quad g_{11} = 1 - \cos^2(t - x); \quad g_{22} = g_{33} = -1;$$

$$g_{01} = g_{10} = -\cos(t - x)(1 + \cos(t - x));$$

describes gravitational planar waves.

Proof We have the nonzero perturbations

$$h_{00} = \cos(t - x)\,(2 + \cos(t - x)); \quad h_{11} = -\cos^2(t - x);$$

$$h_{01} = h_{10} = -\cos(t - x)(1 + \cos(t - x)).$$

It is easy to see that

$$\Box \cos(t - x) = (\partial^0)^2 \cos(t - x) - (\partial^1)^2 \cos(t - x) = 0$$

and

$$\Box \cos(t - x) = (\partial^0)^2 \cos^2(t - x) - (\partial^1)^2 \cos^2(t - x) = 0,$$

therefore the conditions

$$\Box h_{ij} = 0$$

are fulfilled.

Now it seems we have a lot of difficult computations to do in order to prove $R_{ij} = 0$.

It is easy to provide a coordinate transformation for which the Ricci tensor can be computed. Let us consider the Minkowski metric

$$ds^2 = d\bar{t}^2 - d\bar{x}^2 - d\bar{y}^2 - d\bar{z}^2$$

and the transformation

$$\bar{t} = t + \sin(t - x), \quad \bar{x} = x, \quad \bar{y} = y, \quad \bar{z} = z.$$

Since

$$d\bar{t} = (1 + \cos(t - x))dt - \cos(t - x)dx, \quad d\bar{x} = dx, \quad d\bar{y} = dy, \quad d\bar{z} = dz,$$

the Minkowski metric turns into our metric

$$ds^2 = (1 + \cos(t - x)[2 + \cos(t - x)])dt^2 - (1 - \cos^2(t - x))dx^2 - dy^2 - dz^2 -$$

$$-2\cos(t - x)(1 + \cos(t - x))dtdx.$$

Therefore $\bar{R}_{ij} = 0$ transforms into the desired $R_{ij} = 0$.   □

Gravitational waves were among the early predictions of Einstein's General Relativity. Their discovery a century later can be considered one of the greatest achievements of modern Science.

9.15   The Gödel Universe

Another interesting metric is the one describing the so-called Gödel Universe [72] published[7] in 1949.

First, let us show that the metric

$$ds^2 = (dx^0)^2 - (dx^1)^2 + \frac{e^{2x^1}}{2}(dx^2)^2 - (dx^3)^2 + 2e^{x^1}dx^0dx^2$$

written in geometric coordinates, satisfies Einstein's field equations in the case when the cosmological constant is $\Lambda = \dfrac{1}{2}$ and the stress-energy tensor describes dust with constant density $\rho = \dfrac{1}{8\pi G}$. This is called Gödel's first metric . The coefficients involved in computations are

$$(g_{ij}) = \begin{pmatrix} 1 & 0 & e^{x^1} & 0 \\ 0 & -1 & 0 & 0 \\ e^{x^1} & 0 & \dfrac{e^{2x^1}}{2} & 0 \\ 0 & 0 & 0 & -1 \end{pmatrix}, \quad (g^{ij}) = \begin{pmatrix} -1 & 0 & 2e^{-x^1} & 0 \\ 0 & -1 & 0 & 0 \\ 2e^{-x^1} & 0 & -2e^{-2x^1} & 0 \\ 0 & 0 & 0 & -1 \end{pmatrix}.$$

Since only the derivative with respect to x^1 of the metric coefficients can be non-zero, the first type Christoffel symbols are

$$\Gamma_{12,2} = \Gamma_{21,2} = \frac{1}{2}e^{2x^1}, \ \Gamma_{01,2} = \Gamma_{10,2} = \frac{1}{2}e^{x^1};$$

---

[7]The story of this solution is very nice. Kurt Gödel gave it to Albert Einstein as a gift for his 70th birthday when they both lived in Princeton.

$$\Gamma_{02,1} = \Gamma_{20,1} = -\frac{1}{2}e^{x^1}; \ \Gamma_{22,1} = -\frac{1}{2}e^{2x^1};$$

$$\Gamma_{12,0} = \Gamma_{21,0} = \frac{1}{2}e^{x^1}.$$

The non-zero second type Christoffel symbols are

$$\Gamma^0_{10} = \Gamma^0_{01} = 1, \ \Gamma^0_{12} = \Gamma^0_{21} = \frac{1}{2}e^{x^1};$$

$$\Gamma^1_{02} = \Gamma^1_{20} = \frac{1}{2}e^{x^1}; \ \Gamma^1_{22} = \frac{1}{2}e^{2x^1};$$

$$\Gamma^2_{01} = \Gamma^2_{10} = -e^{-x^1}.$$

Then

$$R_{00} = R^i_{0i0} = \frac{\partial \Gamma^i_{00}}{\partial x^i} - \frac{\partial \Gamma^i_{0i}}{\partial x^0} + \Gamma^i_{si}\Gamma^s_{00} - \Gamma^i_{s0}\Gamma^s_{0i} = -\Gamma^i_{10}\Gamma^1_{0i} - \Gamma^i_{20}\Gamma^2_{0i} = -\Gamma^2_{10}\Gamma^1_{02} - \Gamma^1_{20}\Gamma^2_{01} = 1$$

$$R_{22} = R^i_{2i2} = \frac{\partial \Gamma^i_{22}}{\partial x^i} - \frac{\partial \Gamma^i_{2i}}{\partial x^2} + \Gamma^i_{si}\Gamma^s_{22} - \Gamma^i_{s2}\Gamma^s_{2i} = \frac{\partial \Gamma^1_{22}}{\partial x^1} + \Gamma^i_{1i}\Gamma^1_{22} - \Gamma^1_{02}\Gamma^0_{21} - \Gamma^0_{12}\Gamma^1_{20} = e^{2x^1}$$

$$R_{02} = R_{20} = R^i_{2i0} = \frac{\partial \Gamma^i_{20}}{\partial x^i} - \frac{\partial \Gamma^i_{2i}}{\partial x^0} + \Gamma^i_{si}\Gamma^s_{20} - \Gamma^i_{s0}\Gamma^s_{2i} = \frac{\partial \Gamma^1_{20}}{\partial x^1} + \Gamma^i_{1i}\Gamma^1_{20} = e^{x^1}$$

The others R_{ij} are null. Now,

$$R^0_0 = g^{0s}R_{s0} = g^{00}R_{00} + g^{02}R_{20} = 1,$$

$$R^2_2 = g^{2s}R_{s2} = g^{20}R_{02} + g^{22}R_{22} = 0,$$

i.e. the trace is

$$R = 1.$$

Consider the contravariant vector $u^i := (1, 0, 0, 0) = (u^0, u^1, u^2, u^3)$. The corresponding covariant vector is $u_i := g_{is}u^s = (1, 0, e^{x^1}, 0) = (u_0, u_1, u_2, u_3)$.

Let us observe $R_{00} = 1 = u_0u_0$; $R_{22} = e^{2x^1} = u_2u_2$; $R_{02} = R_{20} = e^{x^1} = u_0u_2$, that is, in our case, for $\Lambda = \frac{1}{2}$ and $T_{ij} = \rho u_i u_j = \frac{1}{8\pi G}u_i u_j$ we have

$$R_{ij} - \frac{1}{2}Rg_{ij} + \Lambda g_{ij} = u_i u_j = 8\pi G T_{ij}.$$

Therefore Gödel's metric is a solution for Einstein's field equations when the cosmological constant is $\Lambda = \dfrac{1}{2}$ and $T_{ij} = \dfrac{1}{8\pi G} u_i u_j$.

Consider Gödel's change of coordinates $(t, r, \phi, y) \rightarrow (x^0, x^1, x^2, x^3)$

$$\begin{cases} x^0 = 2t - \phi\sqrt{2} + 2\sqrt{2}\arctan\left(\tan\left(\dfrac{\phi}{2}\right) e^{-2r}\right), \phi \neq \pi;\ x^0 = 2t \ if\ \phi = \pi \\ x^1 = \ln\left[\cosh(2r) + \cos\phi\sinh(2r)\right] \\ x^2 = \dfrac{\sqrt{2}\sin\phi\sinh(2r)}{\cosh(2r) + \cos\phi\sinh(2r)} \\ x^3 = 2y. \end{cases}$$

It can be written in the form

$$\begin{cases} \tan\left(\dfrac{\phi}{2} + \dfrac{x^0 - 2t}{2\sqrt{2}}\right) = \tan\left(\dfrac{\phi}{2}\right) e^{-2r} \\ e^{x^1} = \cosh(2r) + \cos\phi\sinh(2r) \\ x^2 e^{x^1} = \sqrt{2}\sin\phi\sinh(2r) \\ x^3 = 2y. \end{cases}$$

Let us look at the coordinates x^1 and x^2 when $r \geq 0$; $0 \leq \phi \leq \pi$. It can be seen a 2π periodicity of x^1 and x^2 when r is fixed. These coordinates can be called cylindrical coordinates for the manifold M. Computing Gödel's metric in the new coordinates, we find another form of the previous solution of the Einstein field equations, that is

$$ds^2 = 4\left[dt^2 - dr^2 - dy^2 + (\sinh^4 r - \sinh^2 r)d\phi^2 + 2\sqrt{2}\sinh^2 r d\phi dt\right],$$

called second Gödel's second metric.

We do not present the computations because they are heavy to be reported. We leave the calculations as an exercise for the reader. We can prove:

Theorem 9.15.1 *Denote by $M := \mathbb{R}^4$ the set having the coordinates (x^0, x^1, x^2, x^3). Then*

1. For any two events A and B there is a transformation on M carrying A into B, that is there are not privileged points. From the physical point of view, it means that M is homogeneous.

2. M has rotational symmetry, i.e. there exists a transformation of coordinates depending on one parameter only such that A is carried into A.

Proof 1. Consider the transformation

$$\begin{cases} \bar{x}^0 = x^0 + a \\ \bar{x}^1 = x^1 \\ \bar{x}^2 = x^2 \\ \bar{x}^3 = x^3 \end{cases}$$

Let us check that this is an isometry of M. We observe $d\bar{x}^k = dx^k$ and $d\bar{s}^2 = ds^2$. Then, consider two points $A_0(x^0(\lambda_0), x^1(\lambda_0), x^2(\lambda_0), x^3(\lambda_0))$ and $A_1(x^0(\lambda_1), x^1(\lambda_1), x^2(\lambda_1), x^3(\lambda_1))$ joined by the curve $c(\lambda) = (x^0(\lambda), x^1(\lambda), x^2(\lambda), x^3(\lambda))$, $\lambda \in [\lambda_0, \lambda_1]$ and their images $\bar{A}_0(\bar{x}^0(\lambda_0), \bar{x}^1(\lambda_0), \bar{x}^2(\lambda_0), \bar{x}^3(\lambda_0))$, $\bar{A}_1(\bar{x}^0(\lambda_1), \bar{x}^1(\lambda_1), \bar{x}^2(\lambda_1), \bar{x}^3(\lambda_1))$ joined by the curve $\bar{c}(\lambda) = (\bar{x}^0(\lambda), \bar{x}^1(\lambda), \bar{x}^2(\lambda), \bar{x}^3(\lambda))$, $\lambda \in [\lambda_0, \lambda_1]$. It results

$$l_c(A_0, A_1) = \int_{\lambda_0}^{\lambda_1} ||\dot{c}(\lambda)|| d\lambda,$$

where the norm is expressed with respect ds^2. The same,

$$l_{\bar{c}}(\bar{A}_0, \bar{A}_1) = \int_{\lambda_0}^{\lambda_1} ||\dot{\bar{c}}(\lambda)||_1 d\lambda,$$

where this second norm is expressed with respect $d\bar{s}^2$. Since $||\dot{c}(\lambda)|| = ||\dot{\bar{c}}(\lambda)||_1$ it results $l_c(A_0, A_1) = l_{\bar{c}}(\bar{A}_0, \bar{A}_1)$, that is we deal with an isometry of M. Three other isometries can be highlighted:

$$\begin{cases} \bar{x}^0 = x^0 \\ \bar{x}^1 = x^1 + b \\ \bar{x}^2 = e^{-b} x^2 \\ \bar{x}^3 = x^3 \end{cases}, \quad \begin{cases} \bar{x}^0 = x^0 \\ \bar{x}^1 = x^1 \\ \bar{x}^2 = x^2 + c \\ \bar{x}^3 = x^3 \end{cases}, \quad \begin{cases} \bar{x}^0 = x^0 \\ \bar{x}^1 = x^1 \\ \bar{x}^2 = x^2 \\ \bar{x}^3 = x^3 + d \end{cases}.$$

Combining all four previous transformations, any point of M can be mapped into any point of M without changing metric properties of M, i.e. any point can be seen as an origin. Therefore M is homogeneous.

2. The previous discussion allows us to consider any point A as the origin of M. Therefore in the new coordinates (t, r, ϕ, y), $r_A = 0$. Consider the group of transformations with respect to the parameter $K \in \mathbb{R}$,

$$(t, r, \phi, y) \rightarrow (t, r, \phi + K, y).$$

The point A is a fixed point of this group, and according to the previous observation, it exists a 2π periodicity experienced by any other point. Therefore M allows rotations with respect any given point of it. □

A spacetime is time orientable, if the time-like and null vectors can be classified into two classes, the future pointing and the past pointing vectors as we did it in the case of the 2-Minkowski space (that is with respect to a given vector).

Let us remember: in the case of the Minkowski metric $ds^2 = dt^2 - dx^2$, provided by the Minkowski product $< p, q >_M = t_p t_q - x_p x_q$, the vector $e_1 = (1, 0)$ is a time-like vector because $< e_1, e_1 >_M > 0$ and the time-like vector $v = (3, 2)$ becomes a future-pointing time-like vector because $< v, e_1 >_M > 0$. The vector $-v$ becomes a past-pointing time-like vector. The vector $w = (1, 1)$ is a null future-pointing vector, etc.

A curve ψ is called time-like if the tangent vectors $\dot{\psi}$ are time-like future-pointing vectors.

If we choose two events $E_0(\lambda_0)$ and $E_1(\lambda_1)$ connected by a time-like curve, we say that E_0 is in the past of E_1, (or equivalently, E_1 is in the future of E_0) if $\lambda_0 < \lambda_1$.

In the case of Gödel's second metric, the vector $u^i = (1, 0, 0, 0)$ has the property $g_{ij} u^i u^j = 4 > 0$ that is u^i is a time-like vector. If v^j is a time-like vector, that is $g_{ij} v^i v^j > 0$, we say that v^j is future pointing if $g_{ij} u^i v^j > 0$. If $g_{ij} u^i v^j < 0$ the vector v^j is called past pointing. The same, if w^k is a null vector, we can define past pointing and future pointing null vectors according to the sign of $g_{ij} u^i w^k$. Therefore M becomes time orientable.

Theorem 9.15.2 *The time orientable Gödel's universe allows:*
1. closed time-like curves;
2. time-like loops, i.e. any two events connected by a time-like curve can be connected by a closed time-like curve.

Proof 1. Consider the curve $\alpha(s) := (0, R, bs, 0)$.

Its velocity vector is $\dot{\alpha}(s) = v^j = (0, 0, b, 0)$. The norm of this vector depends on

$$g_{ij} v^i v^j = (\sinh^4 R - \sinh^2 R) b^2.$$

If we choose, from the beginning $R > \ln(1 + \sqrt{2})$, i.e. $\sinh R > 1$, this vector is a time-like one. The chosen curve is a time-like curve according to the second statement of the previous theorem and $\alpha(0) = \alpha(2\pi)$. We have obtained a closed time-like curve in M.

2. First, if we look only at the coordinates (r, ϕ, t), we observe that they determine completely the coordinates (x^1, x^2, x^3). More precisely, for particular given (r, ϕ, t), we have particular corresponding (x^1, x^2, x^3). It remains the coordinate x^0 which depends on t; therefore the t−lines of matter, in cylindrical coordinates, are x^0−lines of matter.

Consider the point B_{t_1} with the coordinates $(t_1, R, 0, 0)$. The curve

$$\gamma(s) = \left(t_2 + \frac{t_1 - t_2}{2\pi n} s, R, bs, 0 \right)$$

is time-like because the vector

$$\dot{\gamma}(s) = v^i = \left(\frac{t_1 - t_2}{2\pi n}, 0, b, 0 \right)$$

is time-like. Indeed, for a chosen n, big enough, and $R > \ln(1 + \sqrt{2})$, we have

$$g_{ij}v^i v^j = 4\left[\frac{(t_1 - t_2)^2}{4\pi^2 n^2} + (\sinh^4 R - \sinh^2 R)b^2 - \sinh^2 R \cdot \left(\frac{t_1 - t_2}{2\pi n}\right)b\right] > 0.$$

We observe that $\gamma(0) = B_{t_2}$ and $\gamma(2n\pi) = B_{t_1}$.

Using the same idea, we can derive time-like curves between B_{t_2} and B_{t_3} and between B_{t_3} and B_{t_1}. The concatenation of the three time-like curves is a time-like loop starting from B_{t_1}, passing through B_{t_2}, then to B_{t_3}, to finally reach B_{t_1}. We can conclude that t is not a proper time coordinate, because if it is so, moving forward in time we return in our past. Therefore no global time-coordinate exists in Gödel's universe. □

More about this very nice and difficult subject can be found in [45, 61, 72]. An exhaustive discussion on closed time-like curves and their physical implications can be found in [74]. Here we developed this Universe model because it can be easily framed in our geometric picture.

9.16 Black Holes: A Mathematical Introduction

Black hole physics is undoubtedly one of the more fascinating topic of General Relativity. Here we give only a short mathematical introduction and refer the interested reader to the book by Frolov and Novikov *Black Hole Physics* [74] for a detailed discussion.

9.16.1 Escape Velocity and Black Holes

Suppose we stay on the surface of the Earth imagined as a sphere and we vertically throw a ball. Depending on the speed of throwing, the ball can be higher and higher throwed, but after it reaches a maximum altitude it falls down attracted by the Earth. Which is the speed necessary such that the ball never return?

So, the gravitational force acts between the two involved bodies, the Earth and the ball. If the ball is at distance r from the center of the Earth, we have

$$F = \frac{GMm}{r^2}; \quad K_E = \frac{mv^2}{2}; \quad P_E = -\frac{GMm}{r}.$$

Consider, on the surface of the Earth, the escape velocity v_e. The ball goes higher and higher loosing in time its speed. At infinity, its kinetic energy is 0, the same for the potential energy $-\dfrac{GMm}{r}$. This means that, at infinity, the total energy is 0. At each point of this rectilinear trajectory, the total energy, which is a constant, has to be 0, that is the *escape velocity* can be computed from the condition

$$\frac{mv^2}{2} - \frac{GMm}{r} = 0.$$

It results the escape velocity formula

$$v_e^2 = \frac{2GM}{R_E},$$

where R_E is the radius of the Earth. If someone replace the values, the escape velocity from the Earth gravitational field is almost $11(km)/(s)$.

By definition, a black hole is a "cosmic body" having the escape velocity $> c$, where c is the speed of light in vacuum. According to the fact that there are no speeds greater than the speed of light in vacuum, let us compute how small should be the Earth such that $v_e = c$. We obtain

$$r = \frac{2GM}{c^2} \approx \frac{2 \cdot 6.67 \cdot 10^{-11} 6 \cdot 10^{24}}{9 \cdot 10^{16}} \approx 8.8(mm).$$

Therefore, if all the mass of the Earth is concentrated in a sphere with $8.8\,(mm)$ radius, the Earth should be a black hole and not even photons can leave its surface.

Let us see the difference between the Earth, as we know, and the Earth as a black hole. We have to compute in both cases the gravitational force exerted to a $1\,(kg)$ body.

For the "usual Earth":

$$F = \frac{6.67 \cdot 10^{-11} \cdot 6 \cdot 10^{24} \cdot 1}{(24 \cdot 10^6)^2} \approx 9.8(kg \cdot m)/(s^2) \approx g$$

where g is the constant gravitational acceleration as we expected.

For the "black hole Earth", we have:

$$F = \frac{6.67 \cdot 10^{-11} \cdot 6 \cdot 10^{24} \cdot 1}{(8 \cdot 10^{-3})^2} \approx \frac{1}{2} \cdot 10^{19}(kg \cdot m)/(s^2),$$

that is a tremendous huge force exerted by the black hole to all the bodies on its surface.

Suppose now the Earth transformed instantaneously into a black hole. Is the Moon trajectory affected? This is only a mathematical discussion, of course. Let us look at the formula

$$F = \frac{GMm}{r^2} = \frac{mv^2}{r}.$$

It is the same in both cases, because r is measured from the center of the Earth. We deduce that the Moon continues to orbit the black hole Earth such it does now.

Some other considerations about black holes can be seen when we study them using metrics.

9.16.2 Rindler's Metric and Pseudo-Singularities

Let us define the *Rindler metric* as

$$ds^2 = \frac{(\bar{x}^1)^2}{b^2}(\bar{x}^0)^2 - (\bar{x}^1)^2.$$

What is happening with this metric when $\bar{x}^1 \to 0$? If we are looking only at the first term, the first coefficient approaches 0. We can think that the metric fails to exist. A singularity seems to be highlighted. However, we will show that a suitable change of coordinates transforms the Rindler metric into the ordinary Minkowski metric. We may conclude that $\bar{x}^1 = 0$ is not a physical singularity, but a *pseudo-singularity* (or a *geometric singularity*), that is one which can be removed by a convenient change of coordinates.

Consider the change of coordinates

$$C : \begin{cases} \bar{x}^0(x^0, x^1) = b \tanh^{-1} \dfrac{x^0}{x^1} \\ \bar{x}^1(x^0, x^1) = \sqrt{(x^1)^2 - (x^0)^2}. \end{cases}$$

where $\tanh^{-1}(y) = \dfrac{1}{2} \ln \dfrac{1+y}{1-y}$. If we compute

$$dC = \begin{pmatrix} \dfrac{\partial \bar{x}^0}{\partial x^0} & \dfrac{\partial \bar{x}^0}{\partial x^1} \\ \dfrac{\partial \bar{x}^1}{\partial x^0} & \dfrac{\partial \bar{x}^1}{\partial x^1} \end{pmatrix}$$

the four components are

$$\frac{\partial \bar{x}^0}{\partial x^0} = \frac{bx^1}{(x^1)^2 - (x^0)^2}; \quad \frac{\partial \bar{x}^0}{\partial x^1} = \frac{-bx^0}{(x^1)^2 - (x^0)^2};$$

$$\frac{\partial \bar{x}^1}{\partial x^0} = \frac{-x^0}{\sqrt{(x^1)^2 - (x^0)^2}}; \quad \frac{\partial \bar{x}^1}{\partial x^1} = \frac{x^1}{\sqrt{(x^1)^2 - (x^0)^2}}.$$

Exercise 9.16.1 Compute $dC^t \cdot R \, dC$.

Solution. We have to compute

$$
\begin{pmatrix} \frac{bx^1}{(x^1)^2-(x^0)^2} & \frac{-x^0}{\sqrt{(x^1)^2-(x^0)^2}} \\ \frac{-bx^0}{(x^1)^2-(x^0)^2} & \frac{x^1}{\sqrt{(x^1)^2-(x^0)^2}} \end{pmatrix}
\begin{pmatrix} \frac{(\bar{x}^1)^2}{b^2} & 0 \\ 0 & -1 \end{pmatrix}
\begin{pmatrix} \frac{bx^1}{(x^1)^2-(x^0)^2} & \frac{-bx^0}{(x^1)^2-(x^0)^2} \\ \frac{-x^0}{\sqrt{(x^1)^2-(x^0)^2}} & \frac{x^1}{\sqrt{(x^1)^2-(x^0)^2}} \end{pmatrix},
$$

that is

$$
\begin{pmatrix} \frac{bx^1}{(x^1)^2-(x^0)^2} & \frac{-x^0}{\sqrt{(x^1)^2-(x^0)^2}} \\ \frac{-bx^0}{(x^1)^2-(x^0)^2} & \frac{x^1}{\sqrt{(x^1)^2-(x^0)^2}} \end{pmatrix}
\begin{pmatrix} \frac{(x^1)^2-(x^0)^2}{b^2} & 0 \\ 0 & -1 \end{pmatrix}
\begin{pmatrix} \frac{bx^1}{(x^1)^2-(x^0)^2} & \frac{-bx^0}{(x^1)^2-(x^0)^2} \\ \frac{-x^0}{\sqrt{(x^1)^2-(x^0)^2}} & \frac{x^1}{\sqrt{(x^1)^2-(x^0)^2}} \end{pmatrix}
$$

$$
= \begin{pmatrix} \frac{bx^1}{(x^1)^2-(x^0)^2} & \frac{-x^0}{\sqrt{(x^1)^2-(x^0)^2}} \\ \frac{-bx^0}{(x^1)^2-(x^0)^2} & \frac{x^1}{\sqrt{(x^1)^2-(x^0)^2}} \end{pmatrix}
\begin{pmatrix} \frac{x^1}{b} & -\frac{x^0}{b} \\ \frac{x^0}{\sqrt{(x^1)^2-(x^0)^2}} & \frac{-x^1}{\sqrt{(x^1)^2-(x^0)^2}} \end{pmatrix}
= \begin{pmatrix} 1 & 0 \\ 0 & -1 \end{pmatrix}.
$$

□

The metric in coordinates (x^0, x^1) becomes

$$
ds^2 = (dx^0)^2 - (dx^1)^2.
$$

Therefore we have proved

Theorem 9.16.2 *The change of coordinates*

$$
C : \begin{cases} \bar{x}^0(x^0, x^1) = b \tanh^{-1} \dfrac{x^0}{x^1} \\ \bar{x}^1(x^0, x^1) = \sqrt{(x^1)^2 - (x^0)^2} \end{cases}
$$

transforms the Rindler metric

$$
ds^2 = \frac{(\bar{x}^1)^2}{b^2} (\bar{x}^0)^2 - (\bar{x}^1)^2
$$

into the Minkowski metric

$$
ds^2 = (dx^0)^2 - (dx^1)^2.
$$

As we discussed, $\bar{x}^1 = 0$ is not a physical singularity, but a pseudo-singularity. The lines $x^1 = x^0$ and $x^1 = -x^0$ are called the *horizon of the geometric singularity* $\bar{x}^1 = 0$. Removing of geometric singularities is part of the mathematical theory of black holes we present.

9.16.3 Black Holes Studied Through the Schwarzschild Metric

When we studied the vacuum field equations $R_{ij} = 0$, we started from the Schwarzschild intuition to look for a spherical symmetric solution which describes the relativistic field outside of a non-rotating, massive body.

In the coordinate system $(x^0 = ct, r, \varphi, \theta)$ Schwarzschild chose the form of the solution as

$$ds^2 = c^2 \cdot e^T dt^2 - e^Q dr^2 - r^2 d\varphi^2 - r^2 \sin^2 \varphi d\theta^2,$$

with $T := T(r)$, $Q := Q(r)$ two real functions we need to determine. (In [40], a more general approach is presented considering $T := T(r, t)$ and $Q := Q(r, t)$).

The non-zero Christoffel symbols are

$$\Gamma^0_{01} = \Gamma^0_{10} = \frac{T'}{2}, \quad \Gamma^1_{00} = -\frac{T'}{2} e^{T-Q}, \quad \Gamma^1_{11} = \frac{Q'}{2}, \quad \Gamma^1_{22} = -re^{-Q}, \quad \Gamma^1_{33} = -re^{-Q} \sin^2 \varphi,$$

$$\Gamma^2_{21} = \Gamma^2_{12} = \frac{1}{r}, \quad \Gamma^2_{33} = -\sin \varphi \cos \varphi, \quad \Gamma^3_{31} = \Gamma^3_{13} = \frac{1}{r}, \quad \Gamma^3_{23} = \Gamma^3_{32} = \cot \varphi,$$

where we denote by $'$ the derivative with respect to r. The computations lead to

$$T = -Q = \ln\left(1 + \frac{B}{r}\right).$$

The obtained Schwarzschild metric is

$$ds^2 = c^2 \cdot \left(1 + \frac{B}{r}\right) dt^2 - \frac{1}{\left(1 + \dfrac{B}{r}\right)} dr^2 - r^2 d\varphi^2 - r^2 \sin^2 \varphi d\theta^2.$$

Then we find $B = \dfrac{-2GM}{c^2}$, that is the gravitational Newtonian potential $\phi(x, y, z) = -\dfrac{GM}{r}$ is involved in the coefficients of Schwarzschild metric. We have remembered here these results for two reasons. The first one is the following exercise we need to understand the behavior of the Riemann curvature tensor at the surface of a black hole.

Exercise 9.16.3 Compute R^0_{101}.

Solution. Replacing in R^i_{jkl} formula (in the case $i = k = 0$, $j = l = 1$) the above corresponding Christoffel symbols, we find

$$R^0_{101} = -\frac{1}{2}T'' + \frac{1}{4}T'Q' - \frac{1}{4}(Q')^2,$$

that is

$$R^0_{101} = \frac{-B}{r^3}\left(\frac{1}{1 + \dfrac{B}{r}}\right).$$

□

The second reason is related to the quantity $r_s := \dfrac{2GM}{c^2}$, called the Schwarzschild radius, which gives Schwarzschild metric in the form

$$ds^2 = c^2 \cdot \left(1 - \frac{r_s}{r}\right)dt^2 - \frac{1}{\left(1 - \dfrac{r_s}{r}\right)}dr^2 - r^2 d\varphi^2 - r^2 \sin^2\varphi d\theta^2.$$

If we remember the nature of the Schwarzschild metric, the first term is positive, the others three are negative. If we look at the first term in the case of the Sun, if we replace the gravitational constant G, $M = M_{sun}$, the speed of light in vacuum c and $r = r_{sun}$ we have $1 - \dfrac{r_s}{r} > 0$. If r approaches $r_s = \dfrac{2GM}{c^2}$ the first two terms of the metric have the properties $g_{00} \to 0$ and $g_{11} \to -\infty$. So, the Schwarzschild metric becomes singular. If we compute r_s in the case of our Sun, we find $r_s \approx 3\,km$. So, the anomaly appears when the entire mass of our Sun is concentrated in a sphere with a radius as r_s.

Such a sphere is a *black hole*. The interior of the sphere is called an *interior of the black hole* and it is characterized by the condition $r < r_s$. The surface of the sphere is called the *event horizon* and it is characterized by the condition $r = r_s$. The *exterior of the black hole* is characterized by the condition $r > r_s$.

A clock at $r = r_s$ has its proper time $d\tau = \sqrt{1 - \dfrac{r_s}{r}}dt \to 0$. Which means that the clock is slowed down at maximum; a clock outside the black hole works faster.

What is going on in the interior of the black hole?

We obseve $g_{00} = c^2\left(1 - \dfrac{r_s}{r}\right) < 0$ and $g_{11} = -\dfrac{1}{1 - \dfrac{1}{r_s}} > 0$. Therefore the signs are opposite with respect to the standard ones. It results that t becomes a spatial coordinate and r becomes a temporal coordinate inverting their roles!

However we can prove that the singularity $r = r_s$ is in fact a pseudo-singularity. First, let us see what is happening to Riemann curvature tensor at $r = r_s$.

If we denote $t := x^0$; $r := x^1$; $\varphi := x^2$; $\theta := x^3$ the old coordinates, we may construct the new coordinates:

$$\bar{x}^0 = (x^0 - t_0)\sqrt{1 - \frac{r_s}{r}}; \quad \bar{x}^1 = \frac{x^1 - r_0}{1 - \dfrac{r_s}{r}}; \quad \bar{x}^2 = r_0\left(x^2 - \frac{\pi}{2}\right); \quad \bar{x}^3 = r_0\left(x^3 - \theta_0\right).$$

It results

$$\frac{\partial x^0}{\partial \bar{x}^0} = \frac{1}{\sqrt{1 - \dfrac{r_s}{r}}}; \quad \frac{\partial x^1}{\partial \bar{x}^1} = \sqrt{1 - \frac{r_s}{r}}; \quad \frac{\partial x^2}{\partial \bar{x}^2} = \frac{1}{r_0}; \quad \frac{\partial x^3}{\partial \bar{x}^3} = \frac{1}{r_0};$$

all the other possible partial derivatives are null.

The new Riemann curvature tensor $\bar{R}^0_{101}$ is

$$\bar{R}^0_{101} = \frac{\partial \bar{x}^0}{\partial \bar{x}^i}\frac{\partial x^j}{\partial \bar{x}^1}\frac{\partial x^k}{\partial \bar{x}^0}\frac{\partial x^l}{\partial \bar{x}^1} R^i_{jkl} = \frac{\partial \bar{x}^0}{\partial x^0}\frac{\partial x^1}{\partial \bar{x}^1}\frac{\partial x^0}{\partial \bar{x}^0}\frac{\partial x^1}{\partial \bar{x}^1} R^0_{101} = \left(1 - \frac{r_s}{r}\right)\frac{r_s}{r_0^3}\frac{1}{1 - \dfrac{r_s}{r}} = \frac{r_s}{r_0^3}.$$

Therefore, at the surface of the black hole, that is when $r = r_s$, the Riemann curvature tensor is well defined. The surface of a black hole is not a physical singularity.

At the surface of a black hole, that is when $r = r_s$, the Kruskal–Szekeres metric gives more information than Schwarzschild metric.

We act only on the first two coordinates of the Schwarzschild metric, the other two remain unchanged. The Kruskal–Szeres coordinates look different inside the black hole comparing to the case of the exterior of the black hole.

In the interior of the black hole the Kruskal–Szekeres coordinates are:

$$KS(r < r_s): \begin{cases} V(t,r) = \sqrt{1 - \dfrac{r}{r_s}}\, e^{r/2r_s}\cosh\dfrac{t}{2r_s} \\[3mm] U(t,r) = \sqrt{1 - \dfrac{r}{r_s}}\, e^{r/2r_s}\sinh\dfrac{t}{2r_s} \end{cases}$$

$$U^2 - V^2 = \left(\frac{r}{r_s} - 1\right)e^{r/r_s} < 0; \quad \frac{U}{V} = \tanh\frac{t}{2r_s}, \text{ i.e. } t = 2r_s\tanh^{-1}\frac{V}{U}.$$

When $r < r_s$, that is in the interior of the black hole, we have

$$\frac{\partial V}{\partial t} = A\cdot\sinh\frac{t}{2r_s} \text{ and } \frac{\partial U}{\partial t} = A\cdot\cosh\frac{t}{2r_s} \text{ where } A = \sqrt{1 - \frac{r}{r_s}}\, e^{r/r_s}\frac{t}{2r_s}.$$

$$\frac{\partial V}{\partial r} = B\cdot\cosh\frac{t}{2r_s} \text{ and } \frac{\partial U}{\partial r} = B\cdot\sinh\frac{t}{2r_s} \text{ where } B = \frac{-\dfrac{1}{r_s}}{2\sqrt{1 - \dfrac{r}{r_s}}}e^{r/2r_s} + \frac{1}{2r_s}\sqrt{1 - \frac{r}{r_s}}\, e^{r/2r_s}.$$

$$\begin{pmatrix} A\sinh\dfrac{t}{2r_s} & A\cosh\dfrac{t}{2r_s} \\[3mm] B\cosh\dfrac{t}{2r_s} & B\sinh\dfrac{t}{2r_s} \end{pmatrix} \begin{pmatrix} \dfrac{4r_s^3}{r}e^{-r/r_s} & 0 \\[3mm] 0 & -\dfrac{4r_s^3}{r}e^{-r/r_s} \end{pmatrix} \begin{pmatrix} A\sinh\dfrac{t}{2r_s} & B\cosh\dfrac{t}{2r_s} \\[3mm] A\cosh\dfrac{t}{2r_s} & B\sinh\dfrac{t}{2r_s} \end{pmatrix}$$

$$= \frac{4r_s^3}{r}e^{-r/r_s}\cdot\begin{pmatrix} -A^2 & 0 \\ 0 & B^2 \end{pmatrix} = \begin{pmatrix} 1 - \dfrac{r_s}{r} & 0 \\[3mm] 0 & -\dfrac{1}{1 - \dfrac{r_s}{r}} \end{pmatrix},$$

the last matrix equality because

$$-\frac{4r_s^3}{r}e^{-r/r_s} \cdot A^2 = 1 - \frac{r_s}{r}$$

and

$$B^2 \cdot \frac{4r_s^3}{r}e^{-r/r_s} = -\frac{1}{1 - \dfrac{r_s}{r}}.$$

We have proved the following

Theorem 9.16.4 *When the metric is the Schwarzschild one, in the interior of a black hole described by the condition $r < r_s$, the Kruskal–Szekeres coordinates*

$$KS(r < r_s) : \begin{cases} V(t,r) = \sqrt{1 - \dfrac{r}{r_s}e^{r/2r_s}} \cosh\dfrac{t}{2r_s} \\[2ex] U(t,r) = \sqrt{1 - \dfrac{r}{r_s}e^{r/2r_s}} \sinh\dfrac{t}{2r_s} \end{cases}$$

transforms the Schwarzschild metric into the Kruskal–Szekeres metric

$$ds^2 = \frac{4r_s^3}{r}e^{-r/r_s}\left(dV^2 - dU^2\right) - r^2 d\varphi^2 - r^2\sin^2\varphi d\theta^2.$$

In the case of the exterior of the black hole, the Kruskal–Szekeres coordinates are:

$$KS(r > r_s) : \begin{cases} V(t,r) = \sqrt{\dfrac{r}{r_s} - 1}\,e^{r/2r_s} \cosh\dfrac{t}{2r_s} \\[2ex] U(t,r) = \sqrt{\dfrac{r}{r_s} - 1}\,e^{r/2r_s} \sinh\dfrac{t}{2r_s} \end{cases}$$

$$U^2 - V^2 = \left(\frac{r}{r_s} - 1\right)e^{r/r_s} > 0; \quad \frac{U}{V} = \tanh\frac{t}{2r_s} \text{ i.e. } t = 2r_s\tanh^{-1}\frac{V}{U}. \text{ Similar}$$
computations lead to

Theorem 9.16.5 *When the metric is the Schwarzschild one, in the exterior of a black hole described by the condition $r > r_s$, the Kruskal–Szekeres coordinates*

$$KS(r > r_s) : \begin{cases} V(t,r) = \sqrt{\dfrac{r}{r_s} - 1}\,e^{r/2r_s} \cosh\dfrac{t}{2r_s} \\[2ex] U(t,r) = \sqrt{\dfrac{r}{r_s} - 1}\,e^{r/2r_s} \sinh\dfrac{t}{2r_s} \end{cases}$$

transforms the Schwarzschild metric into the same Kruskal–Szekeres metric

$$ds^2 = \frac{4r_s^3}{r} e^{-r/r_s} \left(dV^2 - dU^2 \right) - r^2 d\varphi^2 - r^2 \sin^2 \varphi d\theta^2.$$

In both cases at $r = r_s$ the singularity has been removed. The only physical singularity corresponds to $r = 0$.

Corollary 9.16.6 *The event horizon of a black hole is a geometric singularity only.*

Let us discuss now the entropy of a black hole. From the Special Relativity chapter we know the following formulas and their meaning:

$$E = hf = h\frac{c}{\lambda}; \ E = mc^2.$$

From this section, we know that the dimension of a black hole is related to its Schwarzschild radius r_s. If a photon is captured by a black hole of mass M and energy E, the black hole changes its mass and energy. This fact can be represented by the formula

$$\Delta M = \frac{\Delta E}{c^2} = \frac{h}{\lambda c} = \frac{h}{r_s c},$$

because λ becomes r_s.

But $\lambda = r_s = \dfrac{2MG}{c^2}$ suggests a variation of the radius described by the formula

$$\Delta r = \frac{2\Delta M G}{c^2} = \frac{2G}{c^2}\Delta M = \frac{2G}{c^2}\frac{h}{r_s c},$$

that is

$$r_s \Delta r = \frac{2Gh}{c^3}.$$

It is worth noticing that the right member is a constant. Furthermore, it is easy to see that the area of a black hole is the area of a sphere of radius r_s,

$$A = 4\pi r_s^2.$$

The derivative with respect to r leads to

$$\frac{dA}{dr} = 8\pi r_s,$$

i.e.

$$dA = 8\pi r_s dr = \frac{16\pi Gh}{c^3}.$$

Therefore, if a photon is captured by a black hole, the black hole area is increasing by the quantity

$$dA = \frac{16\pi G h}{c^3}.$$

If dS is a unitary entropy, the variation of the area becomes

$$dA = \frac{16\pi G h}{c^3} dS,$$

therefore

$$S = \frac{c^3}{16\pi G h} A.$$

This formula is known as *Bekenstein–Hawking formula* for the *black hole entropy*. See [74] for a detailed discussion.

Whit these considerations in mind, it is straightforward to define the *black hole temperature*. The formula which connects the variation of energy, the temperature and the variation in entropy is

$$dE = T dS.$$

If we consider only "one unit" of variation for entropy, i.e. a single photon which changes the black hole energy, we have that

$$dE = T = \frac{hc}{\lambda} = \frac{hc}{r_s} = \frac{hc^3}{2MG}.$$

That is, a single photon changes the temperature of the black hole such that T is proportional to $\frac{1}{M}$. We deduce that smaller black holes are warmer than the massive ones.

9.16.4 The Light Cone in the Schwarzschild Metric

In Minkowski metric, the trajectories of light-rays are determined by the condition $ds^2 = 0$. Since the Minkowski metric in geometric coordinates is

$$ds^2 = (dx^0)^2 - (dx^1)^2,$$

the previous condition becomes

$$x^0 = x^1, \quad x^0 = -x^1,$$

therefore the light cone is highlighted.

In (t, r) coordinates, let us consider the 3-plane $\varphi = 0$ and the corresponding Schwarzschild metric in geometric coordinates

$$ds^2 = \left(1 - \frac{r_s}{r}\right) dt^2 - \frac{1}{\left(1 - \frac{r_s}{r}\right)} dr^2.$$

The condition $ds^2 = 0$ leads to

$$\left(1 - \frac{r_s}{r}\right)^2 dt^2 = dr^2,$$

i.e.

$$\frac{dr}{dt} = \pm \left(1 - \frac{r_s}{r}\right).$$

It is obvious that this is different with respect to the Minkowski condition

$$\frac{dr}{dt} = \pm 1.$$

We can write the equivalent formula

$$\pm \frac{d\left(\frac{r}{r_s}\right)}{\left(1 - \frac{r_s}{r}\right)} = d\left(\frac{t}{r_s}\right).$$

Let us denote $X := \left(\frac{r}{r_s}\right)$ and $Y := \left(\frac{t}{r_s}\right)$. It remains to solve

$$\pm \int \frac{X \, dX}{X + 1} = \int dY.$$

The solutions are $Y = X + \ln |x - 1|$ and $Y = -X - \ln |x - 1|$.

The graphs of the functions $f(X) = X + \ln(1 - X)$ and $g(X) = -X + \ln(1 - X)$, both defined on $(0, 1)$, highlight the light cone in the interior of the black hole.

Suppose $x_0 \in (0, 1)$ and the tangent lines at $(x_0, f(x_0))$ and $(x_0, f(x_0))$. The parallel lines to the tangents in $(x_0, 0)$ show how the light cone looks like in the interior of the black hole.

The same, the graphs of the functions $h(X) = X + \ln(X - 1)$ and $l(X) = -X + \ln(X - 1)$, both defined when $x \in (1, \infty)$ highlight the light cone outside the black hole.

In summary, we gave only some main features of black holes but the physics and the mathematics of these gravitational systems is extremely reach and deserve to be explored in details. Furthermore, after the direct detection of the black hole shadow by the Event Horizon Telescope collaboration [75], a new era started in this fascinating sector of Physics. These gravitational objects, considered only exotic theoretical objects until recently, have become an amazing arena for observational astrophysics and cosmology.

Chapter 10
A Geometric Realization of Relativity: The Affine Universe and de Sitter Spacetime

Rem tene, verba sequentur.

Cato

We want to conclude this book considering a gravity theory without masses which can be constructed in Minkowski spaces using a geometric Minkowski potential. From the point of view of this book, this can be considered a full geometric realization of the relativistic approach. The affine space-like spheres can be seen as the regions of the Minkowski space-like vectors characterized by a constant Minkowski gravitational potential. They highlight, for each dimension $n \geq 3$, a model of spacetime, the de Sitter one, which satisfies Einstein's field equations in the absence of matter, and it is now intuitive why. This chapter is based on results that can be found in [16, 20, 26, 76].

10.1 About the Minkowski Geometric Gravitational Force

Denote by $\mathbb{M}^n$ the Minkowski n-dimensional space, $n \geq 3$, endowed by the Minkowski product

$$\langle a, b \rangle_M := a_0 b_0 - \sum_{\alpha=1}^{n-1} a_\alpha b_\alpha$$

With respect to given $b = (b_0, b_1, \ldots, b_{n-1})$, we consider all vectors $x = (x_0, x_1, \ldots, x_{n-1})$ such that $x - b$ is a space-like vector, that is $\langle x - b, x - b \rangle_M < 0$. We denote by

$$r := \sqrt{-(x_0 - b_0)^2 + \sum_{\alpha=1}^{n-1} (x_\alpha - b_\alpha)^2}$$

W. Boskoff and S. Capozziello, *A Mathematical Journey to Relativity*, UNITEXT for Physics, https://doi.org/10.1007/978-3-030-47894-0_10

the Minkowski "length" of the space-like vector $x - b$ and by

$$u = -\frac{1}{r}(x_0 - b_0, x_1 - b_1, \ldots, x_{n-1} - b_{n-1})$$

the unit vector of $b - x$. We can define the *Minkowski geometric gravitational* force as

$$F_M^n := \frac{1}{n-1} \frac{1}{r^{n-1}} u.$$

If

$$A_M^n := \frac{n-2}{r^{n-1}} u$$

is by definition the *Minkowski geometric gravitational field*, we have the following "*Minkowski-Newton second principle*":

$$F_M^n = \frac{1}{(n-1)(n-2)} A_M^n.$$

Let us define the *Minkowski gradient* and the *Minkowski Laplacian*:

$$\nabla_M := \left(-\frac{\partial}{\partial x_0}, \frac{\partial}{\partial x_1}, \ldots, \frac{\partial}{\partial x_{n-1}} \right)$$

$$\nabla_M^2 := \langle \nabla_M, \nabla_M \rangle_M = \frac{\partial^2}{\partial x_0^2} - \frac{\partial^2}{\partial x_1^2} - \cdots - \frac{\partial^2}{\partial x_{n-1}^2}.$$

For each dimension n, we can define the *Minkowski gravitational potential*

$$\Phi_M^n := -\frac{1}{r^{n-2}}.$$

The following computations

$$\frac{\partial \Phi_M^n}{\partial x_0} = (2-n)\frac{x_0 - b_0}{r^n}; \quad \frac{\partial \Phi_M^n}{\partial x_\alpha} = (n-2)\frac{x_\alpha - b_\alpha}{r^n}, \quad \alpha \in \{1, 2, \ldots, n-1\};$$

$$\frac{\partial^2 \Phi_M^n}{\partial x_0^2} = (2-n)\frac{r^2 + n(x_0 - b_0)^2}{r^{n+2}}; \quad \frac{\partial^2 \Phi_M^n}{\partial x_\alpha^2} = (n-2)\frac{r^2 - n(x_\alpha - b_\alpha)^2}{r^{n+2}}, \quad \alpha \in \{1, 2, \ldots, n-1\}.$$

Lead us to the following two theorems.

Theorem 10.1.1 *The Minkowski gradient of the Minkowski gravitational potential is the opposite of the Minkowski gravitational field.*

Proof It is easy to check the equality $\nabla_M \Phi_M^n = -A_M^n$.   □

Theorem 10.1.2 *The Minkowski Laplacian of the Minkowski gravitational potential is null.*

Proof The same, it is easy to check $\nabla_M^2 \Phi_M^n = 0$.   □

The last relation is the equation of the Minkowski geometric gravitational field.

In the case when b is the origin of the Minkowski space, the Minkowski unitary space-like sphere can be thought as the set of points of $\mathbb{M}^n$ described by the constant gravitational Minkowski potential $\Phi_M^n = -1$. In this theory, at each dimension n, the Minkowski geometrical gravitational force and the Minkowski geometric gravitational field have the physical dimension $\dfrac{1}{(l)^{n-1}}$. The Minkowski gravitational potential has the physical dimension $\dfrac{1}{(l)^{n-2}}$ where (l) is a length.

We may conclude: For each dimension, in the Minkowski space-like vectors region, a natural geometric Minkowski gravity appears in the absence of matter. An equivalent of the Newton gravity theory can be constructed starting from the Minkowski geometric gravitational potential. The affine space-like spheres can be seen as the regions of the Minkowski space-like vectors characterized by a constant Minkowski gravitational potential. They highlight, at each dimension $n \geq 3$, a model of spacetime, the de Sitter one, which satisfies Einstein's field equations in the absence of matter, and it is now intuitive why.

10.2 The de Sitter Spacetime and Its Cosmological Constant

In the case $n = 3$, we choose to represent the 2-surface as

$$X_0^2 - X_1^2 - X_2^2 = -a^2,$$

in the form $f : \mathbb{R} \times (-\pi, \pi) \longrightarrow \mathbb{M}^3$,

$$f(t, x_1) = (a \sinh t, a \cosh t \cos x_1, a \cosh t \sin x_1).$$

Some computations leads to the metric

$$ds_2^2 = a^2 dt^2 - a^2 \cosh^2 t \, dx_1^2.$$

The non-zero Christoffel symbols are

$$\Gamma_{01}^1 = \Gamma_{10}^1 = \tanh t, \quad \Gamma_{11}^0 = \cosh t \sinh t$$

and

$$R^0_{101} = \frac{\partial \Gamma^0_{11}}{\partial t} - \frac{\partial \Gamma^0_{10}}{\partial x_1} + \Gamma^0_{s0}\Gamma^s_{11} - \Gamma^0_{s1}\Gamma^s_{10} = \cosh^2 t.$$

It results $R_{0101} = g_{00}R^0_{101} = a^2 \cosh^2 t$, that is $K^M_f = -\dfrac{1}{a^2}$.

For this 2-de Sitter spacetime, according to the Einstein theorem for surfaces $R_{ij} = K^M_f\, g_{ij}$, we have

$$R_{ij} + \frac{1}{a^2}\, g_{ij} = 0.$$

This last equation can be written also as $R_{ij} - \dfrac{1}{2}R\, g_{ij} = 0$, that is $\Lambda = 0$ and $T_{ij} = 0$.

In the case $n = 4$, the 3-de Sitter spacetime is the Minkowski space-like sphere of $\mathbb{M}^4$ given by the equation

$$X_0^2 - X_1^2 - X_2^2 - X_3^2 = -a^2.$$

The standard parameterization is

$$f(t, x_1, x_2) = (a \sinh t \cos x_2,\, a \cosh t \cos x_1 \cos x_2,\, a \cosh t \sin x_1 \cos x_2,\, a \sin x_2).$$

The metric is

$$ds_3^2 = a^2 \cos^2 x_2\, dt^2 - a^2 \cosh^2 t \cos^2 x_2\, dx_1^2 - a^2 dx_2^2.$$

We observe

$$ds_3^2 = \cos^2 x_2(\, a^2 dt^2 - a^2 \cosh^2 t\, dx_1^2) - a^2 dx_2^2.$$

therefore

$$ds_3^2 = \cos^2 x_2\, ds_2^2 - a^2 dx_2^2.$$

The non-zero Christoffel symbols are

$$\Gamma^0_{02} = \Gamma^0_{20} = -\tan x_2,\ \ \Gamma^0_{11} = \cosh t \sinh t,$$

$$\Gamma^1_{01} = \Gamma^1_{10} = \tanh t,\ \ \Gamma^1_{12} = \Gamma^1_{21} = -\tan x_2,$$

$$\Gamma^2_{00} = -\sin x_2 \cos x_2,\ \ \Gamma^2_{11} = \cosh^2 t \cos x_2 \sin x_2.$$

Now, if we compute

$$R_{ii} = R^s_{isi} = \frac{\partial \Gamma^s_{ii}}{\partial x^s} - \frac{\partial \Gamma^s_{is}}{\partial x^i} + \Gamma^h_{ii}\Gamma^s_{hi} - \Gamma^h_{is}\Gamma^s_{hi},$$

we find

$$R_{00} = -2\cos^2 x_2; \quad R_{11} = 2\cosh^2 t \cos^2 x_2; \quad R_{22} = 2.$$

The other Rici symbols are null, $R_{ij} = 0, i \neq j$. Therefore

$$R_{ij} + \frac{2}{a^2} g_{ij} = 0.$$

If we compute $R := R^i_i$, taking into account $R^i_j = g^{is} R_{sj}$, it results $R = -\frac{6}{a^2}$.
The left hand of Einstein's field equations become

$$R_{ij} - \frac{1}{2}\left(-\frac{6}{a^2}\right) g_{ij} + \Lambda g_{ij}.$$

If we choose $\Lambda = -\frac{1}{a^2}$, the left hand becomes

$$R_{ij} + \frac{2}{a^2} g_{ij},$$

that is the left hand becomes 0. The de Sitter spacetime presented above satisfies the
Einstein field equations

$$R_{ij} - \frac{1}{2} R\, g_{ij} + \Lambda\, g_{ij} = \frac{8\pi G}{c^4} T_{ik}$$

for $R = -\frac{6}{a^2}$, $\Lambda = -\frac{1}{a^2}$ and $T_{ij} = 0$. A spacetime without matter appears as we
expected.

In the case $n = 5$, the parameterization is

$$\begin{cases} X_0 = a\sinh t \cos x_2 \cos x_3 \\ X_1 = a\cosh t \cos x_1 \cos x_2 \cos x_3 \\ X_2 = a\cosh t \sin x_1 \cos x_2 \cos x_3 \\ X_3 = a\sin x_2 \cos x_3 \\ X_4 = a\sin x_3 \end{cases}$$

The metric related to this parameterization is

$$ds^2_4 = a^2 \cos^2 x_2 \cos^2 x_3 dt^2 - a^2 \cosh^2 t \cos^2 x_2 \cos^2 x_3 dx^2_1 - a^2 \cos^2 x_3 dx^2_2 - a^2 dx^2_3.$$

In the same way, as above, it is

$$ds_4^2 = \cos^2 x_3 (a^2 \cos^2 x_2 dt^2 - a^2 \cosh^2 t \cos^2 x_2 dx_1^2 - a^2 dx_2^2) - a^2 dx_3^2,$$

therefore

$$ds_4^2 = \cos^2 x_3 \, ds_3^2 - a^2 dx_3^2.$$

Again, if we compute

$$R_{ii} = R_{isi}^s = \frac{\partial \Gamma_{ii}^s}{\partial x^s} - \frac{\partial \Gamma_{is}^s}{\partial x^i} + \Gamma_{ii}^h \Gamma_{hi}^s - \Gamma_{is}^h \Gamma_{hi}^s,$$

we find

$$R_{00} = -3 \cos^2 x_2 \cos^2 x_3, \quad R_{11} = 3 \cosh^2 t \cos^2 x_2 \cos^2 x_3, \quad R_{22} = 3 \cos^2 x_3, \quad R_{33} = 3,$$

that is

$$R_{ij} + \frac{3}{a^2} g_{ij} = 0,$$

which leads to

$$R = -\frac{12}{a^2}, \quad \Lambda = -\frac{3}{a^2}, \quad T_{ij} = 0,$$

for Einstein's field equations.

In the general case, the $(n-1)$-*de Sitter spacetime* is the Minkowski $(n-1)$-sphere determined by the ends of all the space-like vectors with Minkowski length a.

This is a hipersurface of the Minkowski n-dimensional space $\mathbb{M}^n$ having the algebraic equation

$$X_0^2 - X_1^2 - \cdots - X_{n-1}^2 = -a^2.$$

The related parameterization is

$$\begin{cases}
X_0 = a \sinh t \cos x_2 \cos x_3 \ldots \cos x_{n-2} \\
X_1 = a \cosh t \cos x_1 \cos x_2 \cos x_3 \ldots \cos x_{n-2} \\
X_2 = a \cosh t \sin x_1 \cos x_2 \cos x_3 \ldots \cos x_{n-2} \\
X_3 = a \sin x_2 \cos x_3 \cos x_4 \ldots \cos x_{n-2} \\
X_4 = a \sin x_3 \cos x_4 \ldots \cos x_{n-2} \\
\ldots\ldots\ldots\ldots\ldots\ldots\ldots\ldots\ldots\ldots \\
X_{n-2} = a \sin x_{n-3} \cos x_{n-2} \\
X_{n-1} = a \sin x_{n-2}.
\end{cases}$$

This parameterization makes sense for $n \geq 5$. For $n \geq 6$ we can denote $X_{0,n} := X_0; \ X_{1,n} := X_1; \ X_{n-1,n} := X_{n-1}$ and we can write

$$\begin{cases} X_{0,n} = X_{0,n-1} \cos x_{n-2} \\ X_{1,n} = X_{1,n-1} \cos x_{n-2} \\ \cdots\cdots\cdots\cdots\cdots \\ X_{n-2,n} = X_{n-2,n-1} \cos x_{n-2} \\ X_{n-1,n} = a \sin x_{n-2} \end{cases}$$

with

$$X_{0,n-1}^2 - X_{1,n-1}^2 - \cdots - X_{n-2,n-1}^2 = -a^2.$$

A direct consequence is

$$X_{0,n-1}dX_{0,n-1} - X_{1,n-1}dX_{1,n-1} - \cdots - X_{n-2,n-1}dX_{n-2,n-1} = 0.$$

Using

$$\begin{cases} dX_{0,n} = dX_{0,n-1} \cos x_{n-2} - X_{0,n-1} \sin x_{n-2} dx_{n-2} \\ \cdots\cdots\cdots\cdots\cdots\cdots\cdots\cdots\cdots\cdots\cdots\cdots\cdots\cdots\cdots\cdots\cdots \\ dX_{n-2,n} = dX_{n-2,n-1} \cos x_{n-2} - X_{n-2,n-1} \sin x_{n-2} dx_{n-2} \\ dX_{n-1,n} = a \cos x_{n-2} dx_{n-2} \end{cases}$$

and denoting by

$$ds_k^2 = dX_{0,k+1}^2 - dX_{1,k+1}^2 - \cdots - dX_{k,k+1}^2$$

we obtain

$$ds_{n-1}^2 = a^2 \cos^2 x_{n-2}\, ds_{n-2}^2 - a^2 dx_{n-2}^2, \ n \geq 6,$$

a formula which is the generalization of the formulas obtained for the previous cases $n = 4$ and $n = 5$.

Therefore, in all cases, we proved that the metric is a diagonal one and we have a recursive method to obtain it. Finally it looks like:

$$ds_{n-1}^2 = a^2 \cos^2 x_{n-2}\, ds_{n-2}^2 - a^2 dx_{n-2}^2, \ n \geq 4$$

and

$$ds_2^2 = a^2 dt^2 - a^2 \cosh^2 t\, dx_1^2, \ n = 3.$$

Now, other considerations are in order. If

$$f(t, x_1, x_2, .., x_{n-2}) = \left(X_{0,n-1} \cos x_{n-2}, \ldots, X_{n-2,n-1} \cos x_{n-2}, a \sin x_{n-2}\right)$$

the direct consequence of above results is $\left\langle f, \dfrac{\partial f}{\partial t} \right\rangle_M = 0.$

Another computation leads to $\left\langle f, \dfrac{\partial f}{\partial x_k} \right\rangle_M = 0$, while $\left\langle \dfrac{\partial f}{\partial x_k}, \dfrac{\partial f}{\partial t} \right\rangle_M = 0$,
$\left\langle \dfrac{\partial f}{\partial x_k}, \dfrac{\partial f}{\partial x_j} \right\rangle_M = 0$ are the consequences of the diagonal form of the metric and highlight the orthogonal frame of the tangent space at each point.

Finally, the Minkowski normal to the hypersurface is

$$N(t, x_1, \ldots, x_{n-2}) = \frac{1}{a} f(t, x_1, \ldots, x_{n-2}),$$

that is the Minkowski distance from the origin to the tangent hyperplane at a given point of the hypersurface is a and all the coefficients of the second fundamental form are computed with the formula established for the case $n = 2$,

$$h_{ij} = \left\langle \frac{\partial N}{\partial x_i}, \frac{\partial f}{\partial x_j} \right\rangle_M,$$

therefore

$$h_{ij} = \frac{1}{a} g_{ij}.$$

Since $\langle N, N \rangle_M = -1 < 0$, we have

$$R_{ijkl} = -\left(h_{ik} h_{jl} - h_{il} h_{jk} \right).$$

It results

$$R_{ijkl} = -\frac{1}{a^2} \left(g_{ik} g_{jl} - g_{il} g_{jk} \right), \ i, j, k, l \in \{0, 1, \ldots, n-2\}.$$

Therefore each sectional curvature is

$$K = -\frac{1}{a^2}.$$

From

$$R_{ijij} = -\frac{1}{a^2} \left(g_{ii} g_{jj} - g_{ij} g_{ji} \right),$$

it results

$$g^{mi} R_{ijij} = -\frac{1}{a^2} \left(g^{mi} g_{ii} g_{jj} - g^{mi} g_{ij} g_{ji} \right),$$

that is

$$R^m_{jij} = -\frac{1}{a^2} \left(\delta^m_i g_{jj} - \delta^m_j g_{ji} \right).$$

For $m = i$, it remains

$$R^m_{jmj} = -\frac{1}{a^2} g_{jj},$$

for each $m \neq j$. Finally,

$$R_{jj} = \sum_{m=0,\ m\neq j}^{n-2} R^m_{jmj} = -\frac{n-2}{a^2} g_{jj}.$$

If we start from

$$R_{ijkl} = -\frac{1}{a^2} \left(g_{ik}g_{jl} - g_{il}g_{jk} \right),$$

the same reasoning leads to

$$R^m_{jml} = -\frac{1}{a^2} g_{jl},$$

i.e. for $j \neq l$, we have $R_{jl} = 0$. Therefore

$$R_{ij} + \frac{n-2}{a^2} g_{ij} = 0$$

for all i and j. From this formula, we obtain

$$R = -(n-1)(n-2)\frac{1}{a^2}.$$

Since

$$R_{ij} + \frac{1}{2}(n-1)(n-2)\frac{1}{a^2} g_{ij} - \frac{(n-2)(n-3)}{2}\frac{1}{a^2} g_{ij} = R_{ij} + \frac{n-2}{a^2} g_{ij} = 0$$

it results that, if we choose

$$\Lambda = -\frac{(n-2)(n-3)}{2}\frac{1}{a^2},$$

the previous metric satisfies the Einstein field equations

$$R_{ij} - \frac{1}{2}R\,g_{ij} + \Lambda\,g_{ij} = \frac{8\pi G}{c^4} T_{ij}$$

in the absence of matter, that is with $T_{ij} = 0$.

10.3   Some Physical Considerations

Let us return to the 4th dimensional de Sitter spacetime described by the parameterization

$$
\begin{cases}
X_0 = \sinh t \cos x_2 \cos x_3 \\
X_1 = \cosh t \cos x_1 \cos x_2 \cos x_3 \\
X_2 = \cosh t \sin x_1 \cos x_2 \cos x_3 \\
X_3 = \sin x_2 \cos x_3 \\
X_4 = \sin x_3
\end{cases}
$$

with the metric

$$ds_4^2 = \cos^2 x_2 \cos^2 x_3 dt^2 - \cosh^2 t \cos^2 x_2 \cos^2 x_3 dx_1^2 - \cos^2 x_3 dx_2^2 - dx_3^2.$$

It is difficult to talk about photons traveling in this Universe but if we consider a slice in the previous de Sitter space, determined by $x_2 = x_3 = 0$, we obviously highlight the 2-de Sitter spacetime

$$
\begin{cases}
X_0 = \sinh t \\
X_1 = \cosh t \cos x_1 \\
X_2 = \cosh t \sin x_1
\end{cases}
$$

denoted here by

$$f(t, x_1) = (X_0, X_1, X_2) = (\sinh t, \cosh t \cos x_1, \cosh t \sin x_1),$$

$f : \mathbb{R} \times (-\pi, \pi) \longrightarrow \mathbb{M}^3$, with the metric

$$ds_2^2 = dt^2 - \cosh^2 t dx_1^2$$

and we can hope for a simpler approach of the problem.

It exists two coordinate curves at each given point (t^0, x_1^0). The first one is

$$c_0(t) = f(t, x_1^0)$$

where x_1^0 is a constant. Since

$$\dot{c}_0(t) = \frac{\partial f}{\partial t} = \left(\cosh t, \sinh t \cos x_1^0, \sinh t \sin x_1^0\right)$$

is a time-like vector, i.e. $\left\langle \dfrac{\partial f}{\partial t}, \dfrac{\partial f}{\partial t} \right\rangle = 1$, the curve c_0 is a world line for an observer, that is, we are talking about the evolution in time of an event. The relation

$$\|\dot{c}_0(t)\|_M = 1$$

shows that the parameter t is the proper time because

$$\tau(t) = \int_0^t ||\dot{c}_0(q)||_M dq = t.$$

The other possible curve

$$c_1(x_1) = f(t^0, x_1)$$

where t^0 is a constant, is a "circle" which cuts the "Euclidean hyperboloid," and, at the same time, a space-like curve because $\left\langle \dfrac{\partial f}{\partial x_1}, \dfrac{\partial f}{\partial x_1} \right\rangle = -\cosh^2 t = -\cosh^2 \tau < 0.$

Let us analyze the "circumnavigation problem", that is the possibility to go around the "hyperboloid" in a finite amount of time. The length of c_1 is

$$\int_{-\pi}^{\pi} ||\dot{c}_1(\tau)||_M dx_1 = \int_{-\pi}^{\pi} \cosh \tau dx_1 = 2\pi \cosh \tau.$$

The limit, as τ approaches to ∞, is infinite, therefore this spacetime is unbounded in both given directions.

We are interested in understanding how photons travel in this de Sitter spacetime. Firstly, the metric

$$ds_2^2 = dt^2 - \cosh^2 t \; dx_1^2$$

is described by the metric tensor

$$\begin{pmatrix} 1 & 0 \\ 0 & -\cosh^2 t \end{pmatrix}$$

whose light-cone vectors, in the (t, x) plane, are $L^+ = \begin{pmatrix} x \\ \dfrac{1}{x} \dfrac{}{\cosh t} \end{pmatrix}$ and

$L^- = \begin{pmatrix} x \\ -x \dfrac{1}{\cosh t} \end{pmatrix}$, $x \in \mathbb{R}$.

These vectors were deduced in the same way we deduced them in a Minkowski space, whose metric

$$ds^2 = dt^2 - dx^2$$

is described by the metric tensor

$$\begin{pmatrix} 1 & 0 \\ 0 & -1 \end{pmatrix}.$$

The light cone vectors are $E^+ = \begin{pmatrix} x \\ x \end{pmatrix}$ and $E^- = \begin{pmatrix} x \\ -x \end{pmatrix}$, $x \in \mathbb{R}$.

If we are looking at the vectors L^+, L^- in their transposed form on $\mathbb{M}^3$, from

$$\left(x, \pm\frac{x}{\cosh t}\right) = x \cdot (1, 0) \pm \frac{x}{\cosh t} \cdot (0, 1),$$

the formulas

$$df_x\left(L^+\right) = x \cdot \frac{\partial f}{\partial t} + \frac{x}{\cosh t}\frac{\partial f}{\partial x_1}; \quad df_x\left(L^-\right) = x \cdot \frac{\partial f}{\partial t} - \frac{x}{\cosh t}\frac{\partial f}{\partial x_1},$$

results. This means that the velocity of photons is the ratio (with a sign) between the norms of the spatial vector $\dfrac{x}{\cosh t}\dfrac{\partial f}{\partial x_1}$ and the temporal vector $x\dfrac{\partial f}{\partial t}$, i.e.

$$\pm\frac{1}{\cosh t} \cdot \frac{\|\dot{c}_0(t)\|_M}{\|\dot{c}_1(x_1)\|_M} = \pm 1$$

as we expected.

Let us look again at the tx_1 plane and suppose we have the trajectory of a photon described by a function $x_1(t)$ which is $x_1(\tau)$. In fact, taking into account L^+, we have

$$x_1(q) = \int \frac{2}{e^q + e^{-q}}dq = \int \frac{2e^q}{e^{2q} + 1} = 2\arctan(e^q) + C_1,$$

where C_1 is a constant. This function is increasing. The limit as q approaches $-\infty$ is C_1 and the limit as q approaches $+\infty$ is $\pi + C_1$, therefore a photon image curve in tx_1 plane is completely included in a strip with width π. The same happens for the photon described by

$$x_1(q) = -\int \frac{2}{e^q + e^{-q}}dq = -\int \frac{2e^q}{e^{2q} + 1} = -2\arctan e^q + C_2,$$

where C_2 is a constant.

If we ask for photons having, at the origin of the tx_1 plane the vectors $L^+(0)$, $L^-(0)$ as tangent vectors respectively, we obtain the curves $x_1(\tau) = 2\arctan e^\tau - \dfrac{\pi}{2}$, and $x_1(\tau) = -2\arctan e^\tau + \dfrac{\pi}{2}$, respectively. The images of this two curves are the trajectories of photons in the de Sitter spacetime. Therefore, if we choose the first curve, its image in the de Sitter spacetime is

$$\begin{cases} X_0(\tau) = \sinh \tau \\ X_1(\tau) = \cosh \tau \cos\left(2\arctan e^\tau - \dfrac{\pi}{2}\right) \\ X_2(\tau) = \cosh t \sin\left(2\arctan e^\tau - \dfrac{\pi}{2}\right) \end{cases}.$$

Being

$$\cos\left(2\arctan e^{\tau} - \frac{\pi}{2}\right) = \frac{1}{\cosh \tau}$$

and

$$\sin\left(2\arctan e^{\tau} - \frac{\pi}{2}\right) = \tanh \tau,$$

therefore it is

$$\begin{cases} X_0(\tau) = \sinh \tau \\ X_1(\tau) = 1 \\ X_2(\tau) = \sinh \tau. \end{cases}$$

We leave, as an exercise for the reader, to prove that the second curve is

$$\begin{cases} X_0(\tau) = \sinh \tau \\ X_1(\tau) = 1 \\ X_2(\tau) = -\sinh \tau. \end{cases}$$

Finally, the trajectories of photons in the de Sitter spacetime are lines with slopes 1 and -1 (as we expected) which belong, in this case, to the plane $X_1 = 1$.

Even if it is just the investigation of a light cone at a single point, the reader has to imagine that, at each point of the de Sitter spacetime, the situation is the same: the Euclidean hyperboloid have, at each point, a pair of straight lines embedded into its surface.

10.4 A FLRW Metric for de Sitter Spacetime Given by the Flat Slicing Coordinates Attached to the Affine Sphere

In [26] it is presented a very interesting parameterization of the affine sphere

$$X_0^2 - X_1^2 - \cdots - X_{n-1}^2 = -a^2,$$

using the *flat slicing coordinates*:

$$f: \begin{cases} X_0 = a \sinh \dfrac{t}{a} + \dfrac{r^2}{2a} \cdot e^{t/a} \\[2mm] X_1 = a \cosh \dfrac{t}{a} - \dfrac{r^2}{2a} \cdot e^{t/a} \\[1mm] X_2 = y_1\, e^{t/a} \\ X_3 = y_2\, e^{t/a} \\ \cdots\cdots\cdots\cdots \\ X_{n-1} = y_{n-2}\, e^{t/a} \end{cases}$$

with

$$y_1^2 + y_2^2 + \cdots + y_{n-2}^2 = r^2.$$

It results the relation

$$y_1 dy_1 + y_2 dy_2 + \cdots + y_{n-2} dy_{n-2} = r dr$$

which helps us to find the corresponding metric.

Now,

$$dX_0 = \left[\cosh \frac{t}{a} + \frac{r^2}{2a^2} e^{t/a} \right] dt + \frac{r}{a} e^{t/a} dr$$

$$dX_1 = \left[\sinh \frac{t}{a} - \frac{r^2}{2a^2} e^{t/a} \right] dt - \frac{r}{a} e^{t/a} dr$$

$$dX_k = e^{t/a} \left[dy_{k-1} + \frac{1}{a} y_{k-1} dt \right], \quad k \in \{2, 3, \ldots, n-1\}.$$

If we compute the metric, firstly we obtain

$$dX_0^2 - dX_1^2 = \left[1 + \frac{r^2}{a^2} e^{2t/a} \right] dt^2 + \frac{2r}{a} e^{2t/a} dt dr.$$

Since

$$\sum_{k=2}^{n-1} dX_k^2 = e^{2t/a} \sum_{k=2}^{n-1} \left[dy_{k-1}^2 + \frac{2}{a} y_{k-1} dy_{k-1} dt + \frac{1}{a^2} y_{k-1}^2 dt^2 \right] =$$

$$= e^{2t/a} \sum_{k=2}^{n-1} dy_{k-1}^2 + e^{2t/a} \frac{2r}{a} dr dt + e^{2t/a} \frac{r^2}{a^2} dt^2,$$

finally we find

$$ds^2 = dt^2 - e^{2t/a} \left(dy_1^2 + dy_2^2 + \cdots + dy_{n-2}^2 \right).$$

For this metric we have

$$\begin{cases} \Gamma_{0\alpha}^{\alpha} = \Gamma_{\alpha 0}^{\alpha} = \dfrac{1}{a} \\ \Gamma_{\alpha\alpha}^{0} = \dfrac{1}{a} e^{2t/a}, \quad \alpha \in \{1, 2, \ldots, n-2\}, \end{cases}$$

all the other Christoffel symbols are null. Then,

$$R_{00} = \cancel{R^0_{000}} + R^\beta_{0\beta0} = \sum_{\beta=1}^{n-2} R^\beta_{0\beta0} = \sum_{\beta=1}^{n-2}\left[\frac{\partial\Gamma^\beta_{00}}{\cancel{\partial x^\beta}} - \frac{\cancel{\partial\Gamma^\beta_{0\beta}}}{\partial x^0} + \Gamma^m_{\cancel{00}}\Gamma^\beta_{m\beta} - \Gamma^m_{0\beta}\Gamma^\beta_{m0}\right] =$$

$$= -\frac{n-2}{a^2} = -\frac{n-2}{a^2}g_{00};$$

$$R_{\alpha\alpha} = R^s_{\alpha s\alpha} = \cancel{R^\alpha_{\alpha\alpha\alpha}} + \sum_{s=0,s\neq\alpha}^{n-2} R^s_{\alpha s\alpha} = \sum_{s=0,s\neq\alpha}^{n-2}\left[\frac{\partial\Gamma^s_{\alpha\alpha}}{\partial x^s} - \frac{\cancel{\partial\Gamma^s_{\alpha s}}}{\partial x^\alpha} + \Gamma^m_{\alpha\alpha}\Gamma^s_{sm} - \Gamma^m_{\alpha s}\Gamma^s_{m\alpha}\right] =$$

$$= \frac{2}{a^2}e^{2t/a} + (\Gamma^1_{01} + \Gamma^2_{02} + \cdots + \Gamma^{n-2}_{0n-2})\Gamma^0_{\alpha\alpha} - \Gamma^0_{0\alpha}\Gamma^0_{\alpha\alpha} - \Gamma^0_{\alpha\alpha}\Gamma^\alpha_{0\alpha} = -\frac{n-2}{a^2}e^{2t/a} = -\frac{n-2}{a^2}g_{\alpha\alpha}.$$

Therefore

$$R = -(n-1)(n-2)\frac{1}{a^2},$$

that is choosing

$$\Lambda = -\frac{(n-2)(n-3)}{2}\frac{1}{a^2},$$

the previous metric satisfies the Einstein field equations in absence of matter,

$$R_{ij} - \frac{1}{2}R\,g_{ij} + \Lambda g_{ij} = 0.$$

This metric can be written in the form

$$ds^2 = dt^2 - e^{2t/a}dy^2$$

where

$$dy^2 = dy_1^2 + dy_2^2 + \cdots + dy_{n-2}^2$$

is the flat metric in the y_k coordinates, which explains the name.

This is an example of a *FLRW metric for de Sitter spacetime*.

Example 10.4.1 Consider de Sitter spacetime with one of the previous metric g_{ij}.
It is

$$R_{ij} = -\frac{n-2}{a^2}g_{ij}$$

and

$$R = -(n-1)(n-2)\frac{1}{a^2}.$$

Therefore, in the case of $f(R) = R$ gravity, we obtain Einstein's field equations

$$R_{ij} - \frac{1}{2}R\, g_{ij} + \Lambda\, g_{ij} = 0$$

with the cosmological constant

$$\Lambda = -\frac{(n-2)(n-3)}{2}\frac{1}{a^2},$$

because we have

$$R_{ij} + \frac{1}{2}(n-1)(n-2)\frac{1}{a^2}\, g_{ij} - \frac{(n-2)(n-3)}{2}\frac{1}{a^2}\, g_{ij} = R_{ij} + \frac{n-2}{a^2}\, g_{ij} = 0.$$

If we consider the case of $f(R) = R^2$ gravity, we have

$$f'(R)R_{ij} - \frac{1}{2}f(R)g_{ij} + \Lambda_f\, g_{ij} = 0$$

if

$$\Lambda_f = \frac{(n-1)(n-2)^2(n-5)}{2a^4} = -R\,\frac{(n-2)(n-5)}{2a^2}.$$

Indeed,

$$2R\, R_{ij} - \frac{1}{2}R^2\, g_{ij} - R\,\frac{(n-2)(n-5)}{2a^2}\, g_{ij} = 2R\left(R_{ij} - \frac{1}{4}R\, g_{ij} - \frac{(n-2)(n-5)}{4a^2}\, g_{ij}\right) =$$

$$= 2R\left(R_{ij} + \frac{(n-1)(n-2)}{4a^2}\, g_{ij} - \frac{(n-2)(n-5)}{4a^2}\, g_{ij}\right) = 2R\left(R_{ij} + \frac{n-2}{a^2}\, g_{ij}\right) = 0.$$

Therefore we have the following statement:
The $f(R) = R^2$ gravity equations in absence of matter

$$f'(R)R_{ij} - \frac{1}{2}f(R)g_{ij} + \Lambda_f\, g_{ij} = 0$$

in the case of the cosmological constant

$$\Lambda_f = \frac{(n-1)(n-2)^2(n-5)}{2a^4}$$

are satisfied by any metric

$$ds^2 = g_{ij}dx^i dx^j, \ i, j \in \{0, 1, \ldots, n-2\}$$

having the property

$$R_{ij} = -\frac{n-2}{a^2} g_{ij}.$$

An example is de Sitter spacetime metric

$$ds^2 = dt^2 - e^{2t/a} \left(dy_1^2 + dy_2^2 + \cdots + dy_{n-2}^2 \right).$$

A particular situation happens when $n = 5$. The $f(R)$ equations in vacuum are satisfied in the their original form

$$f'(R)R_{ij} - \frac{1}{2} f(R)g_{ij} = 0,$$

that is no cosmological constant is needed.

Exercise 10.4.2 For the unit space-like affine sphere

$$X_0^2 - X_1^2 - \cdots - X_{n-1}^2 = -1$$

consider the parameterization

$$f : \begin{cases} X_0 = a \sinh t \\ X_1 = y_1 \cosh t \\ \cdots\cdots\cdots\cdots \\ X_{n-1} = y_{n-1} \cosh t \end{cases}$$

with

$$y_1^2 + y_2^2 + \cdots + y_{n-1}^2 = 1,$$

that is

$$y_1 dy_1 + y_2 dy_2 + \cdots + y_{n-2} dy_{n-2} = 0.$$

(i) Show that the corresponding metric is

$$ds^2 = dt^2 - \cosh^2 t \left(dy_1^2 + dy_2^2 + \cdots + dy_{n-1}^2 \right).$$

(ii) Try to understand why the metric can be written in the form

$$ds^2 = dt^2 - \cosh^2 t \, d\Omega_{n-2}^2,$$

where $d\Omega_{n-2}^2$ is the metric of the S^{n-2} sphere.

10.5   Deriving Cosmological Singularities in the Context of de Sitter Spacetime

Let us consider the case $n = 3$. The Euclidean one-sheet hyperboloid, which is in fact a Minkowki sphere, has the algebraic equation

$$X_0^2 - X_1^2 - X_2^2 = -1.$$

Using the flat slicing coordinates

$$f : \begin{cases} X_0 = \sinh t + \dfrac{r^2}{2} \cdot e^t \\[2mm] X_1 = \cosh t - \dfrac{r^2}{2} \cdot e^t \\[2mm] X_2 = r \cdot e^t. \end{cases}$$

we find, as a particular case of our previous discussion, the metric

$$ds^2 = dt^2 - e^{2t} dr^2.$$

For this metric, we have

$$\begin{cases} \Gamma_{01}^1 = \Gamma_{10}^1 = 1 \\ \Gamma_{11}^0 = e^{2t}, \end{cases}$$

all the other Christoffel symbols are null.

Let us consider the Minkowski sphere. At $t = 0$, consider the curve

$$m(r) = \left(\frac{r^2}{2}, 1 - \frac{r^2}{2}, r \right)$$

obtained by replacing $t = 0$ in the parameterization of f. This curve is the intersection between the plane

$$X_0 + X_1 = 1$$

and the Minkowski sphere

$$X_0^2 - X_1^2 - X_2^2 = -1.$$

We can conceive this curve as at the initial singularity at the origin of the de Sitter spacetime. Let us follow the evolution in time of it and choose a point of the singularity,

$$m(r_0) = \left(\frac{r_0^2}{2}, 1 - \frac{r_0^2}{2}, r_0 \right).$$

The evolution in time of this point is the line

$$c(t) = \left(\sinh t + \frac{r_0^2}{2} \cdot e^t, \ \cosh t + \frac{r_0^2}{2} \cdot e^t, \ r_0 \cdot e^t \right).$$

Theorem 10.5.1 *1. $c(t)$ is the intersection between the plane $X_0 + X_1 - \dfrac{1}{r_0} X_2 = 0$
and the previous "hyperboloid" $X_0^2 - X_1^2 - X_2^2 = -1$.
2. The Minkowski product $< \dot{c}(t), \dot{c}(t) >_M$ is 1, i.e. the tangent vector, is a time-like vector.
3. $c(t)$ is a time-like geodesic of de Sitter spacetime.*

Proof We leave for the reader the proof of the first two points which are simple exercises.
The equations of the geodesics are

$$\begin{cases} \dfrac{d^2 t}{d\tau^2} + \Gamma_{11}^0 \dfrac{dr}{d\tau} \dfrac{dr}{d\tau} = 0 \\[3mm] \dfrac{d^2 r}{d\tau^2} + 2\Gamma_{01}^1 \dfrac{dt}{d\tau} \dfrac{dr}{d\tau} = 0, \end{cases}$$

i.e.

$$\begin{cases} \dfrac{d^2 t}{d\tau^2} + e^{2t} \dfrac{dr}{d\tau} \dfrac{dr}{d\tau} = 0 \\[3mm] \dfrac{d^2 r}{d\tau^2} + 2 \dfrac{dt}{d\tau} \dfrac{dr}{d\tau} = 0. \end{cases}$$

The solution $t = \tau$, $r = r_0$ corresponds to the curve c. □

Therefore, the line is the evolution in time of the point and it is the first line we considered in the part of de Sitter spacetime out of the singularity curve.
Now, from each point of

$$m(r) = \left(\frac{r^2}{2}, \ 1 - \frac{r^2}{2}, \ r \right)$$

consider the corresponding curve

$$c_r(t) = \left(\sinh t + \frac{r^2}{2} \cdot e^t, \ \cosh t + \frac{r^2}{2} \cdot e^t, \ r \cdot e^t \right).$$

All these time-like geodesics for $t = \tau > 0$ starting from the initial singularity are part of the texture of the de Sitter spacetime. Observe that not all the Minkowski sphere is the texture of the de Sitter spacetime. Furthermore the parameterization makes sense also for $t = \tau < 0$. We have the image of an evolution of a singularity corresponding to the de Sitter spacetime.

As we said at the beginning of this chapter, gravitational interaction can be obtained starting from Minkowski spacetime according to a slicing procedure. This is only a possible geometric realization of Relativity. Other interesting geometric realizations of spacetimes can be found in [77].

Chapter 11
Conclusions

This book is an "experiment" to demonstrate that, starting from simple arguments of Euclidean Geometry, it is possible to arrive at the geometric formulation of physical theories: in the specific case, Special and General Relativity and, consequently, up to Relativistic Cosmology. Our attempt was aimed, above all, at undergraduate students, in particular those of our university courses, to demonstrate that, by a rigorous and extended mathematical development, theories deemed "difficult", such as General Relativity, can be understood and operationally used. We turned to students, and not to colleagues, to avoid falling into unnecessary technicalities that would have made the text unsuitable for a truly "basic" reading of Special and General Relativity. During the discussion, however, we introduced some advanced topics with the aim of stimulating the reader to further deepen and personal research. Our hope is not to have bored the reader and to have contributed something useful to the vast literature on the subject. Ours was a humble attempt, with no claim to completeness. We hope that our efforts have proved useful to someone eager to understand the wonderful book of Nature with the beautiful language of Mathematics which, as said by Leibniz, is "*the honor of the human spirit*".

© The Editor(s) (if applicable) and The Author(s), under exclusive license to Springer Nature Switzerland AG 2020
W. Boskoff and S. Capozziello, *A Mathematical Journey to Relativity*,
UNITEXT for Physics, https://doi.org/10.1007/978-3-030-47894-0_11

References

1. Bunge, M.: Quantum Theory and Reality. Springer, Berlin (1967)
2. Bunge, M.: Mach's critique of Newtonian mechanics. Amer. J. Phys. **34**, 585 (1966)
3. Lorentz, H.A., Einstein, A., Minkowski, H., Weyl, H.: The Principle of Relativity: A Collection of Original Memoirs on the Special and General Theory of Relativity. Dover Inc., New York (1952)
4. Callahan, J.: The Geometry of Spacetime: Special and General Relativity. Springer, New York (2000)
5. Varićak, V.: Über die nichteuklidische Interpretation der Relativtheorie. Jahreisbericht der Deutschen Mathematiker-Vereinigung **21**, 103 (1912)
6. Carlip, S.: General Relativity: A Concise Introduction. Oxford University Press, Oxford (2019)
7. Boskoff, W.G., Vajiac, A.: Discovering Geometry: An Axiomatic Approach. Matrix Rom Publishing House, Bucharest (2012)
8. Boskoff, W.G., Vajiac, A., Suceava, B.D.: An Exploration of Hilbert's Neutral Geometry. Gazeta Matematica A **26**, 1 (2008)
9. Martin, G.E.: The Foundations of Geometry and the Non-Euclidean Plane. Springer, New York (1975)
10. Torretti, B.: Philosophy of Geometry from Riemann to Poincaré. Reidel Pub, Dordrecht (1978)
11. Boskoff, W.G., Suceava, B.D.: Distances obtained by Barbilian's metrization procedure. Houston J. Math. **33**, 709 (2007)
12. Boskoff, W.G., Suceava, B.D., Ciuca, M.: Revisiting the foundations of Barbilian's metrization procedure. Differ. Geom. Appl. **29**, 577 (2011)
13. Poincaré, H.: La Science et l' Hypothése. E. Flammarion Editeur, Paris (1917)
14. Capuzzo-Dolcetta, R.: Classical Newtonian Gravity. Springer, Dordrecht (2019)
15. Zwicky, F.: On the masses of nebulae and of clusters of nebulae. Astrophys. J. **86**, 217 (1937)
16. Boskoff, W.G., Capozziello, S.: Recovering the Cosmological Constant from Affine Geometry. Int. J. Geom. Meth. Mod. Phys. **16**, 1950161 (2019)
17. Burke, W.: Spacetime. Geometry. Cosmology. University Science Books, Mill Valley (1980)
18. Capozziello, S., Faraoni, V.: Beyond Einstein Gravity. Springer, Dordrecht (2011)
19. Capozziello, S., Funaro, M.: Introduzione alla Relativitá Generale con Applicazioni all'Astrofisica Relativistica e alla Cosmologia. Liguori, Napoli (2005)
20. Coxeter, H.S.M.: A geometrical background for De Sitter's world. Amer. Math. Mon. **50**, 217 (1943)

W. Boskoff and S. Capozziello, *A Mathematical Journey to Relativity*,
UNITEXT for Physics, https://doi.org/10.1007/978-3-030-47894-0

21. Cheng, T.-P.: Relativity, Gravitation, and Cosmology: A Basic Introduction. Oxford, New York (2005)
22. Clarke, C.: Elementary General Relativity. Edward Arnold, London (1979); Wiley, New York (1980)
23. Droste, J.: The field of a single centre in Einstein's theory of the gravitation and the motion of a particle in that field. Proc. R. Neth. Acad. Arts Sci. **19**(1), 197 (1917)
24. Feynman, R.P., Leighton, R.B., Sands, M.: The Feynman Lectures on Physics. Addison-Wesley, Reading (1963)
25. Buchdahl, H.A.: Isotropic coordinates and Schwarzschild metric. Int. J. Theor. Phys. **24**(7), 731 (1985)
26. Hartman, T.: Lecture Notes on Classical de Sitter Space. www.harmanhep.net/GR2017/desitter-lectures-v2.pdf
27. Klingenberg, W.: A Course in Differential Geometry. Springer, New-York (1978)
28. Lawden, D.F.: An Introduction to Tensor Calculus. Relativity and Cosmology. Willey, New York (1982)
29. Eisenhart, L.P.: An Introduction to Differential Geometry. Princeton University Press, Princeton (1947)
30. Lopez, R.: Differential geometry of curves and surfaces in Lorentz-Minkowski space. Int. El. J. Geom. **7**, 44 (2014)
31. Gauss, C.F.: Disquisitiones generales circa superficies curvas. Typis Dieterichianis, Göttingen (1828)
32. Surhone, L.M., Timpledon, M.T., Marseken, S.F.: Poincaré Disk Model. Betascript Pub, Mauritius (2010)
33. Manasse, F.K., Misner, C.W.: Fermi normal coordinates and some basic concepts in differential geometry. J. Math. Phys. **4**(6), 735 (1963)
34. Misner, C., Thorne, K., Wheeler, J.A.: Gravitation. W.H. Freeman, San Francisco (1973)
35. Moore, T.: A General Relativity Workbook. University Science Books, Mill Valley (2013)
36. Arnold, V.I.: Mathematical Methods of Classical Mechanics. Springer, New York (1997)
37. Faraoni, V.: Special Relativity. Springer, Dordrecht (2013)
38. Cohen-Tonnoudji, C., Dui, B., Laloe, F.: Quantum Mechanics. Wiley-VCH, New York (2006)
39. Banks, E.C.: Ernst Mach's World Elements. Kluwer Academic Pub, Dordrecht (2003)
40. Ohanian, H.C., Ruffini, R.: Gravitation and Spacetime. Cambridge University Press, Cambridge (2013)
41. Palatini, A.: Deduzione invariantiva delle equazioni gravitazionali dal principio di Hamilton. Rend. Circ. Mat. Palermo **43**, 203 (1919)
42. Rosen, N.: General relativity and flat space. Phys. Rev. **57**, 147–150 (1940)
43. Schwarzschild, K.: Ubër das Gravitationsfeld eines Massenpunktes nach der Einsteinschen Theorie. Sitzungsberichte der Königlich Preussischen Akademie der Wissenschaften **7**, 189 (1916)
44. Sormani, C., Denson Hill, C., Nurowski, P., Bieri, L., Garnfinkle, D., Yunes, N.: The Mathematics of Gravitational Waves. Notices of AMS (64), **7**, 684 (2017)
45. Stephani, H., Kramer, D., MacCallum, M., Hoenselaers, C., Herlt, E.: Exact Solutions of Einstein's Field Equations, 2nd edn. Cambridge University Press, Cambridge (2003)
46. Will, C.M.: Theory and Experiment in Gravitational Physics. Cambridge University Press, Cambridge (1993)
47. Ferraris, M., Francaviglia, M., Reina, C.: Gen. Rel. Grav. **14**, 243 (1982)
48. Capozziello, S., Lambiase, G.: New Adv. Phys. **7**, 13 (2013)
49. Capozziello, S., De Laurentis, M.: Phys. Rept. **509**, 167 (2011)
50. Tino, G.M., Cacciapuoti, L., Capozziello, S., Lambiase, G., Sorrentino, F.: Precision Gravity Tests and the Einstein Equivalence Principle. Prog. Part. Nucl. Phys. **103772**, (2020)
51. Tino, G.M., et al.: SAGE: a proposal for a space atomic gravity explorer. Eur. Phys. J. D **73**, 228 (2019)
52. Altschul, B., et al.: Quantum tests of the Einstein equivalence principle with the STE-QUEST space mission. Adv. Space Res. **55**, 501 (2015)

53. Starobinsky, A.A.: A new type of isotropic cosmological models without singularity. Phys. Lett. **91B**, 99 (1980)
54. Kolb, E.W., Turner, M.S.: The Early Universe. Addison-Wesley Pub. Co., Inc, Redwood City (1990)
55. Capozziello, S.: Curvature quintessence. Int. J. Mod. Phys. D **11**, 483 (2002)
56. Capozziello, S., Francaviglia, M.: Extended theories of gravity and their cosmological and astrophysical applications. Gen. Relat. Gravit. **40**, 357 (2008)
57. Nojiri, S., Odintsov, S.D.: Introduction to modified gravity and gravitational alternative for dark energy. Int. J. Geom. Meth. Mod. Phys. **4**, 115 (2007)
58. Capozziello, S., De Laurentis, M.: The dark matter problem from f(R) gravity viewpoint. Ann. Phys. **524**, 545 (2012)
59. Capozziello, S., Darabi, F., Vernieri, D.: Equivalence between Palatini and metric formalisms of $f(R)$ gravity by divergence free current. Mod. Phys. Lett. A **26**, 65 (2011)
60. Weinberg, S.: The cosmological constant problem. Rev. Mod. Phys. **61**, 1 (1989)
61. Hawking, S.W., Ellis, G.F.: The Large Scale Structure of Space-time. Cambridge University Press, Cambridge (1973)
62. Peebles, P.J.E.: Principles of Physical Cosmology. Princeton University Press, Princeton (1993)
63. Zaroubi, S.: The epoch of reionization. In: Wiklind, T., Mobasher, B., Bromm, V. (eds.) The First Galaxies. Springer, Berlin (2013)
64. Peacock, J.A.: Physical Cosmology. Cambridge University Press, Cambridge (1999)
65. Amendola, L., Tsujikawa, S.: Dark Energy: Theory and Observations. Cambridge University Press, Cambridge (2015)
66. Zel'dovich, Y.B., Novikov, I.D.: The Structure and Evolution of the Universe. University of Chicago Pr, Chicago (1983)
67. Fermi, E.: Sopra i fenomeni che avvengono in vicinanza di una linea oraria. Rendiconti dell'Accademia dei Lincei **31**, 21 (1922)
68. A. Einstein, *Kosmologische betrachtungen zur allgemeine Relativitätstheorie*, Sitzungsberichte der königlich Preussischen Akademie der Wissenschaften, pp. 142-152 (1917)
69. Maggiore, M.: Gravitational Waves. Oxford University Press, Oxford (2018)
70. Straumann, N.: General Relativity. Springer, Dordrecht (2012)
71. Abbott, B.P., et al.: [LIGO-VIRGO collaboration], GW151226: Observation of gravitational waves from a 22-solar-mass binary black hole coalescence. Phys. Rev. Lett. **116**, 241103 (2016)
72. Gödel, K.: An example of a new type of cosmological solution of Einstein's field equations of gravitation. Rev. Mod. Phys. **21**, 447 (1949)
73. Stephani, H.: General Relativity. Cambridge University Press, Cambridge (1990)
74. Frolov, V.P., Novikov, I.D.: Black Hole Physics. Kluwer Academic Pub, Dordrecht (1998)
75. Akiyama, K., et al., [Event Horizon Telescope Collaboration]: First M87 Event Horizon Telescope Results. I. The Shadow of the Supermassive Black Hole. Astrophys. J. **875**, L1 (2019)
76. Spradlin, M., Strominger, A., Volovich, A.: Les Houches Lectures on de Sitter Space. In: Bachas, C.P., Bilal, A., Douglas, M.R., Nekrasov, N.A., David, F. (eds.) Unity from Duality: Gravity, Gauge Theory and Strings, Les Houches Session LXXVI, July 30 - August 31, 2001. Springer, Berlin (2002)
77. Ryan, M.P., Shepley, L.C.: Homogeneous Relativistic Cosmologies. Princeton University Press, Princeton (1975)

Index

A

Absolute geometry, 24
Action, 201
Acute angle, 18
Additivity of angles, 13
Altitude, 18
Angle, 8
Angle between curves, 74
Angle between n-lines, 64
Angle betwen vectors, 66
Aphelion, 195
Area of a region, 74
Axiom of Archimedes, 20
Axiom of Cantor, 20
Axiom Pash, 4
Axioms of congruence, 8
Axioms of continuity, 20
Axioms of incidence, 2
Axioms of order, 4
Axis of a projective map, 58

B

Barbilian's theorem, 142
Basis of tangent plane, 68
Bekenstein–Hawking formula, 364
Beltrami's identity, 205
Bending of light in the Schwarzschild metric, 299
Betweenness, 4
Bianchi's first formula, 133
Bianchi's first identity, 86
Bianchi's second formula, 133
Black hole, 360
Black hole's entropy, 364
Black hole's temperature, 364
Border, 7

C

Canonically parameterized curve, 123
Catenary problem, 206
Cauchy–Bunyakowsky–Schwartz inequality, 70
Causality Principle, 260
Cayley's theorem, 223
Center of the ellipse, 189
Centripetal acceleration, 196, 197
Centripetal force, 196, 197
Change of coordinates, 110
Christoffel symbols of first kind, 80, 113
Christoffel symbols of second kind, 80, 113
Circle of inversion, 39
Classical momentum, 170, 231
Codazzi–Mainardi equations, 84
Collinear, 2
Complementary half-line, 7
Complementary half-plane, 7
Conformal map, 43
Coplanar, 3
Cosmic string, 345
Cosmological constant, 286
Cosmological principle, 309
Covariance principle, 260
Covariant derivative for contravariant vector $V^k(t)$, 120
Covariant derivative for contravariant vector $V^k(x)$, 119
Covariant divergence, 179
Crossbar theorem, 8

Crossproduct of two vectors, 67
Cross ratio of four ordered lines in a pencil, 47
Cross ratio of four ordered points on a circle, 49
Cross ratio of four ordered points on a line, 45
Curve on a surface, 72

D

D'Alembert operator, 348
Dark energy, 321, 324
Dark matter, 196, 198
Defect of a triangle, 22
Density of mass for the swarm, 333
Density of the incompressible fluid, 176
Density of the relativistic energy of the swarm, 334
De Sitter, 372
Direct geometric inversion, 39
Divergence, 177
Doppler's effect, 244

E

Eccentricity of the ellipse, 189
Einstein equivalence principle, 262
Einstein field equations in vacuum, 279
Einstein–Hilbert action, 276
Einstein notation, 65
Einstein's field equations in vacuum, 277
Einstein's field equations with cosmological constant, 287
Einstein's landscape on the constant gravitational field, 267
Einstein's metric, 304
Einstein's metric in Cartesian coordinates, 307
Einstein's Principle of General Covariance, 266
Einstein's static universe, 343, 344
Einstein's static universe metric, 344
Einstein's tensor, 135
Einstein summation convention, 65
Einstein theorem, 89
Ellipse, 189
Energy-momentum tensor of a swarm of particles, 333, 334
Entrylength of the vector, 66
Equation of a circle, 35
Equivalence principle, 260
Escape velocity, 356
Euclidean distance, 33, 67

Euclidean geometry, 25
Euclidean inner product, 30, 66
Euclidean norm, 30
Euclidean parallelism axiom, 24
Euclidean perpendicular, 30
Euclidean perpendicular vectors, 66
Euclidean rotation matrix, 30
Euler–Lagrange equation, 204, 210
Event, 220
Event horizon, 360
Exterior angle, 17
Exterior angle theorem, 17
Exterior of a black hole, 360
Exterior of an angle, 8
Exterior of the disk model, 164

F

Fermi's coordinates, 327
First fundamental form, 70
First order variation of the action, 202
Flat slicing coordinates, 379
FLRW metric for de Sitter spacetime, 381
FLRW models of universe, 309
Foci of the ellipse, 189
4-momentum vector, 333
4-velocity of a particle moving at speed v, 333
Frequency of light, 244
$f(R)$ field equations in vacuum, 281
$f(R)$ generalized field equations, 281
$f(R)$ gravity, 279
Future-pointing space-like vector, 37
Future-pointing time-like vector, 37

G

Galilean equivalence principle, 172
Galilean transformations, 224
Gauss' equations, 84
Gauss' formulas, 81
Gaussian curvature, 77
Gaussian curvature geometric interpretation, 78
Gauss map, 69
General Einstein–Hilbert action, 277
General Einstein's field equations, 278, 335
General gravitational field equation, 180
Generalized Minkowski-Pythagoras theorem, 38
Generalized Pythagoras theorem, 31
Geodesic, 113, 120
Geodesic of a surface, 92
Geodesics of the Poincaré half-plane, 212

Geometric coordinates, 225
Geometric inversion, 39
Geometric inversion properties, 40
Geometric involution, 53
Geometric singularity, 357
Geometric tidal acceleration equations, 187, 189
Geometric units, 225
Gödel's first metric, 350
Gödel's second metric, 352
Gödel's universe, 350
Gradient, 173
Gradient notation ∇, 173
Gradient operator, 238
Gravitational acceleration, 172
Gravitational constant, 171
Gravitational field, 172
Gravitational force, 171
Gravitational mass, 171
Gravitational planar waves, 349
Gravitational potential, 172, 200
Gravitational potential energy $P_E(r)$, 200
Gravitational radius, 294
Gravitational weak equivalence principle, 263
Group of linear transformations, 111

H
Half-line, 6
Half-plane, 7
Hamilton's stationary action principle, 202
Harmony law, 196
Height of a triangle, 18
Hemisphere metric, 161
Hemisphere model, 160, 164
Hessian matrix, 175
Hessian matrix notation $d^2\Phi$, 175
Homologous points, 50
Horizon of the geometric singularity, 358
Hubble's constant, 318
Hubble's law, 317
Hyperbolic rotation, 229
Hyperbolic rotation matrix, 36
Hyperboloid metric, 163
Hyperboloid model, 164

I
Incompressible fluid flow, 176
Inertial mass, 172
Inflation, xvii
∞-points, 64
Interferometer, 218

Interior alternate angles, 19
Interior of a black hole, 360
Interior of an angle, 8
Interior of a triangle, 8
Intrinsic geometric properties, 74
Invariance principle, 246
Isometry, 71

J
Jacobian determinant, 111
Jacobian matrix, 111

K
Kepler's first law, 194
Kepler's second law, 195
Kepler's third law, 196
Kinetic energy, 200
Kinetic relativistic energy, 237
Kruskal–Szekeres coordinates, 361
Kruskal–Szekeres metric, 362

L
Laplace equation for gravity, 176
Laplace operator, 173, 182, 238
Laplace operator notation ∇^2, 173
Laplacian, 173
Laue's scalar, 278
Legendre's theorem, 22
Length contraction, 230
Length of a curve, 73
Light cone, 221
Lightlike vectors, 94
Linearized metric, 349
Line at infinity, 57
Lines, 2
Lorentz frame, 263
Lorentz geometric transformations, 233
Lorentz physical transformations, 233
Lorentz transformation, 221
Lorentz transformations in geometric coordinates, 226

M
Major axis of the ellipse, 189
Maxwell's equations, 240
Mechanical Lagrangian, 201
Metric induced by the Minkowski 3-space structure, 98
Metric of a hemisphere, 161
Metric of a surface, 72
Metric of the exterior of the disk, 157

Michelson–Morley experiment, 218
Midpoint of a segment, 18
Minkowski crossproduct, 95
Minkowski distance, 95
Minkowski first fundamental form, 97
Minkowski–Gauss curvature, 101, 103
Minkowski–Gauss' equations, 101
Minkowski–Gauss frame, 97
Minkowski geometric gravitational field,
     368
Minkowski geometric gravitational force,
     368
Minkowski gradient, 368
Minkowski gravitational potential, 368
Minkowski hypotenuse, 38
Minkowski Laplacian, 368
Minkowski legs, 38
Minkowski metric in spherical spatial coor-
     dinates, 289
Minkowski-Newton second principle, 368
Minkowski orthogonal, 35
Minkowski perpendicular vectors, 95
Minkowski product, 35, 94
Minkowski-Pythagoras theorem, 37
Minkowski right triangle, 38
Minor axis of the ellipse, 189
Missing matter, 198

N
N-collinear, 64
Net outflow, 177
Newton's vacuum field equation, 174
N-line, 64
Non-colinear, 2
Non-coplanar, 3
Non-Euclidean geometry, 27
Non-Riemannian metric, 112
Non-secant lines, 17
N-point, 64
N-triangle, 64
Null cone, 37
Null vectors, 35, 94

O
Observer, 220
Obtuse angle, 18
Orbital period of a planet, 196
Orbit of a planet in Schwarzschild metric,
     291
Orbit's equation solution in Schwarzschild
     metric, 295
Oriented hyperbolic angle, 36

P
Palatini's formula, 274
Parallel line, 25
Past-pointing space-like vector, 37
Past-pointing time-like vector, 37
Pencils of lines, 47
Perihelion, 195
Period corresponding to a given frequency
     of light, 244
Perspective map, 54
Perspective pencils of lines, 57
Photon energy, 238
Photon frequency, 238
Photon wavelength, 238
Physical coordinates, 225
Planck constant, 238
Poincaré disk model, 163
Poincaré distance of half-plane, 148
Poincaré distance of the disk, 63, 145
Poincaré half-plane model, 164
Poincaré metric of the disk, 146, 150
Poincaré metric of the half-plane, 148, 149,
     151
Point at infinity of a line, 47
Points, 2
Poisson's equation for gravity, 183
Postulates of Special Relativity, 220
Potential, 199
Potential energy, 200
Precedes, 7
Prejective map on a line, 51
Principle of covariance, 332
Projective map between two lines, 50
Projective map of a circle, 50
Projective pencils of lines, 54
Proper time in General Relativity, 294
Proper time in Special Relativity, 235
Pseudo-Riemannian metric, 112
Pseudo-singularity, 357
Pseudosphere, 87
Ptolemy's theorem, 44
Pythagoras' theorem, 33

R
Rate of separation of geodesics, 187
Ray, 6
Reference frame, 220
Relativistic counterparts, 336
Relativistic Doppler's effect, 244
Relativistic field equations in vacuum, 332
Relativistic mass, 231
Relativistic momentum, 231

Relativistic velocities addition, 224
Rest energy, 237
Ricci's equations, 82
Ricci tensor, 113
Riemann symbols, first type, 82
Riemann symbols, second type, 82
Riemannian curvature tensor, covariant, 113
Riemannian curvature tensor, mixed, 113
Riemannian metric, 112
Right angle, 15
Rindler's metric, 357

S

Schwarzschild metric, 290
Schwarzschild radius, 294
Schwarzschild's solution, 288
Secant, 4
Second fundamental form, 75
Segment, 5
Self-homologous point, 54
Semi-Riemannian metric, 112
Sides, 5
Signature of a metric, 112
Simultaneity, 225, 226
Simultaneous, 225
Slope, 34
Slow particle, 338
Space, 2
Space-like vectors, 35, 95
Spacetime, 220
Spatial Poincaré half-plane model, 164
Speed of light in a gravitational field, 299
Starobinsky's inflation, 279
Stationary point of the functional, 203, 210
Steiner's theorem, 54
Strong Equivalence Principle, 263
Surface, 67
Swarm of particles, 333
Symmetric geometric inversion, 40

T

Tachyon, 322
Tangent plane, 68
Tangent space, 112
Tensor contravariant of rank k and covariant of rank p, 110
Tensor (k,p), 110
Thales theorem, 34
Theorema Egregium, 79, 86
Theorem ASA, 12
Theorem SAS, 12
Theorem SSS, 14
Tidal acceleration, 184
Tidal acceleration equations, 185
Tidal acceleration equations in Fermi's coordinates, 329
Tidal effect, 184
Tidal vector, 186
Time dilation, 229
Time-like vectors, 35, 95
Total deflection, 303
Total energy, 201
Tractrix, 87
Translated disk model, 163
Transversal theorem, 8
Triangle inequality, 18
Twins paradox, 230

V

Vector parallel transported along a curve, 91, 120
Vertexes, 5
Violation of speed of light limit, 299

W

Weak equivalence principle, 262
Weak gravitational field, 336
Weingarten's formulas, 77
World line, 220

Printed in the United States
by Baker & Taylor Publisher Services